# Foundations of Rock Mechanics in Oil and Gas Engineering

Yuanfang Cheng · Chuanliang Yan ·
Zhongying Han

# Foundations of Rock Mechanics in Oil and Gas Engineering

中国石油大学出版社
CHINA UNIVERSITY OF PETROLEUM PRESS

Springer

Yuanfang Cheng
School of Petroleum Engineering
China University of Petroleum
Qingdao, China

Chuanliang Yan
School of Petroleum Engineering
China University of Petroleum
Qingdao, China

Zhongying Han
School of Petroleum Engineering
China University of Petroleum
Qingdao, China

ISBN 978-981-99-1416-6 ISBN 978-981-99-1417-3 (eBook)
https://doi.org/10.1007/978-981-99-1417-3

Jointly published with China University of Petroleum Press
The print edition is not for sale in China (Mainland). Customers from China (Mainland) please order the print book from: China University of Petroleum Press.
ISBN of the Co-Publisher's edition: 978-7-5636-7733-7



This Springer imprint is published by the registered company Springer Nature Singapore Pte Ltd.
The registered company address is: 152 Beach Road, #21-01/04 Gateway East, Singapore 189721, Singapore

# Preface

Rock mechanics is an area of applied science that has been recognized as a self-contained engineering discipline for only a few decades. It covers a range of knowledge that includes the mechanical properties of rocks, various methods for the analysis of rock stresses induced by applied disturbances, a set of accepted principles for expressing the response of rock masses to loads, and a set of scientific methods for applying these concepts and methods to the analysis and solution of practical physical problems. For engineering rock mechanics, the main study is rock stability and fragmentation. Rock stability mainly includes the branches of borehole stability, roadway stability, slope stability, quarry mine pressure, rock movement and impact ground pressure. Rock crushing includes branches such as mechanical crushing, blasting rock breaking, hydraulic rock breaking and electrophysical rock breaking.

The principle of rock mechanics has been widely used in the fields of petroleum, mining, human defense, attempted roads and other engineering fields, making the engineering design or process design more reasonable and producing great economic benefits.

The application of rock mechanics in petroleum engineering has been relatively slow to develop. However, in the last three decades, due to the deepening of extraction depth, increasing difficulty of extraction and the emergence of new and complex processes such as extended reach wells, horizontal wells and hydraulic fracturing, the downhole complexities have become more prominent. A large number of excellent rock mechanics experts began to pay attention to the research of rock mechanics in oil and gas engineering, which promoted the development of the subject of rock mechanics application in petroleum engineering.

As the application of rock mechanics in petroleum engineering is a fairly new discipline, petroleum engineers need a book that systematically introduces the basic theory of rock mechanics. The authors deeply realized this when they were engaged in this research work. Based on the analysis and summary of more than 200 documents and books, and the author's achievements in rock mechanics research of oil and gas engineering for a long time, the offset copy of this book was compiled in 1998 and has been used in the teaching of petroleum engineering in China University of Petroleum for more than ten years. In 2015, the rock mechanics in oil and gas engineering was finally published by China University of Petroleum Press on the basis of offset printing manuscript and teaching courseware and was used in the elective courses of petroleum engineering undergraduate students. As the scale of ultra-deep well development of oil and gas fields has increased rapidly in the past decade, the demand for rock mechanics has become more urgent. In this context, China University of Petroleum has adjusted the training program for the undergraduate major of petroleum engineering, making rock mechanics a compulsory course in English. Therefore, this textbook has been rewritten. It is worth mentioning that in recent years, young doctors have joined the course group of rock mechanics, bringing vitality to the study of rock mechanics in oil and gas engineering. Representative figures Dr. Chuanliang Yan and Dr. Zhongying Han participated in the compilation of this textbook.

The first seven chapters of this book introduce the basic knowledge of rock mechanics. This book systematically introduces the basic theory of rock mechanics, including the stress and strain analysis, the composition and basic physical properties of rocks, the characteristics of rock deformation and strength, rock mechanics experimental methods, rock strength failure criteria and in situ stress state. Among them, Chap. 1 is an overview prepared by Prof. Yuanfang Cheng, Chap. 2 is stress and strain prepared by Dr. Zhongying Han, Chap. 3 is rock composition and rock physical properties prepared by Dr. Chuanliang Yan, Chap. 4 is strength and deformation characteristics of rocks prepared by Prof. Yuanfang Cheng, Chap. 5 is strength characterization and laboratory measurement of rocks prepared by Dr. Zhongying Han, Chap. 6 is rock strength failure criteria and Chap. 7 is in situ stress state prepared by Dr. Chuanliang Yan. The last three chapters are petroleum engineering applications, mainly introducing the application methods of rock mechanics theory in the fields of wellbore stability, oil well sand production and hydraulic fracturing. Among them, Dr. Chuanliang Yan wrote Chap. 8 mechanical principle of wellbore stability, Prof. Yuanfang Cheng wrote Chap. 9 mechanical principle of hydraulic fracturing and Dr. Zhongying Han wrote Chap. 10 mechanical principle of oil well sand production. Finally, Prof. Yuanfang Cheng prepared the whole manuscript. This book can be used as a textbook for petroleum engineering undergraduate majors and similar majors and also has a certain reference value for on-site petroleum engineering technicians engaged in rock mechanics.

Qingdao, China
July 2022

Yuanfang Cheng

# Contents

# Chapter 1
# Rock Mechanics and Petroleum Engineering

## 1.1 General Concepts

Rock mechanics has evolved through the social practice of people. Although people have been working with rocks (or soil) for a long time, rock mechanics has become a separate discipline in recent decades. In ancient times, our ancestors lived in caves and used rocks as tools and weapons, and there was a "stone age". The Neolithic megalithic culture was popular in Europe over 4000 B.C. Stonehenge, now world-famous in southern England, is one of these sites.

Around 2700 B.C., the working people of ancient Egypt built the pyramids. In the sixth century B.C., the Babylonians built Hanging Gardens in the mountains. In 613–591 B.C., Chinese people built the first barrage in history on the Pai River in Anhui Province. In 256–251 B.C., the Dujiangyan Water Conservancy Project was built on the Min River in Sichuan. Drilling technology was developed around 254 B.C. (during the time of King Zhao of Qin). In 218 B.C., the Ling Canal, a waterway linking the Yangtze and Pearl River systems, was dug in Guangxi, and a masonry diversion weir was built. The Great Wall of China was built in the northern mountains from 221 to 206 B.C. Zhaozhou Bridge was built in 595–605 A.D. in Sui Dynasty. The stone arch bridge designed and built by Li Chun, a famous craftsman, is the crystallization of the wisdom of the ancient working people and has created a new situation for bridge construction in China. From the 1140s to the 1830s, salt wells were drilled in Sichuan using the percussion drilling technique, with the maximum depth exceeding 1000 m. The drilling process is described in detail in "Tian Gong Kai Wu" (1637 A.D.) by Song Ying Xing of the Ming Dynasty, with various forms of tools for drilling, salvaging, and treating wells.

It is inevitable that some basic knowledge of rock mechanics will be applied in the construction of these projects. However, as a discipline, the study of rock mechanics only began around the 1950s. Due to the increasing scale and complexity of modern geotechnical engineering, it has been difficult to meet the engineering requirements by experience alone. Especially since the middle of the twentieth century, the scale and scope of rock engineering has increased significantly, in which serious accidents



Y. Cheng et al., *Foundations of Rock Mechanics in Oil and Gas Engineering*,
https://doi.org/10.1007/978-981-99-1417-3_1

have occurred, such as the destruction of the Malpasset dam in France in December 1959, the failure of a reservoir caused by a large landslide at the Vajont dam in Italy in October 1960, a major blowout at the Dina 2 well in the Tarim oil field in April 2001, the bedrock landslide in Wulong County, Chongqing in May 2001, etc. all caused great loss of life and property. The occurrence of these accidents has made it clearer the need to study the reflection of rock masses on engineering activities.

The generally recognized definition of rock mechanics was first proposed by the US National Committee on Rock Mechanics in 1964, and then revised in 1974 as: "Rock mechanics is a science that studies the theory and application of mechanical properties of rocks and rock masses, and is a mechanical branch that discusses the response of rocks and rock masses to the force field of their surrounding physical environment." This definition points out the relationship between rock as a natural material and its geological environment, formation history, crustal movement and engineering factors.

Engineering rock mechanics is a discipline that applies the principles of engineering mechanics to the design and construction analysis of geotechnical structures. Obviously this discipline is closely related to the main elements of classical mechanics and mechanics of continuum. But as the reader will appreciate, it is due to a number of special factors that the discipline is regarded as a separate discipline. In order to be able to determine the properties and theoretical calculations of naturally occurring rock masses from a mechanical point of view for the purpose of engineering construction, the research methodology of rock mechanics includes three main components, scientific experiments, theoretical analysis and engineering verification, which are closely linked and mutually reinforcing.

Scientific experiments are the basis for the development of rock mechanics, which includes the determination of mechanical parameters of rocks in the laboratory, model tests, in-situ testing of rocks in the field and monitoring techniques, determination of in-situ stress and determination of rock formations (Zhou 1990; Zhou and Yang 2005).

The classical theories of elasticity and plasticity were the basis of the theoretical analysis. Later on, rheology, viscoelastic-plastic theory, fracture mechanics and damage mechanics were gradually applied. New theories of numerical computation such as finite element, boundary element, discrete element, block mechanics and inverse analysis are also widely used in rock mechanics. As can be seen, the theoretical basis of rock mechanics is quite broad, which leads to the high difficulty of rock mechanics research.

Engineering verification is a means to test the correctness of theoretical analysis and a way to guide engineering practice. In applying knowledge of rock mechanics to solve specific engineering problems, it is necessary to maintain close contact and cooperation with engineering design and construction. An example is the method of convergence measurement, information monitoring and feedback design and construction, which has been developed rapidly recently. It brings theoretical analysis and engineering closer together and greatly improves the quality and progress of construction on site. The introduction of big data and artificial intelligence theory has deepened the integration of theoretical analysis with engineering practice.

The application of rock mechanics is quite wide, and it can be said that rock mechanics problems exist in all geotechnical related engineering, such as hydraulic and hydroelectric power, mining engineering, tunnel engineering, petroleum engineering, nuclear waste disposal, earthquake prediction, etc. This book deals only with the basic principles of rock mechanics and their application in petroleum engineering.

## 1.2 Inherent Complexity of Rock Mechanics

Rock mechanics embodies a set of principles, primary knowledge and various analytical methods related to the general field of applied mechanics. Why then should rock mechanics be considered as a separate engineering discipline? We illustrate this question in the following six ways (Chen 1986).

### *1.2.1 Failure Properties of Rocks*

The failure of engineering materials in general occurs in the tensile stress region, requiring a sound theory to elucidate the properties of these material media before and after damage. In contrast, the stress field in rock mass is generally a compressive stress field. Therefore, the theories established by general engineering mechanics are not directly applicable to the failure of rocks. The unique complex state of compressed rocks is related to the frictional forces generated between the individual microfracture surfaces at the onset of failure. This makes the strength of the rock very sensitive to changes in lateral pressure (confining pressure). Thus, concepts such as the orthogonal principle, the associated flow rule, and the plasticity principle are called into question when analyzing the strength of rocks and the nature of deformation after damage. A related issue is the localized phenomenon of rock rupture, in which the rupture of a rock manifests itself as a sharp shear zone in the rock medium, which separates the unaffected rock material.

### *1.2.2 Size Effect*

Joints and fractures generated due to other geological causes are prevalent in the rock mass, forming macroscopic discontinuity surfaces. The strength and deformation characteristics of the rock mass are influenced by a combination of the properties of the rock block (i.e., the rock continuum media unit) and the various geological structural surfaces. It is generally accepted that during drilling, the characteristics of rock broken by bit are the strength characteristics of intact rock. Excavation tunnels in jointed rock can reflect the nature of the joint system, when the final section of the tunnels depends on the spatial orientation of the joints (Brown 1987). The damage

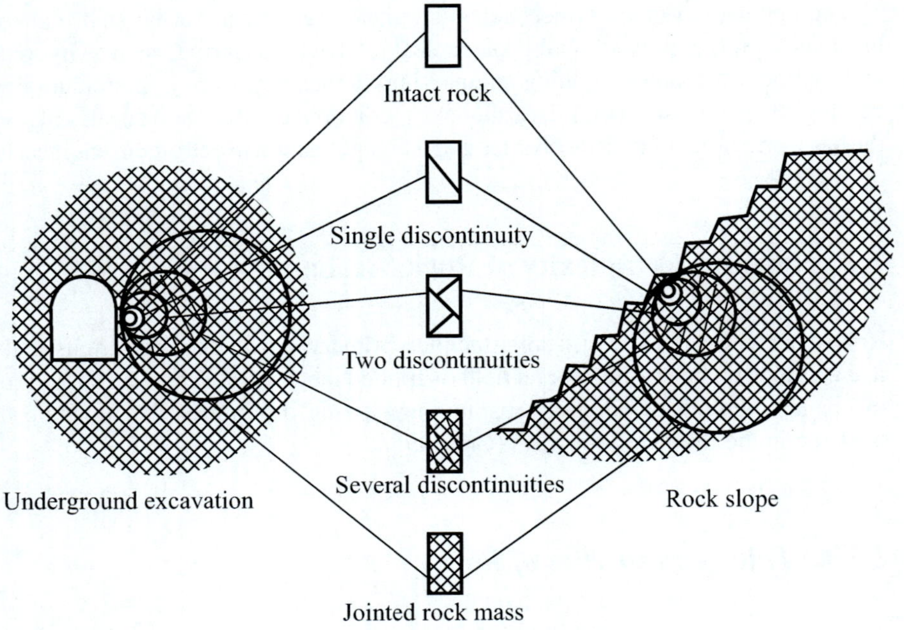


**Fig. 1.1** Rock feature models for rock engineering at different spatial scales

within the bedrock of the Malbasai arch dam in 1959 was the result of being cut by faults and joints into a single rock block that moved. The relationship between rock engineering and rock feature parameters at different spatial scales is shown in Fig. 1.1.

It is evident that the relative dimensions between the engineering object and the rock mass are different, and the treatment methods are also different. In order to describe the mechanical properties of the rock mass accurately, some new methods must be proposed to infer the integrated properties of the rock mass from the properties of its constituent units and to validate these methods.

### *1.2.3 Tensile Strength*

Rocks differ from other common engineering materials (except concrete) due to their low tensile strength. The stress of rock specimen in uniaxial tensile strength test is one order of magnitude lower than that in uniaxial compression test. Due to the presence of joints and other fractures in the rock, which resist only minimal or no tensile stresses, the tensile strength of the rock is assumed to be zero in some projects, and for this reason the rock is referred to as a "non-tensile" material (Li 1983; You 2007). For borehole stability analysis, this property means that if tensile stresses occur in the borehole, a leakage will occur.

### 1.2.4 Impact of Groundwater

Groundwater affects the mechanical properties of rock masses in two ways. The most obvious one is through the principle of effective stress. Pressurized water in joints separating rock masses reduces the effective normal stress between the rock surfaces. Thus reducing the potential resistance to shear that may arise due to friction. Porous rocks reduce the ultimate strength of the rock under these conditions. Another effect of groundwater on the mechanical properties of rocks is the deleterious effect of groundwater on special rocks and minerals. For example, clay layers can soften in the presence of groundwater, reducing the strength of the rock and increasing its deformation. Rocks such as shales and muddy sandstones can exhibit a significant reduction in strength with water intrusion. Caves are subsurface spaces in soluble rocks formed by karstic action and are the result of long-term groundwater dissolution in limestone areas, as shown in Fig. 1.2.

The effect of groundwater on rock properties is particularly important in oil and gas engineering. During the drilling process water-based drilling fluid will gradually infiltrate the muddy rock and reduce its strength, resulting in the phenomenon of borehole spalling and falling blocks; secondly, water injection is one of the main measures to improve the recovery rate, and water injection will change the properties of muddy sandstone such as strength, porosity and permeability, resulting in sand production and lower production rate.

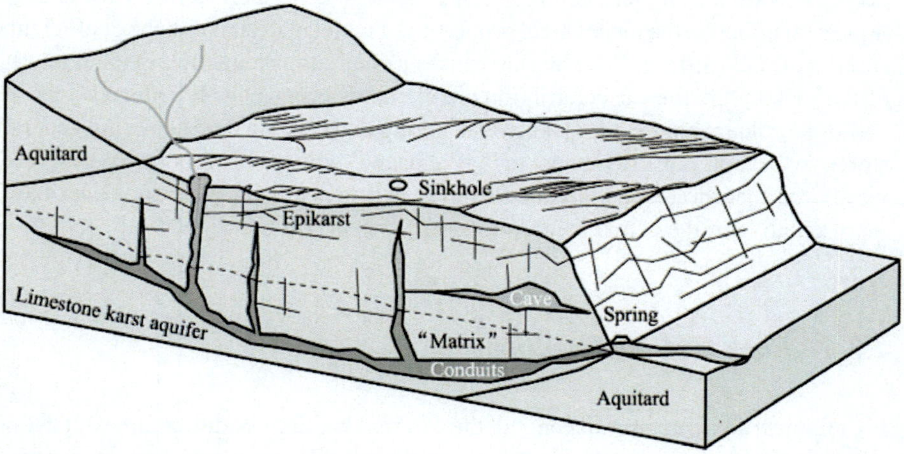


**Fig. 1.2** Karst caves

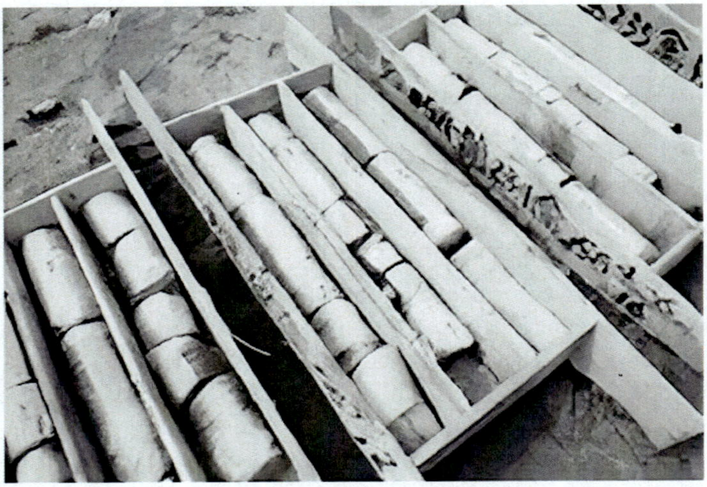

**Fig. 1.3** Weathering of rock cores

### *1.2.5 Weathering*

Weathering is the chemical or physical change of a rock due to the action of atmospheric and water dissolution on its surface, a process similar to the corrosion that occurs on ordinary materials. Weathering has a stake in engineering because of its important effect on the mechanical properties of intact materials and the coefficient of friction of rock surfaces. Weathering causes a continuous reduction in the properties of the rock, while the surface friction coefficient is progressively reduced.

For petroleum engineering, it is worth noting that cores in core banks are generally stored for several years (as shown in Fig. 1.3), and some types of rocks are so severely weathered that the results obtained from experiments using these rocks are hardly representative of the actual conditions of the subsurface rocks.

### *1.2.6 Rock Outside Loading*

For engineering structures in general, the external loading conditions are well defined when performing stress analysis. However, for subsurface rock masses, it is difficult to know exactly the original stress state prior to engineering disturbances. the This original stress state is called the in-situ stress state, determination of it is an important research topic in the discipline of rock mechanics.

The above discussion does not give the full range of special problems to be studied in rock mechanics. It is clear, however, that the discipline of rock mechanics extends beyond the traditional scope of applied mechanics and that it necessarily includes many topics not yet covered by any other engineering discipline. The investigation of these problems constitutes a unique study of rock mechanics.

## 1.3 Rock Mechanics Problems in Oil and Gas Engineering

Oil and gas development is very different compared to coal mining, where oil layers are usually buried at depths of several kilometers. To product oil from the lower formation, it is necessary to drill through its upper layers to form an oil and gas channel (borehole). The borehole may pass through high pressure layers, collapse layers, leaky layers, fracture zones and other formations. The rock mechanics properties of different formations may be different, and their failure mechanisms are also different. In order to drill a well quickly and with high quality, the failure mechanisms of rocks and the drillability of formations should be studied, and different types of drill bits should be designed and selected for the characteristics of different formations. In order to form thousands of meters of borehole, we are faced with the problem of borehole stability. Only by maintaining borehole stability can we ensure the smooth progress of drilling and subsequent oil and gas development. In field practice, in the past, borehole stability was mainly studied from the perspective of mud chemistry, and various mud systems and chemical additives were studied to facilitate borehole stability. These studies have played a role in reducing borehole collapse, but they have not been able to encompass all the mechanisms of borehole collapse, and therefore have not been able to analyze the problem comprehensively. This will be evident through the study of this book. Hydraulic fracturing is an important production enhancement measure in the oil and gas development, and fracturing engineering design is a typical rock mechanics engineering design problem, where the height, propagation length and orientation of the fracture can be determined through rock mechanics analysis. Good or bad fracturing construction plays a decisive role in the effect of increasing production. Sand production is another rock mechanics problem in the oil recovery process. For sandstone reservoirs with low strength, if the production pressure difference is too large, it will lead to shear damage in the reservoir, so that the sand particles which are separated from the rock skeleton will flow into the borehole and bring a series of difficulties to the oil production work. Such analysis of the sand production mechanism of the reservoir is the basic work of sand prevention research. In addition, maintaining the long-term stability of the casing is also closely related to the mechanical properties of the formation. In the salt paste layer, mud or shale layer, and reservoir there have been phenomena of casing deformation, squeezing and even wrong breakage due to the external load of the formation. Water injection is often used to improve recovery, and the flow of water in sandstone formations causes hydration of clay minerals, resulting in changes in rock porosity and permeability, which is also an area of rock mechanics research. In addition to this, there are also fluid-solid coupling problems such as the depletion of formation pressure and ground subsidence during the development of the reservoir.

In conclusion, the subject of application of rock mechanics in petroleum engineering contains very rich connotations and requires the hard work of researchers dedicated to this field of study to solve these problems more satisfactorily.

## 1.4 History of the Development of Rock Mechanics

Rock mechanics is an emerging discipline developed only in the twentieth century, and because of its relevance to many industrial departments, a great deal of human and material resources have been invested in extensive research in various countries. In Europe, it is worth introducing the Austrian rock mechanics Salzburg School. In 1951, J. Stini and L. Muller from Austria, together with experts and scholars from geology, mechanics, engineering and other disciplines, initiated and held the first international rock mechanics academic conference with the theme of rock mechanics in Salzburg, Austria. The relationship between geological factors such as structural surfaces of rock masses and the mechanical properties of rock masses was discussed. In 1962, scholars from all over the world, including those from different schools of thought, joined together with this organization to establish a worldwide organization (International Society of Rock Mechanics).

In the former Soviet Union, people were working on rock mechanics related to mines as early as the 1950s, for example, Schliniel, Baron and others. They generally based their research on the theory of mechanics of continuous media. In addition to universities with special research on rock mechanics and mining physics, research institutes and relevant industrial sectors in other republics also attach great importance to the study of rock mechanics problems.

The early research work of rock mechanics in the United States was carried out in combination with mining work. In 1965, the American Geophysical Union's Institute of Mining and Metallurgy, the Civil Engineering Society, the Materials Society, the Geological Society, and the Mining Society joined together to form the Academic Committee on Rock Mechanics, later renamed the Federal Committee on Rock Mechanics. The exchange and popularization of the achievements in various fields of rock mechanics further promoted the development of rock mechanics. They started with the first national academic conference on rock mechanics held at Colorado School of Mines in 1964, and have been held basically once a year since then.

In China, with the development of national economic construction after the founding of the People's Republic of China, some research institutions on rock mechanics were established one after another, such as the Institute of Rock and Soil Mechanics of Chinese Academy of Sciences, etc. Before 1966, only a few industrial departments paid more attention to the research work on rock mechanics, for example, hydropower, coal mining, metallurgy, etc. With the experience and lessons gained from engineering practice, more and more technical personnel in departments (such as construction, railway, petroleum, national defense, civil air defense, etc.) realize that they must fully understand the mechanical properties of rock in order to solve the technical problems of rock engineering. After 1976, many academic groups resumed their activities and some new academic organizations on rock mechanics were established. In 1978, the Chinese Academy of Sciences and the Ministry of Foreign Affairs jointly submitted to the State Council a report on "Application for Participation in the International Society of Rock Mechanics and Attendance at the Fourth Congress of the Society", which was approved by the State Council. Then, China officially established the National Group China of the International Society of Rock Mechanics, and joined the international society of rock mechanics as a member of the group. In September 1979, a 10-member Chinese delegation headed by Mr. Chen Zongji and Mr. Gu Dezhen attended the 4th International Society of Rock Mechanics in Montreux, Switzerland.

Mr. Chen Zongji gave presentations on the development of the Chinese rock mechanics discipline at the Council of the International Society of Rock Mechanics and at academic conferences, respectively, which had a wide international impact. In addition, under the initiative of Mr. Chen Zongji and Mr. Pan Jiazheng, the Preparatory Committee of the Chinese Society of Rock Mechanics and Engineering was established in 1982. After more than three years of unremitting efforts, a national level society was established in 1985: Chinese Society of Rock Mechanics and Engineering. The new prelude of rock mechanics research in China was unveiled. With the development of domestic rock mechanics discipline technology, China's position in the world rock mechanics community gradually increased, and from 2011 to 2015, Dr. Feng Xiating took over the position of President of the International Society of Rock Mechanics and Engineering, marking the position of China's rock mechanics community in the international community to a new height (FU 1998, 2003; Feng 2002).

In the petroleum engineering field, the scientific research of rock mechanics started late, and the laboratory of rock mechanics was prepared only after 1977. In China, the first laboratory of rock mechanics for petroleum engineering was built in 1980 in the department of Professor Huang Rongzun of East China Petroleum Institute (the predecessor of China University of Petroleum), and the related research work was carried out systematically. Firstly, During the period 1980–1985, the research on formation fracture pressure forecasting technology was carried out. During the period 1986–1990, a series of researches on casing damage mechanism and prevention, as well as the rheological law of compound salt paste layer were carried out. During the period 1991–1995, more extensive research work was carried out, such as in-situ stress determination technology, borehole stability technology, sand production mechanism, hydraulic fracturing and other research. Later, with the support of the national Project 211 construction funds, a systematic laboratory of rock mechanics for oil and gas engineering was set up by Professor Cheng Yuanfang. In addition to the Rock Mechanics Laboratory of China University of Petroleum, many research institutes of CNPC, SINOPEC and others have built laboratories specializing in rock mechanics. Since then, there has been a thriving research scene in the field of rock mechanics of oil and gas engineering in China.

## References

Brown ET. Analytical and computational methods in engineering rock mechanics. Australia: Allen & Unwin Ltd.; 1987.

Chen Z. Mechanical properties of rock and tectonic stress field. Beijing: Geological Publishing Press; 1986.

Feng X. Introduction to intelligent rock mechanics. Beijing: Science Press; 2002.

Fu B. Rock mechanics and engineering towards the new century. In: Zhou G, editor. Discipline development and technical progress. Beijing: Science and Technology Press of China; 1998.

Fu B. Current status and future of rock mechanics research. National Geotechnical and Engineering Academic Conference; 2003; Beijing.

Li X. Mechanical properties of rock block. Beijing: China Coal Industry Press; 1983.

You M. Mechanical properties of rocks. Beijing: Geological Publishing House; 2007.

Zhou W. Advanced rock mechanics. Beijing: Water Resources and Electric Power Press; 1990.

Zhou W, Yang Q. Numerical calculation methods for rock mechanics. Beijing: China Electric Power Press; 2005.

# Chapter 2
# Stress and Strain

## 2.1 Stress

### *2.1.1 Stress Vector*

Before studying the mechanical properties of rocks and their applications, the basic concepts of mechanics such as stress and strain should first be established (Yang 1980; Zhilun 1980).

When examining the external forces acting on a particular rock mass, the external forces can be divided into volume and surface forces, depending on their domain of action. The so-called volume forces are external forces distributed on the internal volume of the object, which, for a rock body, is generally gravity; the so-called surface forces are external forces acting on the surface of the rock body, such as concentrated forces, hydrostatic pressure of liquid, external collapse pressure of the casing, etc.

When a rock body is subjected to an external force, there will be forces interacting between different parts of its interior, i.e. internal forces. For a rock body in equilibrium, to study the internal forces at any point $P$ inside it, suppose that a plane $C$ past the point $P$ is used to truncate it into two parts $A$ and $B$. If part $B$ is removed, the action of $B$ on $A$ should be replaced by the distributed internal force. Examine the tiny area of the plane $C$ including the point $P$, as shown in Fig. 2.1a.

Let the external normal to the plane $C$ be $\boldsymbol{n}$, the area of the microplane be $\Delta S$, and the combined internal force acting on the microplane be $\Delta F$. The average internal force distribution on this microplane is $\Delta F/\Delta S$. It can be seen that the magnitude and distribution of the internal force varies from section to section; even in the same section, the distribution of the internal force at each point varies in magnitude and direction. In order to determine the distributed force at a point on an arbitrary section, the concept of stress will be introduced. Assuming that the rock mass is uniform and continuous, the internal forces should be distributed continuously across the cross section. For a microsurface area $\Delta S$, the average stress is



Y. Cheng et al., *Foundations of Rock Mechanics in Oil and Gas Engineering*,
https://doi.org/10.1007/978-981-99-1417-3_2

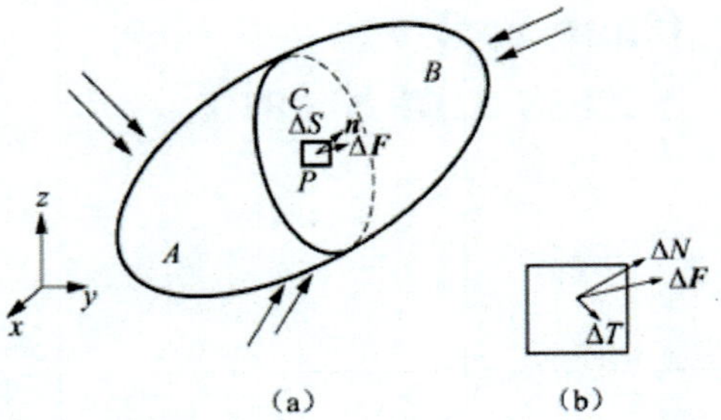


**Fig. 2.1** Definition of stress

$$\overline{F}_m = \frac{\Delta F}{\Delta S} \tag{2.1}$$

The average stress $\overline{F}_m$ is related to the size of the area taken, and in order to eliminate the effect of area to obtain the stress at that point, the concept of limit is used to approximate $\Delta S$ infinitely small to that point, i.e.

$$\overline{F} = \lim_{\Delta s \to 0} \frac{\Delta \overline{F}}{\Delta S} = \frac{\mathrm{d}\overline{F}}{\mathrm{d}S} \tag{2.2}$$

***F*** is called the full stress at point P on section $\Delta S$, which is a vector quantity.

Typically decomposing $\Delta \overline{F}$ into a component force $\Delta N$ perpendicular to the section, along the normal direction, and a component force $\Delta S$ parallel to the section in the tangential direction, as shown in Fig. 2.1b, then

$$\begin{aligned} \sigma_n &= \lim_{\Delta s \to 0} \frac{\Delta N}{\Delta S} \\ \tau_n &= \lim_{\Delta s \to 0} \frac{\Delta T}{\Delta S} \end{aligned} \tag{2.3}$$

$\sigma_n$ and $\tau_n$ are called the normal and shear stresses at point $P$ on section $C$, respectively, and their dimensions are [force][length]$^{-2}$, and the international unit is Pa i.e. N/m$^2$.

The rock is subjected to loading, and the stresses at internal points are generally different, and even at the same point the stress components vary with the orientation of the cross-section through that point. Therefore, the stresses are related not only to the location of the point under examination, but also to the orientation of the section taken at that point. The stress vectors acting on all the different external normal directions on the micro-plane at the same point constitute the stress state at that point.

### 2.1.2 Stress Tensor

For the convenience of the study, a micro hexahedron with one side of length d$x$, d$y$, and d$z$ along the $x$, $y$, and $z$ coordinate axes at any point within the rock mass is usually studied, as shown in Fig. 2.2. Normal and shear stresses exist in each section of the hexahedron. Generally, the shear stress is again decomposed into two components along the coordinate axes. For example, if the outer normal of a section is parallel to the $x$-axis, the normal stress acting on the section is denoted as $\sigma_x$; the two shear stresses on the section are denoted as $\tau_{xy}$ and $\tau_{xz}$, where the first footnote indicates outside normal direction of the section where the shear stress is located and the second footnote indicates the direction of the shear stress. For example, $\tau_{xy}$ means that the shear stress acts on the section normal to the $x$-axis, pointing in the $y$-axis direction. Thus there are nine stress components in the six sections of the micro-unit, three normal stresses $\sigma_x$, $\sigma_y$, $\sigma_z$; and six shear stress components $\tau_{xy}$, $\tau_{xz}$, $\tau_{yx}$, $\tau_{yz}$, $\tau_{zx}$, $\tau_{zy}$, which are not uncorrelated but are two equal to each other. For example, take the line segment connecting the centers of the front and rear sections as the moment axis, and for this axis take the distance $m_x$, which is known from the equilibrium conditions as

$$\sum m_x(\overline{F}) = 0$$

That is,

$$2\tau_{yz} \cdot \mathrm{d}x \cdot \mathrm{d}z \cdot \frac{\mathrm{d}y}{2} - 2\tau_{zy} \cdot \mathrm{d}x \cdot \mathrm{d}y \cdot \frac{\mathrm{d}z}{2} = 0 \tag{2.4}$$

Got it,

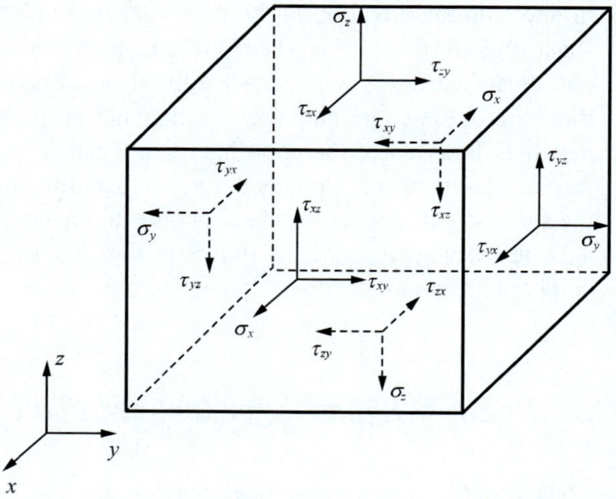


**Fig. 2.2** Stress state at any point

$$\tau_{yz} = \tau_{zy} \tag{2.4a}$$

Similarly,

$$\begin{aligned} \tau_{zx} &= \tau_{xz} \\ \tau_{zy} &= \tau_{yz} \end{aligned} \tag{2.4b}$$

It can be seen that only six of the nine stress components on the micro-unit are independent, namely $\sigma_x$, $\sigma_y$, $\sigma_z$ and $\tau_{xy}$, $\tau_{yz}$, $\tau_{zx}$. If these six stress components are known, then the stress components on any azimuthal section past that point can be found. Therefore, the stress components on the micro-unit near that point can represent **the stress state** at that point.

To express the stress state at a point, the stress components can be arranged according to certain rules so that each row represents the stress component of a section on a micro-unit, i.e.

$$\boldsymbol{\sigma}_{ij} = \begin{vmatrix} \sigma_x & \tau_{xy} & \tau_{xz} \\ \tau_{yx} & \sigma_y & \tau_{yz} \\ \tau_{zx} & \tau_{zy} & \sigma_z \end{vmatrix} \tag{2.5}$$

$\boldsymbol{\sigma}_{ij}$ is called the stress tensor, which is a second-order symmetric tensor, where $i, j = x, y, z$. The stress components on the corresponding cross-sections are then obtained, while the normal stresses $\sigma_{xx}$, $\sigma_{yy}$, $\sigma_{zz}$ are abbreviated as $\sigma_x$, $\sigma_y$, $\sigma_z$; the shear stresses $\sigma_{xy}$, $\sigma_{yz}$, $\sigma_{zx}$ are rewritten as $\tau_{xy}$, $\tau_{yz}$, $\tau_{zx}$.

For rock mechanics, the positive and negative signs of the stress components are specified as follows (Cai et al. 2002): normal stresses are positive for compressive stresses and negative for tensile stresses. The positive and negative signs of shear stress are specified as follows: when the direction of the outer normal of a section is in line with the direction of the coordinate axes (the section is called positive), the shear stress on that face is positive when it points in the negative direction of the other coordinate axes; it is negative when the shear stress on the face points in the positive direction of the coordinate axes. If the outer normal on a section is in line with the negative direction of the coordinate axis (called negative), the shear stress on the face is positive when it points to the positive direction of the other coordinate axes; the shear stress is negative when it points to the negative direction of the coordinate axis on the negative side, as shown in Fig. 2.2, the normal stress and shear stress components are all positive.

### *2.1.3 Stress Equation for Inclined Plane*

The stress components vary from point to point in the rock, even for the same point, the normal stress and shear stress on each azimuth section are also different. The

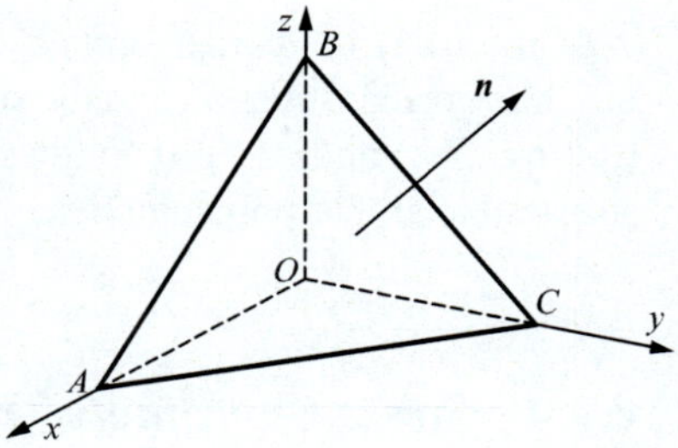


**Fig. 2.3** Relationships between stresses in each section of the differential body of a three-dimensional space unit

relationship between the stress components along the coordinate plane and the stress components in an arbitrary inclined section is studied below.

Assume that the six stress components $\sigma_x$, $\sigma_y$, $\sigma_z$, $\tau_{xy}$, $\tau_{yz}$ and $\tau_{zx}$ are known on the micro-unit body. Suppose an arbitrary inclined section ABC is intercepted in the micro-unit body, where the directional cosines between the outer normal $\boldsymbol{n}$ of the inclined section and the coordinate axes $x$, $y$ and $z$ are $l$, $m$ and $n$, respectively, and the area of the inclined triangular ΔABC is $\mathrm{d}S$, as shown in Fig. 2.3. Then the areas of ΔOBC, ΔOAC, and ΔOAB are $l\mathrm{d}S$, $m\mathrm{d}S$ and $n\mathrm{d}S$, respectively. Assume that the full stress of the inclined section ABC is $\overline{F}$, where the projections on the xyz axis are $F_x$, $F_y$, $F_z$, respectively.

According to the equilibrium condition $\sum F_x = 0, \sum F_y = 0, \sum F_z = 0$, it can be concluded that

$$\begin{aligned} F_x &= \sigma_x l + \tau_{yx} m + \tau_{zx} n \\ F_y &= \tau_{xy} l + \sigma_y m + \tau_{zy} n \\ F_z &= \tau_{xz} l + \tau_{yz} m + \sigma_z n \end{aligned} \tag{2.6}$$

Then projecting $F_x$, $F_y$, $F_z$ onto the $\boldsymbol{n}$-axis normal to the outside of this section gives the normal stress $\sigma_n$ on the inclined section ABC.

$$\sigma_n = F_x l + F_y m + F_z n \tag{2.7}$$

Bringing Eq. (2.6) into Eq. (2.7) yields

$$\sigma_n = \sigma_x l^2 + \sigma_y m^2 + \sigma_z n^2 + 2\tau_{xy} lm + 2\tau_{yz} mn + 2\tau_{zx} nl \tag{2.8}$$

Since the total stress in the inclined section is

$$F = \sqrt{F_x^2 + F_y^2 + F_z^2}$$

So the shear stress in the inclined section $\tau_n$ is

$$\tau_n = \sqrt{F^2 - \sigma_n^2} \tag{2.9}$$

If the six stress components $\sigma_x$, $\sigma_y$, $\sigma_z$, $\tau_{xy}$, $\tau_{yz}$ and $\tau_{zx}$ are known for the micro-unit, the normal stress $\sigma_n$ and the shear stress $\tau_n$ can be found for any inclined section by Eqs. (2.8) and (2.9). It follows that if the six stress components are known, the stress state at that point can be completely determined.

### 2.1.4 Stress Coordinate Transformation

The stress equation on the inclined section is derived in the *xyz* coordinate system with six known stress components of the micro-unit $\sigma_x$, $\sigma_y$, $\sigma_z$, $\tau_{xy}$, $\tau_{yz}$ and $\tau_{zx}$. The following discusses how to characterize the stress state at a point in the new coordinate system $x'y'z'$. For the convenience of the analysis, assume that the new coordinate system $x'$ axis is in line with the direction normal to the outside of the inclined section, and that the $y'$ and $z'$ axes are taken to be on the inclined section and perpendicular to each other, as shown in Fig. 2.4. The cosine of the direction between the new coordinate system and the original coordinate system at this point is shown in Table 2.1.

Then the normal stress in the direction normal to this inclined section $\sigma_{x'} = \sigma_n$, simply by rewriting *l*, *m* and *n* in Eq. (2.8) as $l_1$, $m_1$and, $n_1$, is given by

$$\sigma_{x'} = \sigma_x l_1^2 + \sigma_y m_1^2 + \sigma_z n_1^2 + 2\tau_{xy} l_1 m_1 + 2\tau_{yz} m_1 n_1 + 2\tau_{zx} n_1 l_1 \tag{2.10}$$

And the shear stresses pointing along the $y'$ axis in the inclined section $\tau_{x'y'}$ Simply replace *l*, *m* and *n* in $F_x$, $F_y$, and $F_z$ in Eq. (2.6) with $l_1$, $m_1$, $n_1$ and then decompose along the $y'$ axis to obtain

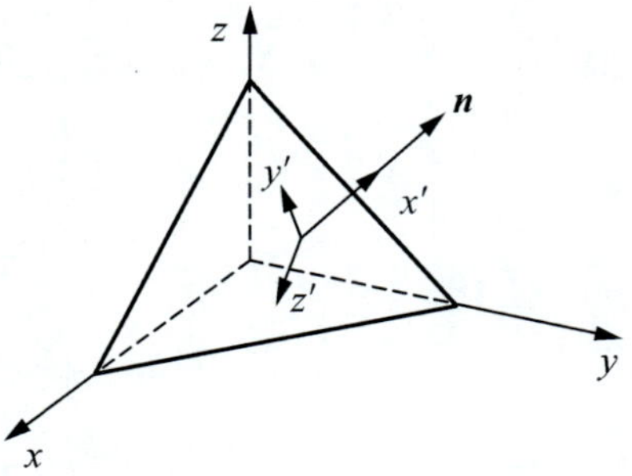


**Fig. 2.4** Relationship between the new coordinates $x'y'z'$ and the original coordinates *xyz*

**Table 2.1** Directional cosine of the new coordinates with respect to the original coordinates

| New coordinates | Original coordinates | | |
|---|---|---|---|
| | $x$ | $y$ | $z$ |
| $x'$ | $l_1$ | $m_1$ | $n_1$ |
| $y'$ | $l_2$ | $m_2$ | $n_2$ |
| $z'$ | $l_3$ | $m_3$ | $n_3$ |

$$\tau_{x'y'} = F_x l_2 + F_y m_2 + F_z n_2$$

or,

$$\begin{aligned}\tau_{x'y'} &= \sigma_x l_1 l_2 + \sigma_y m_1 m_2 + \sigma_z n_1 n_2 + \tau_{xy}(l_1 m_2 + l_2 m_1) \\ &\quad + \tau_{yz}(m_1 n_2 + m_2 n_1) + \tau_{zx}(n_1 l_2 + n_2 l_1)\end{aligned} \tag{2.11}$$

Similarly, decomposing $F_x$, $F_y$, and $F_z$ on the $z'$ axis yields

$$\begin{aligned}\tau_{x'z'} &= \sigma_x l_1 l_3 + \sigma_y m_1 m_3 + \sigma_z n_1 n_3 + \tau_{xy}(l_3 m_1 + l_1 m_3) \\ &\quad + \tau_{yz}(m_3 n_1 + m_1 n_3) + \tau_{zx}(n_3 l_1 + n_1 l_3)\end{aligned} \tag{2.12}$$

Equations (2.10), (2.11) and (2.12) are the stress components in the inclined section and the three stress components in the plane perpendicular to the $x'$ axis in the new coordinate system.

By analogy, the six stress components of the new coordinate system $x'y'z'$ can be found, and the result is written in matrix form as

$$\begin{Bmatrix}\sigma_{x'} \\ \sigma_{y'} \\ \sigma_{z'} \\ \tau_{x'y'} \\ \tau_{y'z'} \\ \tau_{z'x'}\end{Bmatrix} = \begin{bmatrix} l_1^2 & m_1^2 & n_1^2 & 2l_1 m_1 & 2m_1 n_1 & 2n_1 l_1 \\ l_2^2 & m_2^2 & n_2^2 & 2l_2 m_2 & 2m_2 n_2 & 2n_2 l_2 \\ l_3^2 & m_3^2 & n_3^2 & 2l_3 m_3 & 2m_3 n_3 & 2n_3 l_3 \\ l_1 l_2 & m_1 m_2 & n_1 n_2 & l_1 m + l_2 m_1 & m_1 n_2 + m_2 n_1 & n_1 l_2 + n_2 l_1 \\ l_2 l_3 & m_2 m_3 & n_2 n_3 & l_2 m_3 + l_3 m_2 & m_2 n_3 + m_3 n_2 & n_2 l_3 + n_3 l_2 \\ l_3 l_1 & m_3 m_1 & n_3 n_1 & l_3 m_1 + l_1 m_3 & m_3 n_1 + m_1 n_3 & n_3 l_1 + n_1 l_3 \end{bmatrix} \begin{Bmatrix}\sigma_x \\ \sigma_y \\ \sigma_z \\ \tau_{xy} \\ \tau_{yz} \\ \tau_{zx}\end{Bmatrix} \tag{2.13}$$

The above equation is the coordinate transformation equation for the stress component. By adding the first three rows of the above equation according to the directional cosine, we get,

$$\sigma_{x'} + \sigma_{y'} + \sigma_{z'} = \sigma_x + \sigma_y + \sigma_z = \text{constant} \tag{2.14}$$

It follows that the sum of the three normal stresses is not affected by the transformation of the coordinate system and remains constant.

### *2.1.5 Principal Stresses and Principal Directions*

Knowing the six stress components at a point, the stress components on any inclined section through that point can be found according to Eqs. (2.8) and (2.9). It can be proved that through the point there are always three special planes perpendicular to each other and with no shear stress through this point, the section is called the principal plane, the direction of its outer normal is called **the principal direction** of

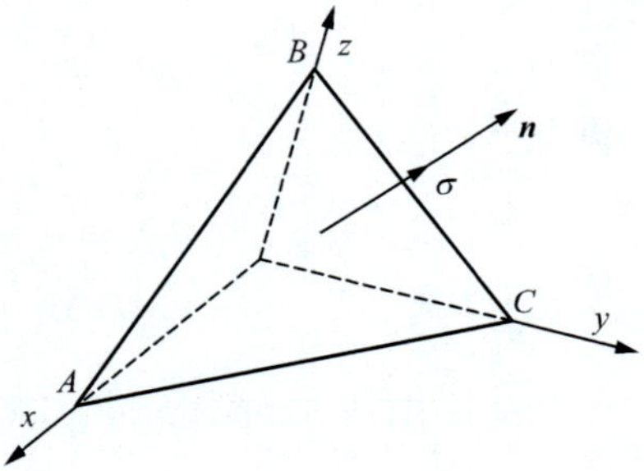


**Fig. 2.5** Stress relationship between micro-unit body and principal plane in 3D space

the point, the normal stress acting on the principal plane is called the principal stress (Sun 1983). The magnitude of the principal stresses on this face and their principal directions are analyzed below.

If the six stress components of the micro-unit are known and the *ABC* inclined section is assumed to be a principal plane, its full stress must be the principal stress $\sigma$, since the shear stress on the principal plane is zero, and the direction of the full stress coincides with the principal direction $\boldsymbol{n}$, as in Fig. 2.5. The projections of the principal stress $\sigma$ on the *x*, *y*, and *z* axes, respectively, are

$$\left.\begin{aligned} F_x &= \sigma l \\ F_y &= \sigma m \\ F_z &= \sigma n \end{aligned}\right\} \tag{2.15}$$

where *l*, *m*, and *n* are the directional cosines of the principal direction with respect to the *x*, *y*, *z* axes, respectively. Bringing Eq. (2.15) into Eq. (2.6) gives,

$$\left.\begin{aligned} \sigma l &= \sigma_x l + \tau_{yz} m + \tau_{zx} n \\ \sigma m &= \tau_{xy} l + \sigma_y m + \tau_{zy} n \\ \sigma n &= \tau_{xz} l + \tau_{yz} m + \sigma_z n \end{aligned}\right\} \tag{2.16}$$

To wit

$$\left.\begin{aligned} (\sigma_x - \sigma) l + \tau_{yz} m + \tau_{zx} n = 0 \\ \tau_{xy} l + (\sigma_y - \sigma) m + \tau_{zy} n = 0 \\ \tau_{xz} l + \tau_{yz} m + (\sigma_z - \sigma) n = 0 \end{aligned}\right\} \tag{2.17}$$

and

$$l^2 + m^2 + n^2 = 1 \tag{2.18}$$

Equations (2.17) and (2.18) are algebraic equations in four variables $\sigma$, *l*, *m*, and *n*. Solving for them yields the principal stress and principal direction.

Based on the knowledge of linear algebra, there are

$$\begin{vmatrix} (\sigma_x - \sigma) & \tau_{yz} & \tau_{zx} \\ \tau_{zy} & (\sigma_y - \sigma) & \tau_{zy} \\ \tau_{xz} & \tau_{yz} & (\sigma_z - \sigma) \end{vmatrix} = 0 \tag{2.19}$$

Expanding this determinant yields a cubic power equation about $\sigma$, i.e.

$$\begin{aligned} &\sigma^3 - (\sigma_x + \sigma_y + \sigma_z)\sigma^2 + (\sigma_x\sigma_y + \sigma_y\sigma_z + \sigma_z\sigma_x - \tau_{xy}^2 - \tau_{yz}^2 - \tau_{zx}^2)\sigma \\ &- (\sigma_x\sigma_y\sigma_z + 2\tau_{xy}\tau_{yz}\tau_{zx} - \sigma_x\tau_{yz}^2 - \sigma_y\tau_{zx}^2 - \sigma_z\tau_{xy}^2) = 0 \end{aligned} \tag{2.20}$$

The above equation can be written in the following form

$$\sigma^3 - I_1\sigma^2 + I_2\sigma - I_3 = 0 \tag{2.21}$$

Of which

$$\begin{aligned} I_1 &= \sigma_x + \sigma_y + \sigma_z \\ I_2 &= \sigma_x\sigma_y + \sigma_y\sigma_z + \sigma_z\sigma_x - \tau_{xy}^2 - \tau_{yz}^2 - \tau_{xz}^2 \\ I_3 &= \sigma_x\sigma_y\sigma_z + 2\tau_{xy}\tau_{yz}\tau_{zx} - \sigma_x\tau_{yz}^2 - \sigma_y\tau_{zx}^2 - \sigma_z\tau_{xy}^2 \end{aligned} \tag{2.22}$$

The three roots $\sigma_1$, $\sigma_2$, and $\sigma_3$ of $\sigma$ can be found from Eq. (2.20) or Eq. (2.21) and can be shown to be all real roots. This shows that there are indeed three principal stresses in the stress state of a point $\sigma_1$, $\sigma_2$ and $\sigma_3$, and $\sigma_1 > \sigma_2 > \sigma_3$. These three principal stresses are called the maximum, intermediate and minimum principal stresses respectively. By substituting the principal stresses back into Eq. (2.17), the cosine of the three directions corresponding to the principal stresses can be found, and the directions of the three principal stresses can be determined. It can be shown that these three principal directions are perpendicular to each other.

From the above analysis, if the stress state at a point has been determined, the principal stress and principal direction obtained by Eqs. (2.17) and (2.21) are always constant, regardless of how the coordinate axes *x*, *y*, and *z* are chosen. In other words, the principal stress magnitude and principal direction are independent of the choice of the coordinate system. Therefore, the coefficient in Eq. (2.21) must be constant, that is, it does not depend on the selection of coordinates, so the coefficients $I_1$, $I_2$, $I_3$ are called the first, second and third invariants of the stress tensor.

If we choose axes that coincide with the three principal directions, the stress tensor invariants can also be expressed in terms of principal stresses as follows

$$\left.\begin{aligned} I_1 &= \sigma_1 + \sigma_2 + \sigma_3 \\ I_2 &= \sigma_1\sigma_2 + \sigma_2\sigma_3 + \sigma_3\sigma_1 \\ I_3 &= \sigma_1\sigma_2\sigma_3 \end{aligned}\right\} \tag{2.23}$$

The physical meaning of the first invariant of the stress tensor $I_1$ is relatively clear, since the average stress at a point $\sigma_m$ is

$$\sigma_m = \frac{1}{3}(\sigma_1 + \sigma_2 + \sigma_3) = \frac{1}{3}\left(\sigma_x + \sigma_y + \sigma_z\right) = \frac{1}{3}I_1 \tag{2.24}$$

Therefore, the first invariant of the stress tensor $I_1$ is three times the average stress $\sigma_m$.

### 2.1.6 Maximum Shear Stress

If the principal stress and principal direction at a point in the rock mass have been determined, choose the axes *x*, *y* and *z* parallel to the principal stresses $\sigma_1$, $\sigma_2$, and $\sigma_3$, respectively, and suppose an inclined section *ABC* is intercepted with area *S* and direction cosine *l*, *m*, and *n*. The full stress in this section is $\overline{F}$, as shown in Fig. 2.6.

The projections of the full stress $\overline{F}$ on the *x*, *y*, and *z* coordinate axes are $F_x$, $F_y$ and $F_z$, respectively, and according to the equilibrium condition $\sum F_x = 0$, $\sum F_y = 0$ $\sum F_z = 0$, we get

$$F_x S = \sigma_1 S l$$
$$F_y S = \sigma_2 S m$$
$$F_z S = \sigma_3 S n$$

Projecting $F_x$, $F_y$ and $F_z$ onto the outer normal $\boldsymbol{n}$ of the inclined section yields the normal stresses $\sigma_n$ in the inclined section as

$$\sigma_n = F_x l + F_y m + F_z n = \sigma_1 l^2 \sigma_2 m^2 + \sigma_3 n^2 \tag{2.25}$$

The total stress in the inclined section is $F = \sqrt{F_x^2 + F_y^2 + F_z^2}$.

Thus, the shear stress in the inclined section $\tau_n$ is

$$\tau_n^2 = F^2 - \sigma_n^2 = F_x^2 + F_y^2 + F_z^2 - \sigma_n^2 \tag{2.26}$$

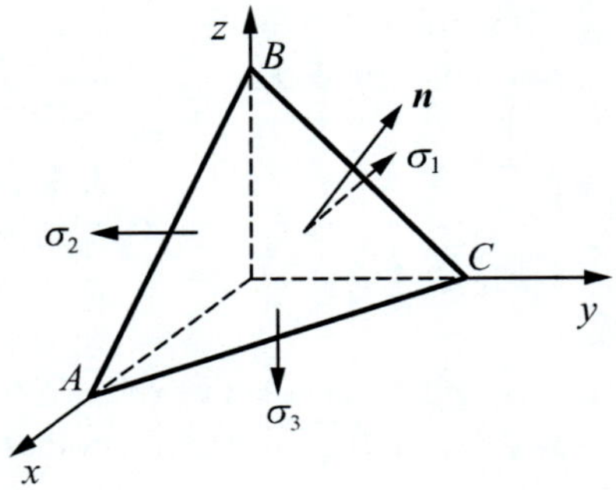


**Fig. 2.6** Plot of principal stress versus maximum shear stress

To wit

$$\tau_n^2 = \sigma_1^2 l^2 + \sigma_2^2 m^2 \sigma_3^2 n^2 - \left(\sigma_1 l^2 + \sigma_2 m^2 + \sigma_3 n^2\right)^2 \tag{2.27}$$

Using $l^2 + m^2 + n^2 = 1$, eliminating one of the three directional cosines, e.g. eliminating n, gives

$$\tau_n^2 = \sigma_1^2 l^2 + \sigma_2^2 m^2 + \sigma_3^2\left(1 - l^2 - m^2\right) - \left[\sigma_1 l^2 + \sigma_2 m^2 + \sigma_3\left(1 - l^2 - m^2\right)\right]^2 \tag{2.28}$$

Taking the partial derivatives of $\tau_n^2$ with respect to $l$ and $m$ respectively and making them zero, the extreme value of the shear stress at any point of the inclined section can be found as

$$\frac{\partial \tau_n^2}{\partial l} = 0, \quad l\left[(\sigma_1 - \sigma_2)l^2 + (\sigma_2 - \sigma_3)m - \frac{1}{2}(\sigma_1 - \sigma_2)\right] = 0$$
$$\frac{\partial \tau_n^2}{\partial m} = 0, \quad m\left[(\sigma_1 - \sigma_3)l^2 + (\sigma_2 - \sigma_3)m^3 - \frac{1}{2}(\sigma_2 - \sigma_3)\right] = 0$$

Based on the two equations above, the three sets $l$ and $m$ can be found, i.e.

- $l = 0, m = 0$
- $l = 0, m = \pm\frac{1}{\sqrt{2}}$
- $l = \pm\frac{1}{\sqrt{2}}, m = 0$

Substitute $l$, $m$ above into $l^2 + m^2 + n^2 = 1$ to find the corresponding $n$. Then substitute $l$, $m$ and $n$ back into Eq. (2.28) to find the extreme values of $\tau_n$. Similarly the solutions for the other three sets of very large and very small shear stresses can be found, as shown in Table 2.2.

From the table it can be seen that the last three groups are the extreme values of shear stress we require, viz.

$$\left.\begin{aligned} \tau_1 &= \pm\tfrac{1}{2}(\sigma_2 - \sigma_3) \\ \tau_2 &= \pm\tfrac{1}{2}(\sigma_3 - \sigma_1) \\ \tau_3 &= \pm\tfrac{1}{2}(\sigma_1 - \sigma_3) \end{aligned}\right\} \tag{2.29}$$

**Table 2.2** Extreme values of $\tau_n$

| | Group I | Group II | Group III | Group IV | Group V | Group VI |
|---|---|---|---|---|---|---|
| $l$ | 0 | 0 | $\pm 1$ | 0 | $\pm 1/\sqrt{2}$ | $\pm 1/\sqrt{2}$ |
| $m$ | 0 | $\pm 1$ | 0 | $\pm 1/\sqrt{2}$ | 0 | $\pm 1/\sqrt{2}$ |
| $n$ | $\pm 1$ | 0 | 0 | $\pm 1/\sqrt{2}$ | $\pm 1/\sqrt{2}$ | 0 |
| $\tau_n$ | 0 | 0 | 0 | $\pm(\sigma_2 - \sigma_3)/2$ | $\pm(\sigma_3 - \sigma_1)/2$ | $\pm(\sigma_1 - \sigma_2)/2$ |

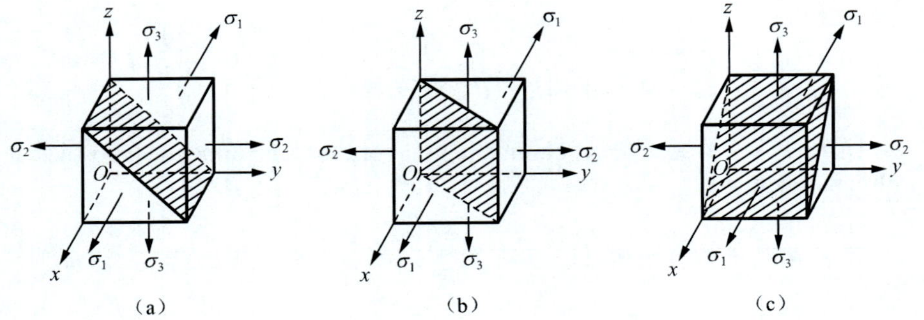


**Fig. 2.7** Cross section for main shear stress

If $\sigma_1 > \sigma_2 > \sigma_3$, then the maximum shear stress is

$$\tau_{\max} = \frac{1}{2}(\sigma_1 - \sigma_3) \tag{2.30}$$

This shows that the maximum shear stress is half the difference between the maximum principal stress and the minimum principal stress, and its direction cosine is $\pm\, 1/\sqrt{2}, 0, \pm\, 1/\sqrt{2}$; the cross section on which it acts is shown in Fig. 2.7c. The other two principal shear stresses have their action planes shown in Fig. 2.7a, b. They are each in a cross section at an angle of 45° in the principal direction. In addition to the main shear stresses, there are also normal stresses in this section, which are given by Eq. (2.25) as, $(\sigma_2 - \sigma_3)/2$, $(\sigma_3 - \sigma_1)/2$ and $(\sigma_1 - \sigma_2)/2$.

When $\sigma_1 = \sigma_2 = \sigma_3 = \sigma$, the principal shear stress is zero regardless of the values of $l$, $m$ and $n$ and the normal stress is $\sigma$. This indicates that there are only equal normal stresses and no shear stresses in the cross section in any direction. This situation is usually referred to as a **hydrostatic stress state**. In this stress state, only volumetric strain can be induced in the micrometeoroid without any change in shape.

### *2.1.7 Stress Mohr Circle*

The stress state at a point in the rock mass in three dimensions can also be expressed in a planar graph using stress circles (Reismann and Pawlik 1980; Goodman 1980). For example, if a micro-unit is taken out of a rock mass and the principal stresses act on each section $\sigma_1, \sigma_2, \sigma_3$, if the $x$, $y$ and $z$ axes are taken to coincide with the three principal stress directions, respectively, as shown in Fig. 2.6. Then the directional cosines of the outer normal of any inclined section are $l$, $m$, and $n$. The normal and shear stresses can be written as

$$\begin{aligned}\sigma_n &= \sigma_1 l^2 \sigma_2 m^2 + \sigma_3 n^2 \\ \tau_n^2 &= \sigma_1^2 l^2 + \sigma_2^2 m^2 + \sigma_3^2 n^2 - \sigma_n^2\end{aligned}$$

Based on the geometric relations, there are

$$l^2 + m^2 + n^2 = 1$$

It can be solved for

$$\begin{aligned} l^2 &= \frac{\tau_n^2 + (\sigma_n - \sigma_2)(\sigma_n - \sigma_3)}{(\sigma_1 - \sigma_2)(\sigma_2 - \sigma_3)} \\ m^2 &= \frac{\tau_n^2 + (\sigma_n - \sigma_1)(\sigma_n - \sigma_3)}{(\sigma_2 - \sigma_1)(\sigma_2 - \sigma_3)} \\ n^2 &= \frac{\tau_n^2 + (\sigma_n - \sigma_1)(\sigma_n - \sigma_2)}{(\sigma_3 - \sigma_1)(\sigma_3 - \sigma_2)} \end{aligned} \tag{2.31}$$

If $l$, $m$, and $n$ are real numbers, the right end of the above equal sign must be positive, and under the conditions $\sigma_1 > \sigma_2 > \sigma_3$, the first equation of Eq. (2.31) has a denominator greater than zero, so its numerator

$$\tau_n^2 + (\sigma_n - \sigma_2)(\sigma_n - \sigma_3) \geq 0 \tag{2.32}$$

If the right-angle coordinate system is taken in the stress plane with $\sigma$ as the horizontal axis and $\tau$ as the vertical axis. When $l = 0$, this is the set of inclined sections parallel to $\sigma_1$.

$$\tau_n^2 + (\sigma_n - \sigma_2)(\sigma_n - \sigma_3) = 0 \tag{2.33}$$

It is also the equation of the stress circle with the center of the circle on the transverse axis. The stress circles can be drawn by $\sigma_2$ and $\sigma_3$. Analyzing the second and third equations of Eq. (2.31) similarly, two stress circles can be obtained, and they represent the stress states on two sets of inclined sections parallel to $\sigma_2$ and $\sigma_3$. These three circles are called Mohr stress circles, so that the stress state is represented by three stress circles, and the coordinates of any point $D$ in the region surrounded by these three stress circles (shaded part in Fig. 2.8) represent the stress states on the micro-unit not parallel to $\sigma_1$, $\sigma_2$ and $\sigma_3$ on any inclined section.

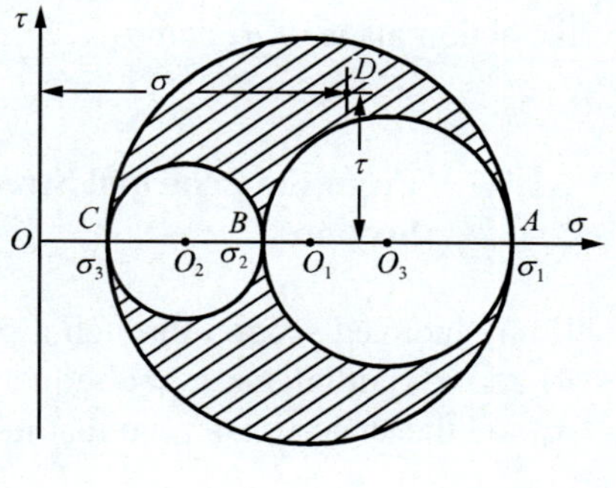


**Fig. 2.8** Mohr stress circle for the 3D stress state

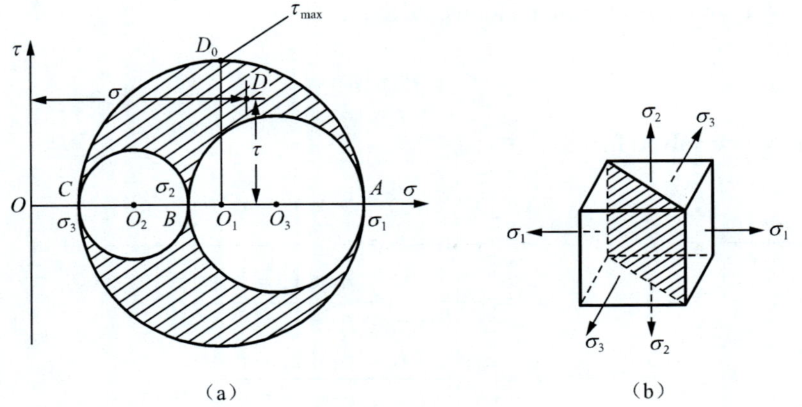


**Fig. 2.9** Mohr stress circle for the three-way stress state

#### 2.1.7.1 Maximum Shear Stress from Principal Stress

If the principal stress meets $\sigma_1 > \sigma_2 > \sigma_3$, first consider the set of inclined sections parallel to $\sigma_2$, such as the shaded section shown in Fig. 2.9b. It is not difficult to understand that since the intermediate principal stress $\sigma_2$ is parallel to that inclined section, it is independent of the normal and shear stresses in that inclined section. It depends only on the maximum principal stress $\sigma_1$ and the minimum principal stress $\sigma_3$, so for this inclined section, it can be reduced to a two dimensional stress state, and the stresses in this set of inclined sections can be determined from the corresponding points on the stress circles drawn by $\sigma_1$ and $\sigma_3$. Similarly, the stress circles parallel to $\sigma_1$ and $\sigma_3$ can be determined from the corresponding points on the stress circles drawn by $\sigma_2$, $\sigma_3$ and $\sigma_1$, $\sigma_2$, resulting in the three stress circles shown in Fig. 2.9a.

It is easy to see from the figure that the vertical coordinate of the highest point $D_0$ is the maximum shear stress in all inclined sections, and its magnitude is equal to the radius of the stress circle drawn by $\sigma_1$ and $\sigma_3$, i.e.

$$\tau_{\max} = \frac{1}{2}(\sigma_1 - \sigma_2) \tag{2.34}$$

The action plane is parallel to the intermediate principal stress and is at $\pm\, 45°$ to the action plane of $\sigma_1$ and $\sigma_3$.

#### 2.1.7.2 From the Principal Stress to Find the Stress in Any Inclined Section

If any inclined section through a point that is neither parallel to $\sigma_1$ nor parallel to $\sigma_2$ and $\sigma_3$, but intersects it obliquely, the stress point representing the normal stress $\sigma_n$ and the shear stress $\tau_n$ in that inclined section must lie within the circumference divided by the three stress circles shown in Fig. 2.9a, as indicated by a point $D$ in the shaded region in Fig. 2.9a.

The coordinates of this point represent the normal and shear stresses in a certain inclined section, which is shown in Fig. 2.10a. The stress circle solution method is now used to find the stresses in this inclined section.

The graphing steps are as follows (Fig. 2.10b)

(1) Select $\sigma$ and $\tau$ as the horizontal and vertical axes of the right-angle coordinate system on the stress plane, respectively, and measure $OA_1 = \sigma_1$, $OA_2 = \sigma_2$, $OA_3 = \sigma_3$ on the $\sigma$ axis at a certain scale to obtain the points $A_1$, $A_2$ and $A_3$, respectively.
(2) Using $A_2A_3$, $A_1A_2$, and $A_1A_3$ as diameters and each with its point $O_1$, $O_2$, $O_3$ as the center, draw three stress circles $K_{23}$, $K_{12}$, and $K_{13}$.
(3) Make a vertical line of $\sigma$ axis at points $A_1$ and $A_3$, i.e. $A_1T_1$, $A_3T_3$ respectively.
(4) A line made at an angle $\alpha$ (measured in a counterclockwise direction) to the line $A_1T_1$ from point $A_1$ intersects the great circle $K_{13}$ at point $A$. A line made at an angle $\gamma$ (measured in a clockwise direction) to $A_3T_3$ from point $A_3$ intersects the great circle $K_{13}$ at point $B$.
(5) With $O_3$ as the center and $O_3B$ as the radius, make a $BC$ arc; with $O_1$ as the center and $O_1A$ as the radius, make an $AC$ arc and intersect the two arcs at point $C$. The coordinates of point $C$ represent the normal and shear stresses in this inclined section, respectively.

If the three-dimensional stress state is simplified to a two-dimensional stress state, either inclined section with the principal stress as the coordinate axis is simplified to a diagonal straight line, as shown in Fig. 2.11. The directional cosine of the external normal $\boldsymbol{n}$ to $\sigma_1$ is $l$ (the directional angle is $\alpha$) and the directional cosine to $\sigma_2$ is $m$. The normal stress acting on the inclined section is $\sigma_n$ and the shear stress is $\tau_n$. The combined force is $\boldsymbol{F}$.

According to the conditions for the equilibrium of forces, $\sum F_x = 0$, $\sum F_y = 0$, resulting in, $F_x = \sigma_1 l$, $F_y = \sigma_2 m$.

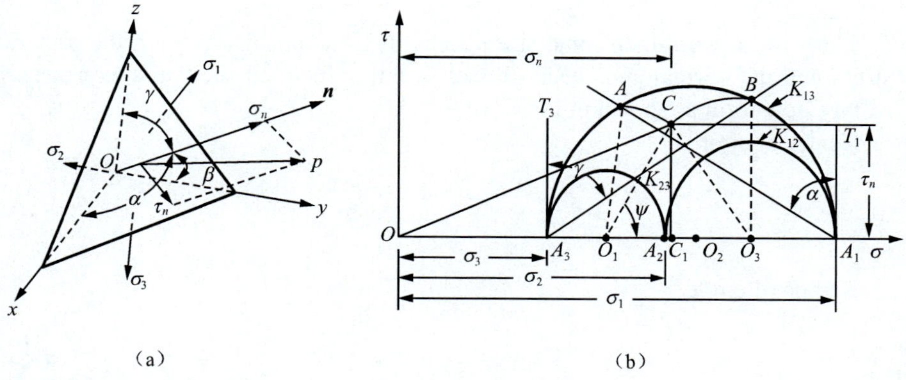


**Fig. 2.10** Using stress circles to find the stress in any inclined section

**Fig. 2.11** Inclined section stresses in a plane stress state

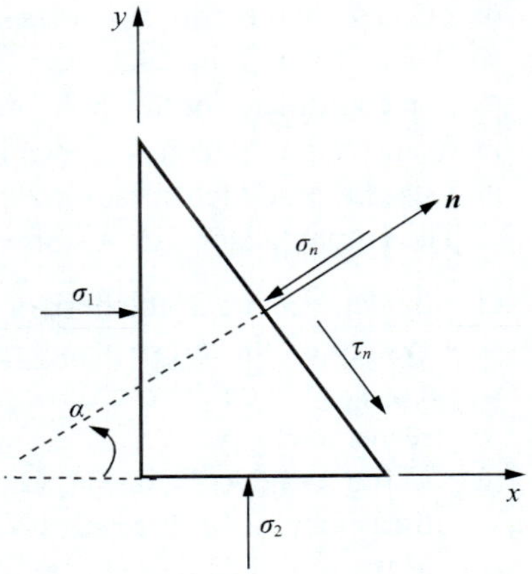


Projecting $F_x$, $F_y$ onto the direction normal to the outer inclined section yields the normal stresses in the inclined section $\sigma_n$ as

$$\sigma_n = F_x l + F_y m = \sigma_1 l^2 + \sigma_2 m^2 = \sigma_1 \cos^2 \alpha + \sigma_2 \sin^2 \alpha \tag{2.35}$$

Therefore, the shear stress in the inclined section is obtained from the relationship between the full force and the partial force as

$$\tau_n^2 = F^2 - \sigma_n^2 = F_x^2 + F_y^2 - \sigma_n^2 \tag{2.36}$$

Further collation gives

$$\sigma_n = \frac{1}{2}(\sigma_1 + \sigma_2) + \frac{1}{2}(\sigma_1 - \sigma_2)\cos 2\alpha \tag{2.37}$$

$$\tau_n = lm(\sigma_1 - \sigma_2) = -\frac{1}{2}(\sigma_1 - \sigma_2)\sin 2\alpha \tag{2.38}$$

Thus, for a two-dimensional stress state, the relationship between the principal stress and the normal and shear stresses in either inclined section is expressed in Mohr circle form as shown in Fig. 2.12.

Center of circle

$$H = \frac{1}{2}(\sigma_1 + \sigma_2) \tag{2.39}$$

Radius of circle

$$R = \frac{1}{2}(\sigma_1 - \sigma_2) \tag{2.40}$$

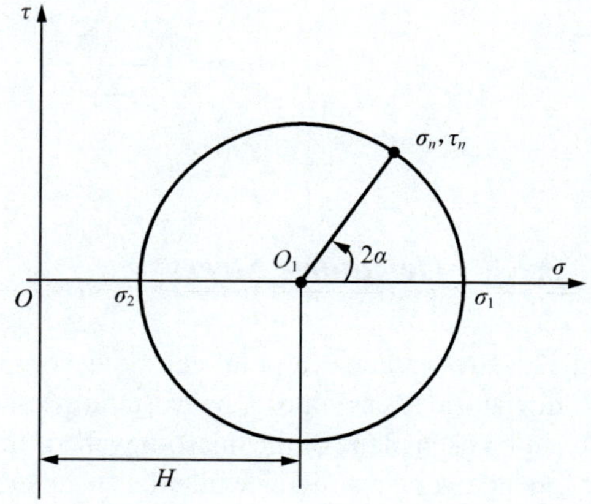


**Fig. 2.12** Mohr circle in two-dimensional stress state

If the general stress states $\sigma_x$, $\sigma_y$, and $\tau_{xy}$ are known, the principal stresses $\sigma_1$ and $\sigma_2$ can be obtained by the graphing method, as shown in Fig. 2.13. The graphing procedure is as follows.

- Select $\sigma$ and $\tau$ as the horizontal and vertical axes of the right-angle coordinate system on the stress plane, respectively, and measure $\sigma_x$ and $\sigma_y$ on the $\sigma$ axis at a certain scale and draw the vertical lines $A_1A_2$ and $B_1B_2$, respectively.
- Measure $\tau_{xy}$ on $A_1A_2$ (line $AA_2$) and $-\tau_{xy}$ on $B_1B_2$ (line $BB_2$).
- Connect the two points $A$ and $B$, which is the diameter of Mohr circle in this stress state, and the intersection $O_1$ of $AB$ and the horizontal axis $\sigma$ is the location of the center of the circle, that is, to obtain this Mohr circle.
- The intersections of the Mohr circle on the transverse axis are the two points where the two principal stresses $\sigma_1$ and $\sigma_2$ are located.

From Fig. 2.13, it can be seen that the maximum shear stress in any stress state is the radius of the Mohr circle and is related to the principal stress as follows

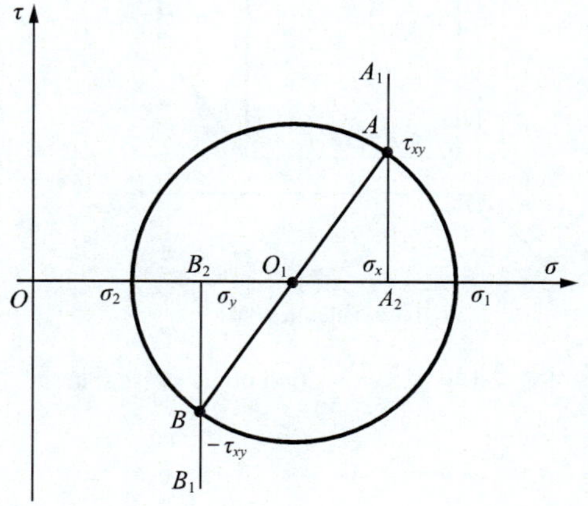


**Fig. 2.13** Moore's circle for any two-dimensional stress state

$$\sigma_{1,2} = \frac{\sigma_x + \sigma_y}{2} \pm \sqrt{\left(\frac{\sigma_x - \sigma_y}{2}\right)^2 + \tau_{xy}^2} \tag{2.41}$$

### 2.1.8 Deviatoric Stress

The stress state at a point can be decomposed into a hydrostatic stress state and a deviatoric stress state. A hydrostatic stress state is one in which only normal stresses act on each plane of the micro-hexahedron, and the magnitude of the normal stresses are all average stresses while the shear stresses are zero, as shown in Fig. 2.14a.

$$\sigma_0 = \frac{1}{3}(\sigma_1 + \sigma_2 + \sigma_3) \tag{2.42}$$

The deviatoric stress state is the portion remaining after hydrostatic stresses are deducted from the stress state, as shown in Fig. 2.14b.

#### 2.1.8.1 Characteristics of Hydrostatic Stress State

For a hydrostatic stress state, each face is the principal plane, and the normal stress $\sigma_0$ on each face is the principal stress, with the corresponding Mohr circle degenerating to a point on the $\sigma$ axis. Thus, hydrostatic stress is a stress state in which the stresses are the same on each face.

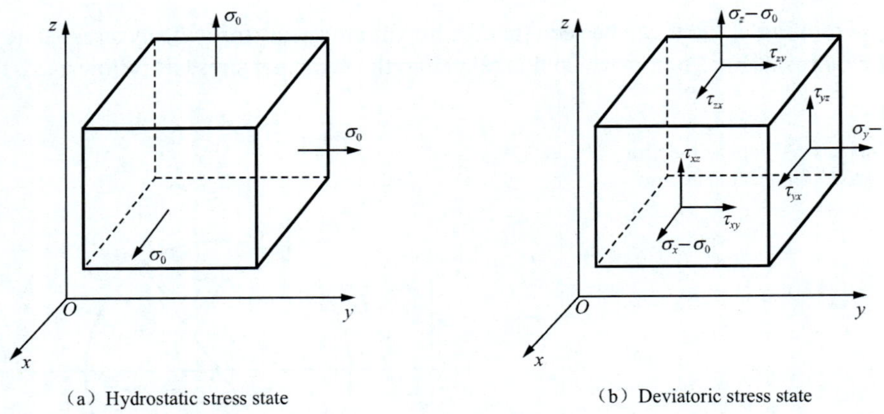


**Fig. 2.14** Decomposition of the stress state

### 2.1.8.2 Characteristics of the Deviatoric Stress State

Replacing the stress principal in Eq. (2.21) with the deviatoric stress principal yields the characteristic equation for the deviatoric stress principal as

$$s^3 - J_1 s^2 - J_2 s - J_3 = 0 \tag{2.43}$$

where

$$\begin{aligned} J_1 &= \sigma_x - \sigma_0 + \sigma_y - \sigma_0 + \sigma_z - \sigma_0 = 0 \\ J_2 &= \frac{1}{6}\left[\left(\sigma_x - \sigma_y\right)^2 + \left(\sigma_y - \sigma_z\right)^2 + (\sigma_z - \sigma_x)^2 + 6\left(\tau_{xy}^2 + \tau_{yz}^2 + \tau_{zx}^2\right)\right] \\ J_3 &= s_x s_y s_z + 2 s_{xy} s_{yz} s_{zx} - s_x s_{yz}^2 - s_y s_{zx}^2 - s_z s_{xy}^2 \end{aligned} \tag{2.44}$$

They are the 3 invariants of the deviatoric stress, where the first invariant is zero and the second invariant is the most used. Some invariants related to $J_2$ will be described later.

Solving the characteristic Eq. (2.43) for the deviatoric stress yields the three principal values of the deviatoric stress, $s_1$, $s_2$, and $s_3$, as

$$\begin{aligned} s_1 &= \frac{2\sqrt{J_2}}{\sqrt{3}} \sin\left(\theta_\sigma + \frac{2\pi}{3}\right) \\ s_2 &= \frac{2\sqrt{J_2}}{\sqrt{3}\sin\theta_\sigma} \\ s_3 &= \frac{2\sqrt{J_2}}{\sqrt{3}} \sin\left(\theta_\sigma - \frac{2\pi}{3}\right) \end{aligned} \tag{2.45}$$

where, $\theta_\sigma$ is called the Lode angle and is

$$\theta_\sigma = \frac{1}{3}\sin^{-1}\left[\frac{-\sqrt{27}J_3}{2(J_2)^{3/2}}\right] \tag{2.46}$$

It is known from the characteristics of the hydrostatic stress state that the principal direction of the deviatoric stress tensor coincides with the principal direction of the stress tensor, and that their principal values have the following relationship

$$\begin{aligned} s_1 &= \sigma_1 - \sigma_0 \\ s_2 &= \sigma_2 - \sigma_0 \\ s_3 &= \sigma_3 - \sigma_0 \end{aligned} \tag{2.47}$$

### 2.1.9 Equilibrium Differential Equations

If the rock mass is considered as a continuous medium, the points within the rock mass will produce different stress states and corresponding displacement and strain states when subjected to loading at the boundary in equilibrium. In other words the stress, displacement and strain components are all single-valued continuous functions of the spatial coordinate *xyz*. Suppose a unit differential is taken out within the rock mass (as in Fig. 2.2), and the axes of Cartesian coordinate system are taken to coincide with its sides, with side lengths d*x*, d*y*, and d*z*, respectively, and the stress components on the $x = 0$ plane of the differential are $\sigma_x$, $\tau_{xy}$, and $\tau_{zx}$; the stress components on the $x = \mathrm{d}x$ plane have a small amount of variation, omitting the higher-order terms, and are as follows,

$$\sigma_x + \frac{\partial \sigma_x}{\partial x}\mathrm{d}x,\ \tau_{xy} + \frac{\partial \tau_{xy}}{\partial x}\mathrm{d}x,\ \tau_{zx} + \frac{\partial \tau_{zx}}{\partial x}\mathrm{d}x.$$

The remaining stress components, all of which can be introduced in a similar way as described above. The unit differential of the stress components shown in Fig. 2.2 is thus obtained. Using *XYZ* to denote the components of the unit volume force vector along the *x*, *y*, and *z* directions, and considering the equilibrium of the unit differential along the *x* direction, we obtain

$$\begin{aligned}&\left(\sigma_x + \frac{\partial \sigma_x}{\partial x}\mathrm{d}x\right)\mathrm{d}y\mathrm{d}z - \sigma_x\mathrm{d}y\mathrm{d}z + \left(\tau_{yx} + \frac{\partial \tau_{yx}}{\partial y}\mathrm{d}y\right)\mathrm{d}x\mathrm{d}z - \tau_{yx}\mathrm{d}x\mathrm{d}z\\&\quad + \left(\tau_{zx} + \frac{\partial \tau_{zx}}{\partial z}\mathrm{d}z\right)\mathrm{d}x\mathrm{d}y - \tau_{zx}\mathrm{d}x\mathrm{d}y + X\mathrm{d}x\mathrm{d}y\mathrm{d}z = 0\end{aligned} \tag{2.48}$$

Rectifying the above equation and dividing by the volume of the unit differential d*x*d*y*d*z* yields

$$\frac{\partial \sigma_x}{\partial x} + \frac{\partial \tau_{yx}}{\partial y} + \frac{\partial \tau_{zx}}{\partial z} + X = 0 \tag{2.49a}$$

Similarly, establishing the equilibrium conditions in the *y* and *z* directions, it follows that

$$\begin{aligned}&\frac{\partial \sigma_y}{\partial y} + \frac{\partial \tau_{xy}}{\partial x} + \frac{\partial \tau_{zy}}{\partial z} + Y = 0\\&\frac{\partial \sigma_z}{\partial z} + \frac{\partial \tau_{xz}}{\partial x} + \frac{\partial \tau_{yz}}{\partial y} + Z = 0\end{aligned} \tag{2.49b}$$

Equations (2.49a, 2.49b) is the equilibrium differential equation for a continuous medium.

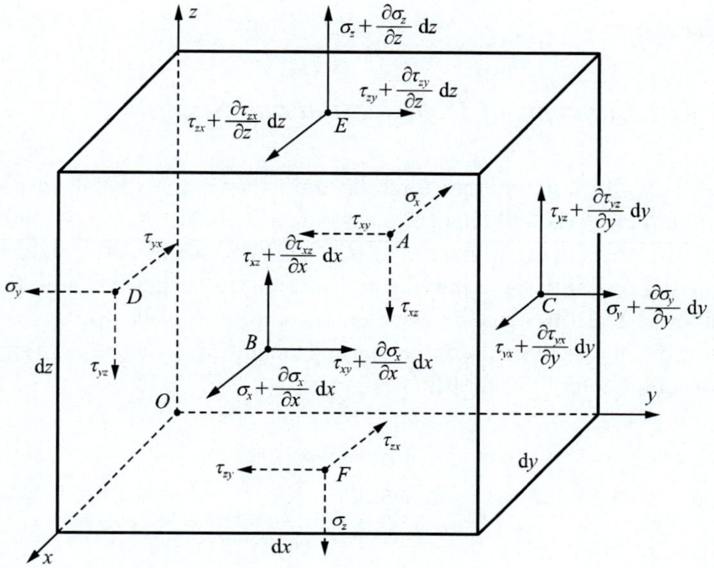


**Fig. 2.15** Equilibrium of a unitary differential body

Examine the moment balance of the unit differential body in Fig. 2.15. Taking the moment for the axis parallel to the z-direction through the form center point, both the stress and body force components along the reaction line through the form center point or parallel to the z-axis have zero moment to that axis, so the moment balance equation is

$$\left(\tau_{xy}\mathrm{d}y\mathrm{d}z\right)\mathrm{d}x-\left(\tau_{yx}\mathrm{d}x\mathrm{d}z\right)\mathrm{d}y=0 \tag{2.50}$$

This leads to:

$$\tau_{xy}=\tau_{yx} \tag{2.51a}$$

Similarly, taking moments for the axes parallel to the $x$ and $y$ directions through the point at the center of the form, we get

$$\begin{aligned}\tau_{yz}&=\tau_{zy}\\ \tau_{xz}&=\tau_{zx}\end{aligned} \tag{2.51b}$$

This is the principle of reciprocal shear stress.

## 2.2 Strain

### 2.2.1 *The Concepts of Deformation and Strain*

Under the action of an external force or temperature, the spatial position of each point in the rock mass will change, that is, displacement. As shown in Fig. 2.16, a fixed set of Cartesian coordinates *xyz* is established. Consider a material point *A*, whose spatial position in the initial state is *A*, and it reaches $A'$ under the action of external force. The difference between the two positions is the displacement of point *A*. Assume that the medium is continuous and that the displacement is a single-valued continuous function of the coordinates *xyz*, i.e.

$$\begin{aligned} u &= u(x, y, z) \\ v &= v(x, y, z) \\ w &= w(x, y, z) \end{aligned} \tag{2.52}$$

In general, the displacements at each point in the rock mass are different, and the set of displacement vectors at each point defines the displacement field of the rock mass. After displacement, the size and shape of the rock mass will change, except for rigid body translation and rotation, because the displacement at each point within it is generally not the same, and this change is called deformation, which includes volume deformation and shape distortion.

To describe the deformation of a point in a rock, an arbitrary tiny line segment past this point, the line element, is examined, and the relative change in length and relative change in direction of the line element before and after deformation are the key quantities. For example, the line element *AB* past point *A* in Fig. 2.16 has length $l_0$ before deformation and length $l$ after deformation, and the relative change in length of the line element before and after deformation is defined as the normal strain.

$$\varepsilon = \frac{l - l_0}{l_0} \tag{2.53a}$$

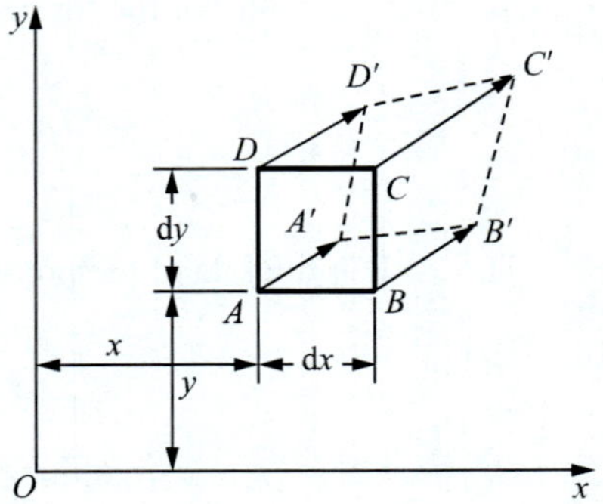


**Fig. 2.16** Deformation of plane differential element

To describe the relative change of direction of line element *AB*, another line element *AD* passing through the same point and perpendicular to line element *AB* is used as a reference, and the angle between the two line elements after deformation is $\alpha$, and the change in angle before and after deformation is defined as the shear strain $\gamma$,

$$\gamma = \frac{\pi}{2} - \alpha \tag{2.53b}$$

The strain state at a given point can be characterized by the normal strain on the three edges of the micro-unit and the shear strain of the included angle change. That is $\varepsilon_x, \varepsilon_y, \varepsilon_z, \gamma_{xy}, \gamma_{yx}, \gamma_{yz}, \gamma_{zy}, \gamma_{zx}, \gamma_{xz}$. It is easily seen from the geometric relationship that $\gamma_{xy} = \gamma_{yx}, \gamma_{yz} = \gamma_{zy}, \gamma_{xz} = \gamma_{zx}$. If the nine strain components of the strain state at a point within the rock mass are known, they can be determined by the strain components in any direction at that point. Similarly to the stress tensor, the strain state at a point can be expressed in terms of the strain tensor as follows

$$\varepsilon_{ij} = \begin{vmatrix} \varepsilon_x & \frac{1}{2}\gamma_{xy} & \frac{1}{2}\gamma_{xz} \\ \frac{1}{2}\gamma_{yx} & \varepsilon_y & \frac{1}{2}\gamma_{yz} \\ \frac{1}{2}\gamma_{zx} & \frac{1}{2}\gamma_{zy} & \varepsilon_z \end{vmatrix} \tag{2.54}$$

If let $\varepsilon_{xy} = \frac{1}{2}\gamma_{xy}$, $\varepsilon_{yz} = \frac{1}{2}\gamma_{yz}$, $\varepsilon_{zx} = \frac{1}{2}\gamma_{zx}$, then the strain tensor can also be expressed as

$$\varepsilon_{ij} = \begin{vmatrix} \varepsilon_x & \varepsilon_{xy} & \varepsilon_{xz} \\ \varepsilon_{yx} & \varepsilon_y & \varepsilon_{yz} \\ \varepsilon_{zx} & \varepsilon_{zy} & \varepsilon_z \end{vmatrix} \tag{2.55}$$

The strain component is caused by the different displacements generated at each point, and the relationship between the displacement component and the strain component is further discussed below.

### 2.2.2 Geometric Equations

If a reference coordinate system *xyz* is chosen, in which the object is properly constrained, and any point *m*, within the object, is moved to *m*′ when a load is applied on the boundary, the displacement *mm*′ at that point can be decomposed into *u*, *v*, and *w* along the *x*, *y*, and *z* axes, as shown in Fig. 2.17. To facilitate the study of the change in its side length and angle, the microelement can be projected onto the three coordinate planes *xOy*, *xOz*, and *yOz*, respectively, as shown in Fig. 2.18. This allows the deformation of the entire hexahedron to be determined from the deformation of these projections. Since the deformations are small, two parallel planes projected on either coordinate plane can be combined into one projection, differing from each other by only one higher-order differential.

Now examine one of the planes, say the $xOy$ plane, as shown in Fig. 2.19, and suppose that the projection on that plane is a micrometer of the rectangle $ABCD$, deformed as a rhombus $A'B'C'D'$, and let the displacements of point $A$ along the $x$ and $y$ axes be $u$ and $v$, respectively. Then the displacements of point $B$ along the $x$ and $y$ axes are respectively $u + \frac{\partial u}{\partial x}\mathrm{d}x$, $v + \frac{\partial v}{\partial x}\mathrm{d}x$.

Similarly, the displacement of point $C$ along the $x$ and $y$ axes is $u + \frac{\partial u}{\partial y}\mathrm{d}y$, $v + \frac{\partial v}{\partial y}\mathrm{d}y$.

According to the definition of normal strain, the normal strain along the $x$-axis is

$$\varepsilon_x = \frac{\left(u + \frac{\partial u}{\partial x}\mathrm{d}x\right) - u}{\mathrm{d}x} = \frac{\partial u}{\partial x} \tag{2.56}$$

Similarly, the normal strain at point $A$ along the $y$-axis is

$$\varepsilon_y = \frac{\partial v}{\partial y} \tag{2.57}$$

Now study the change in the right angle between $AB$ and $BC$ of the microelement, as seen in Fig. 2.19.

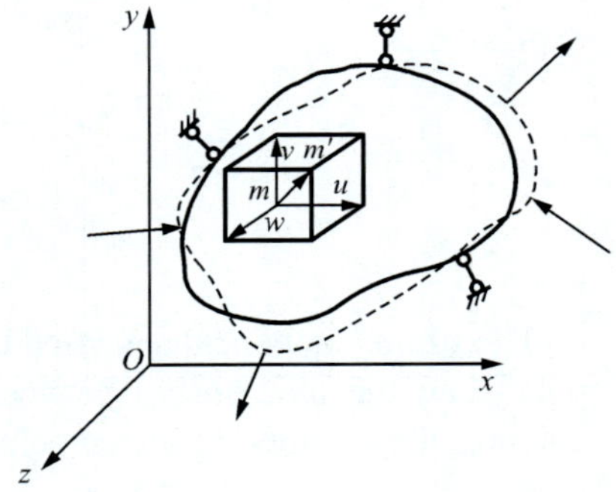


**Fig. 2.17** Displacement decomposition along the coordinate axes

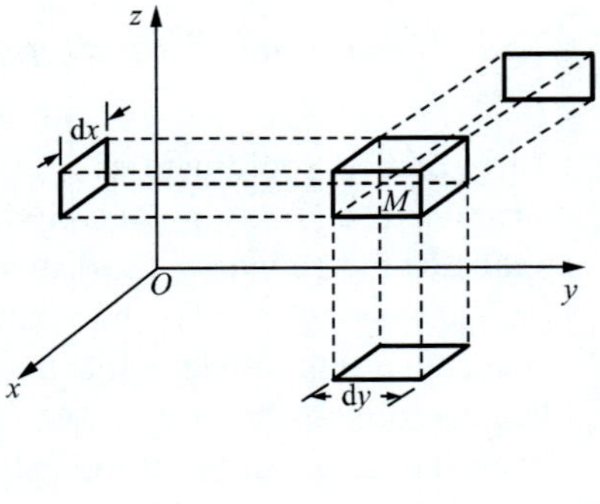


**Fig. 2.18** Projection of the micrometric body

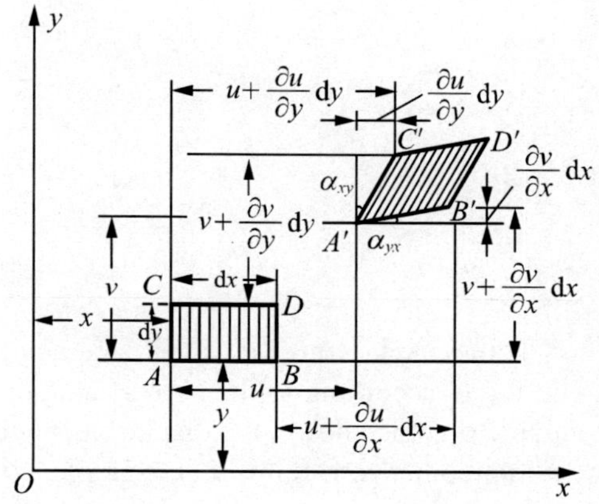


**Fig. 2.19** Relationship between strain and displacement

$$\alpha_{yx} \approx tg\alpha_{yx} = \frac{\left(v + \frac{\partial v}{\partial x}\mathrm{d}x\right) - v}{dx + \frac{\partial u}{\partial x}\mathrm{d}x} = \frac{\frac{\partial v}{\partial x}}{\left(1 + \frac{\partial u}{\partial x}\right)} \tag{2.58}$$

Since the assumptions are small deformations, $\frac{\partial u}{\partial x}$ can be omitted when compared to 1, giving

$$\alpha_{yx} = \frac{\partial v}{\partial x} \tag{2.59}$$

Similarly

$$\alpha_{xy} = \frac{\partial u}{\partial y} \tag{2.60}$$

So the amount of change of the right angle in the *xOy* plane after the deformation is

$$\gamma_{xy} = \alpha_{xy} + \alpha_{yx} \tag{2.61}$$

That is,

$$\gamma_{xy} = \frac{\partial v}{\partial x} + \frac{\partial u}{\partial y} \tag{2.62}$$

The above equation represents the shear strain at point *A* in the *xOy* plane.

The normal and shear strains on the other two planes *xOz* and *yOz* are studied similarly, yielding the corresponding two sets of strain components, which are synthetically summarized as

$$\left.\begin{aligned}\varepsilon_x &= \frac{\partial u}{\partial x}\\ \varepsilon_y &= \frac{\partial v}{\partial y}\\ \varepsilon_z &= \frac{\partial w}{\partial z}\\ \gamma_{xy} &= \frac{\partial u}{\partial y} + \frac{\partial v}{\partial x}\\ \gamma_{yz} &= \frac{\partial v}{\partial z} + \frac{\partial w}{\partial y}\\ \gamma_{zx} &= \frac{\partial u}{\partial z} + \frac{\partial w}{\partial x}\end{aligned}\right\} \tag{2.63}$$

If the displacement $u$, $v$, and $w$ at any point is known, the strain component can be found according to the above equation, and conversely if the strain component at any point is known, the displacement component cannot be uniquely determined without other constraints. For example, if the displacement components at any point are $u'$, $v'$, $w'$ and $u$, $v$, $w$ differ by $u_0$, $v_0$, $w_0$, then the same strain components can be obtained according to Eq. (2.63). It follows that in the absence of other constraints, the displacement component from the strain component is not fully determined because it survives displacements unrelated to the deformation, and in order to eliminate these displacements, appropriate constraints must be added to the boundary of the object.

It should also be suggested here that the micro-unit in the $xOy$ plane, after deformation of the line segments along the $x$ and $y$ axes respectively, turns by an angle $\alpha_{xy}$ and $\alpha_{yx}$, which makes the diagonal $AD$ of the micro-unit produce a turning angle after deformation because of the unequal rotation angle of $\alpha_{xy}$, $\alpha_{yx}$. If we first consider the angle of rotation of the diagonal $AD$ due to $\frac{\partial u}{\partial y}$, as shown in Fig. 2.20a.

$$\omega_x' = -\frac{1}{2}\alpha_{xy} = -\frac{1}{2}\frac{\partial u}{\partial y} \tag{2.64}$$

where, a negative value indicates a clockwise rotation of the $AD$ line.

As seen in Fig. 2.20b.

$$\omega_x'' = \frac{1}{2}\alpha_{yx} = \frac{1}{2}\frac{\partial v}{\partial x} \tag{2.65}$$

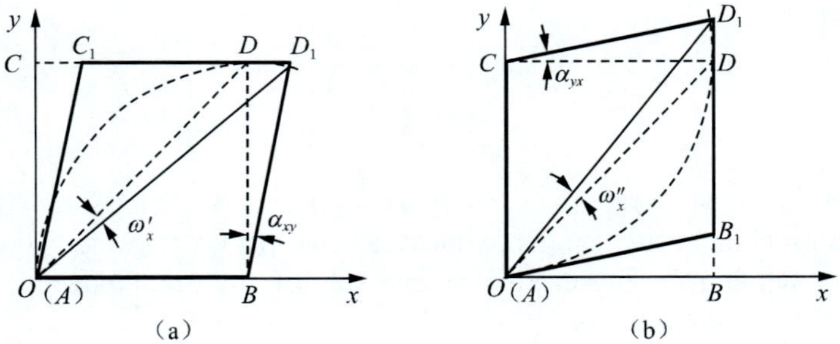


**Fig. 2.20** Rotation of the diagonal of the unitary differential

where, $\omega_x''$ is the angle of clockwise rotation of the line *AD*.

When $\frac{\partial u}{\partial y}$ and $\frac{\partial v}{\partial x}$ is present simultaneously, then the angle of rotation of the diagonal *AD* about the *z*-axis is

$$\omega_x = \frac{1}{2}\left(\frac{\partial v}{\partial x} - \frac{\partial u}{\partial y}\right) \quad (2.66a)$$

Similarly, the angles of the diagonals of the micro elements projected on the *xOz* plane and the *yOz* plane around the *y* axis and the *x* axis are,

$$\omega_y = \frac{1}{2}\left(\frac{\partial u}{\partial z} - \frac{\partial \omega}{\partial x}\right)$$
$$\omega_z = \frac{1}{2}\left(\frac{\partial \omega}{\partial y} - \frac{\partial v}{\partial z}\right) \quad (2.66b)$$

Equations (2.66a, 2.66b) represents the overall rotation of the micro elements in space.

### 2.2.3 *State of Strain*

In general, the strain components at various points in a rock mass are not the same, and even at the same point, the normal and shear strains vary in each direction. For any point where the normal and shear strain components along the *xyz* axis are known, how do you determine the strain components through that point in either direction?

#### 2.2.3.1 Normal Strain in Any Direction

Let the rock be taken as a differential line segment *AB* near point *A* before deformation. Its directional cosines with respect to the coordinate axis *xyz* are *l*, *m*, *n*. And the coordinates of the two points *A* and *B* are *A*(*x*, *y*, *z*) and *B*($x + \mathrm{d}x$, $y + \mathrm{d}y$, $z + \mathrm{d}z$) respectively. As shown in Fig. 2.21, if the position of the deformed *AB* line segment is shifted to the position of $A'B'$, At this time, the cosines of the direction between it and *x*, *y*, *z* are $l_1, m_1, n_1$, respectively. The coordinates of the two points $A'$ and $B'$ are $A'(x + u, y + v, z + w)$, $\mathrm{B}'(x + \mathrm{d}x + u', y + \mathrm{d}y + v', z + \mathrm{d}z + w')$, where *u*, *v*, *w* and $u'$, $v'$, $w'$ are the displacement components at the deformation of points *A* and *B*, respectively. They are all functions of *x*, *y*, *z*. The displacement components of $B'$ neglecting higher order differentials are

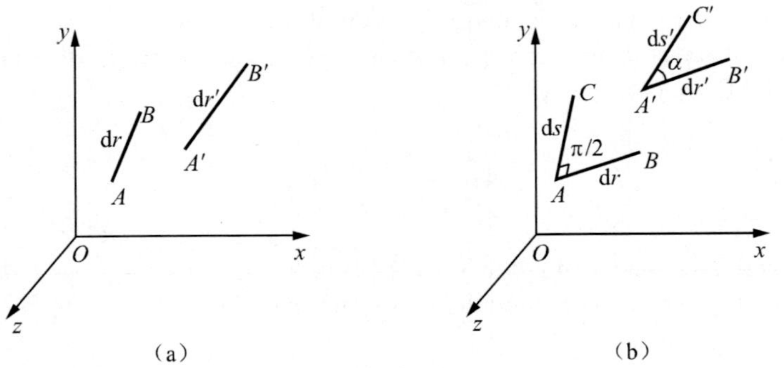


**Fig. 2.21** Strain components of the three-way strain state

$$\left.\begin{aligned} u' &= u + \tfrac{\partial u}{\partial x}\mathrm{d}x + \tfrac{\partial u}{\partial y}\mathrm{d}y + \tfrac{\partial u}{\partial z}\mathrm{d}z \\ v' &= v + \tfrac{\partial v}{\partial x}\mathrm{d}x + \tfrac{\partial v}{\partial y}\mathrm{d}y + \tfrac{\partial v}{\partial z}\mathrm{d}z \\ w' &= w + \tfrac{\partial w}{\partial x}\mathrm{d}x + \tfrac{\partial w}{\partial y}\mathrm{d}y + \tfrac{\partial w}{\partial z}\mathrm{d}z \end{aligned}\right\} \tag{2.67}$$

The $AB$ line segment is deformed into the $A'B'$ line segment, which is projected on the coordinate axis as

$$\left.\begin{aligned} (x + \mathrm{d}x + u') - (x + u) &= \mathrm{d}x + \tfrac{\partial u}{\partial x}\mathrm{d}x + \tfrac{\partial u}{\partial y}\mathrm{d}y + \tfrac{\partial u}{\partial z}\mathrm{d}z \\ (y + \mathrm{d}y + v') - (y + v) &= \mathrm{d}y + \tfrac{\partial v}{\partial x}\mathrm{d}x + \tfrac{\partial v}{\partial y}\mathrm{d}y + \tfrac{\partial v}{\partial z}\mathrm{d}z \\ (z + \mathrm{d}z + w') - (z + w) &= \mathrm{d}z + \tfrac{\partial w}{\partial x}\mathrm{d}x + \tfrac{\partial w}{\partial y}\mathrm{d}y + \tfrac{\partial w}{\partial z}\mathrm{d}z \end{aligned}\right\} \tag{2.68}$$

Let the normal strain of the line segment $AB$ be $\varepsilon_r$, which is deformed to $A'B' = \mathrm{d}r'$ and whose length is $\mathrm{d}r + \varepsilon_r \mathrm{d}r$. By the geometric law, the square of the length of the line segment $A'B'$ is equal to the sum of the squares of the lengths on the projection axis, i.e.

$$\begin{aligned} (\mathrm{d}r + \varepsilon_r \mathrm{d}r)^2 = &\left(\mathrm{d}x + \frac{\partial u}{\partial x}\mathrm{d}x + \frac{\partial u}{\partial y}\mathrm{d}y + \frac{\partial u}{\partial z}\mathrm{d}z\right)^2 + \left(\mathrm{d}y + \frac{\partial v}{\partial x}\mathrm{d}x + \frac{\partial v}{\partial y}\mathrm{d}y + \frac{\partial v}{\partial z}\right)^2 \\ &+ \left(\mathrm{d}z + \frac{\partial w}{\partial x}\mathrm{d}x + \frac{\partial w}{\partial y}\mathrm{d}y + \frac{\partial w}{\partial z}\mathrm{d}z\right)^2 \end{aligned} \tag{2.69}$$

Since $\mathrm{d}x = l\mathrm{d}r$, $\mathrm{d}y = m\mathrm{d}r$, and $\mathrm{d}z = n\mathrm{d}r$, dividing both sides of the equation above by $(\mathrm{d}r)^2$, gives

$$\begin{aligned} (1 + \varepsilon_r)^2 = &\left[l\left(1 + \frac{\partial u}{\partial x}\right) + m\frac{\partial u}{\partial y} + n\frac{\partial u}{\partial z}\right]^2 + \left[l\frac{\partial v}{\partial x} + m\left(1 + \frac{\partial v}{\partial y}\right) + n\frac{\partial v}{\partial z}\right]^2 \\ &+ \left[l\frac{\partial w}{\partial x} + m\frac{\partial w}{\partial y} + n\left(1 + \frac{\partial w}{\partial z}\right)\right]^2 \end{aligned} \tag{2.70}$$

The above equation has $\varepsilon_r < 1$ and $\varepsilon_x < 1,\ \varepsilon_y < 1,\ \varepsilon_z < 1$, so expand it by the binomial theorem and omit the square terms or product terms, i.e.

$$1 + 2\varepsilon_r = l^2\left(1 + 2\frac{\partial u}{\partial x}\right) + 2lm\frac{\partial u}{\partial y} + 2ln\frac{\partial u}{\partial z} + m^2\left(1 + 2\frac{\partial v}{\partial y}\right) + 2mn\frac{\partial v}{\partial z} + 2ml\frac{\partial v}{\partial x}$$
$$+ n^2\left(1 + 2\frac{\partial w}{\partial z}\right) + 2nl\frac{\partial w}{\partial x} 2mn\frac{\partial w}{\partial y} \tag{2.71}$$

Considering $l^2 + m^2 + n^2 = 1$, the above equation can be written as

$$\varepsilon_r = l^2\frac{\partial u}{\partial x} + m^2\frac{\partial v}{\partial y} + n^2\frac{\partial w}{\partial z} + mn\left(\frac{\partial w}{\partial y} + \frac{\partial v}{\partial z}\right) + nl\left(\frac{\partial u}{\partial z} + \frac{\partial w}{\partial x}\right)$$
$$+ lm\left(\frac{\partial v}{\partial x} + \frac{\partial u}{\partial y}\right) \tag{2.72}$$

To wit

$$\varepsilon_r = l^2\varepsilon_x + m^2\varepsilon_y + n^2\varepsilon_z + mn\gamma_{yz} + nl\gamma_{zx} + lm\gamma_{zy} \tag{2.73}$$

The above equation shows that, knowing the six strain components of a point and the directional cosine of any direction, the normal strain along the direction through the point can be obtained.

#### 2.2.3.2 Shear Strain in Any Two Mutually Perpendicular Directions

Two mutually perpendicular line segments $AB$ and $AC$ are provided through point $A$ with lengths d$r$ and d$s$ and direction cosines $l_1, m_1, n_1$ and $l_2, m_2, n_2$, respectively. As shown in Fig. 2.21b, if a deformation occurs, the angle between these two line segments will change. Let the angle between the deformed line segments $A'B'$ and $A'C'$ be $\alpha$, and since small deformation is assumed, by strain definition,

$$\gamma_{rs} = \frac{\pi}{2} - \alpha \approx \sin\left(\frac{\pi}{2} - \alpha\right) = \cos\alpha \tag{2.74}$$

Let the direction cosines of the deformed line segments $A'B'$ and $A'C'$ be $l_1', m_1', n_1'$ and $l_2', m_2', n_2'$ respectively. By the two-vector product relation we get

$$\gamma_{rs} = \cos\alpha = l_1'l_2' + m_1'm_2' + n_1'n_2' \tag{2.75}$$

Then, by finding the relationship between the direction cosines $l_1', m_1', n_1'$ and $l_2', m_2', n_2'$ and the displacement after deformation, brought into the above equation and simplified, the shear strain $\gamma_{rs}$ of the two line segments can be obtained. From Fig. 2.21b it can be seen that after deformation the cosine of the angle between $A'B'$

and the x-axis is

$$l' = \frac{\mathrm{d}x + \mathrm{d}u}{\mathrm{d}r'} = \frac{\mathrm{d}r}{\mathrm{d}r'}\left(\frac{\mathrm{d}x + \mathrm{d}u}{\mathrm{d}r}\right) \tag{2.76}$$

According to Eq. (2.68), we get

$$l_1' = \frac{\mathrm{d}r}{\mathrm{d}r'}\left[\left(1 + \frac{\partial u}{\partial x}\right)l + \frac{\partial u}{\partial y}m + \frac{\partial u}{\partial z}w\right] \tag{2.77a}$$

Similarly

$$\begin{aligned} m_1' &= \frac{\mathrm{d}r}{\mathrm{d}r'}\left[\left(\frac{\partial v}{\partial x}\right)l_1 + \left(1 + \frac{\partial v}{\partial y}\right)m_1 + \frac{\partial v}{\partial z}n_1\right] \\ n_1' &= \frac{\mathrm{d}r}{\mathrm{d}r'}\left[\left(\frac{\partial w}{\partial x}\right)l_1 + \frac{\partial w}{\partial y}m_1 + (1 + \frac{\partial w}{\partial z})n_1\right] \end{aligned} \tag{2.77b}$$

The above equation uses $\mathrm{d}x/\mathrm{d}r = l_1$, $\mathrm{d}y/\mathrm{d}r = m_1$ and $\mathrm{d}z/\mathrm{d}r = n_1$. Substituting $\mathrm{d}r' = (1 + \varepsilon_r)\mathrm{d}r$ into the above equation and omitting the higher order differential yields,

$$\begin{aligned} l_1' &= \left(1 - \varepsilon_r + \frac{\partial u}{\partial x}\right)l_1 + \frac{\partial u}{\partial y}m_1 + \frac{\partial u}{\partial z}n_1 \\ m_1' &= \frac{\partial v}{\partial x}l_1 + \left(1 - \varepsilon_r + \frac{\partial v}{\partial y}\right)m_1 + \frac{\partial v}{\partial z}n_1 \\ n_1' &= \frac{\partial w}{\partial x}l_1 + \frac{\partial w}{\partial y}m_1 + \left(1 - \varepsilon_r + \frac{\partial w}{\partial z}\right)n_1 \end{aligned} \tag{2.78}$$

Similar steps give

$$\begin{aligned} l_2' &= \left(1 - \varepsilon_s + \frac{\partial u}{\partial x}\right)l_2 + \frac{\partial u}{\partial y}m_2 + \frac{\partial u}{\partial z}n_2 \\ m_2' &= \frac{\partial v}{\partial x}l_2 + \left(1 - \varepsilon_s + \frac{\partial v}{\partial y}\right)m_2 + \frac{\partial v}{\partial z}n_2 \\ n_2' &= \frac{\partial w}{\partial x}l_2 + \frac{\partial w}{\partial y}m_2 + \left(1 - \varepsilon_s + \frac{\partial w}{\partial z}\right)n_2 \end{aligned} \tag{2.79}$$

Substituting (2.78) and (2.79) into (2.75), omitting the partial derivatives and the strain quadratic term higher-order differential, and after finishing, we get

$$\gamma_{rs} = (l_1 l_2 + m_1 m_2 + n_1 n_2)(1 - \varepsilon_r - \varepsilon_s) + 2\left(l_1 l_2 \frac{\partial u}{\partial x} + m_1 m_2 \frac{\partial v}{\partial y} + n_1 n_2 \frac{\partial w}{\partial z}\right)$$

$$+(l_1 m_2 + l_2 m_1)\left(\frac{\partial u}{\partial y} + \frac{\partial v}{\partial x}\right) + (m_1 n_2 + m_2 n_1)\left(\frac{\partial v}{\partial z} + \frac{\partial w}{\partial y}\right)$$
$$+(n_1 l_2 + l_1 n_2)\left(\frac{\partial w}{\partial x} + \frac{\partial w}{\partial z}\right) \tag{2.80}$$

Since d$r$ and d$s$ are perpendicular to each other, there is,

$$l_1 l_2 + m_1 m_2 + n_1 n_2 = 0 \tag{2.81}$$

According to the geometric Eq. (2.63) we get

$$\gamma_{rs} = 2l_1 l_2 \varepsilon_x + 2m_1 m_2 \varepsilon_y + 2n_1 n_2 \varepsilon_z + (l_1 m_2 + l_2 m_1)\gamma_{xy} + (m_1 n_2 + m_2 n_1)\gamma_{yz}$$
$$+ (n_1 l_2 + n_2 l_1)\gamma_{zx} \tag{2.82}$$

The above equation is the formula for calculating the shear strain of any two mutually perpendicular direction line segments.

#### 2.2.3.3 Strain Coordinate Transformation Equation

With Eqs. (2.73) and (2.82), it is not difficult to derive the coordinate transformation formula. If the original coordinates of any point in the object are $x$, $y$, and $z$, and its strain components are $\varepsilon_x, \varepsilon_y, \varepsilon_z, \gamma_{xy}, \gamma_{yz}, \gamma_{zx}$, the strain component of the new coordinate system $x'y'z'$ can be obtained. The direction cosine between these two coordinate systems is as the stress component, as shown in Table 1.2. It is easy to write the strain component of the new coordinate system according to Eqs. (2.73) and (2.82) as

$$\begin{aligned}
\varepsilon_{x'} &= \varepsilon_x l_1^2 + \varepsilon_y m_1^2 + \varepsilon_z n_1^2 + \gamma_{xy} l_1 m_1 + \gamma_{yz} m_1 n_1 + \gamma_{zx} n_1 l_1 \\
\varepsilon_{y'} &= \varepsilon l_2^2 + \varepsilon_y m_2^2 + \varepsilon_z n_2^2 + \gamma_{xy} l_2 m_2 + \gamma_{yz} m_2 n_2 + \gamma_{zx} n_2 l_2 \\
\varepsilon_{z'} &= \varepsilon_x l_3^2 + \varepsilon_y m_3^2 + \varepsilon_z n_3^2 + \gamma_{xy} l_3 m_3 + \gamma_{yz} m_3 n_3 + \gamma_{zx} n_3 l_3 \\
\gamma_{x'y'} &= 2\varepsilon_x l_1 l_2 + 2\varepsilon_y m_1 m_2 + 2\varepsilon_z n_1 n_2 + \gamma_{xy}(l_1 m_2 + l_2 m_1) \\
&\quad + \gamma_{yz}(m_1 n_2 + m_2 n_1) + \gamma_{zx}(n_1 l_2 + n_2 l_1) \\
\gamma_{y'z'} &= 2\varepsilon_x l_2 l_3 + 2\varepsilon_y m_2 m_3 + 2\varepsilon_z n_2 n_3 + \gamma_{xy}(l_2 m_3 + l_3 m_2) \\
&\quad + \gamma_{yz}(m_2 n_3 + m_3 n_2) + \gamma_{zx}(n_2 l_3 + n_3 l_2) \\
\gamma_{z'x'} &= 2\varepsilon_x l_3 l_1 + 2\varepsilon_y m_3 m_1 + 2\varepsilon_z n_3 n_1 + \gamma_{xy}(l_3 m_1 + l_1 m_3) \\
&\quad + \gamma_{yz}(m_3 n_1 + m_1 n_3) + \gamma_{zx}(n_3 l_1 + n_1 l_3)
\end{aligned} \tag{2.83}$$

#### 2.2.3.4 Principal Strain and Principal Direction

When the shear strain at a point in a rock mass is zero in a certain direction, that direction is called the principal direction of strain; there are three principal directions past a point, and the normal strain in the principal direction of strain is called the principal strain. To find the maximum, intermediate, and minimum principal strains $\varepsilon_1, \varepsilon_2, \varepsilon_3$, similar to finding the principal stress, the following system of equations can be used.

$$\begin{aligned} &2(\varepsilon_x - \varepsilon)l + \gamma_{xy}m + \gamma_{zx}n = 0 \\ &\gamma_{xy}l + 2(\varepsilon_y - \varepsilon)m + \gamma_{zy}n = 0 \\ &\gamma_{xz}l + \gamma_{yz}m + 2(\varepsilon_z - \varepsilon)n = 0 \end{aligned} \tag{2.84}$$

where, $l^2 + m^2 + n^2 = 1$.

The determinant of the coefficients of the above system of equations is zero, so it follows that

$$\varepsilon^3 - I_1'\varepsilon^2 + I_2'\varepsilon - I_3' = 0 \tag{2.85}$$

among others

$$\begin{aligned} I_1' &= \varepsilon_x + \varepsilon_y + \varepsilon_z \\ I_2' &= \varepsilon_x\varepsilon_y + \varepsilon_y\varepsilon_z + \varepsilon_z\varepsilon_x - \frac{1}{4}\left(\gamma_{xy}^2 + \gamma_{yz}^2 + \gamma_{zx}^2\right) \\ I_3' &= \varepsilon_x\varepsilon_y\varepsilon_z + \frac{1}{4}\gamma_{xy}\gamma_{yz}\gamma_{zx} - \frac{1}{4}\left(\varepsilon_x\gamma_{yz}^2 + \varepsilon_y\gamma_{zx}^2 + \varepsilon_z\gamma_{xy}^2\right) \end{aligned} \tag{2.86}$$

They are called the first, second, and third invariants of the strain tensor, respectively.

If the coordinates are chosen to coincide with the three principal directions of strain, the strain tensor can be expressed in terms of the main strain as

$$\begin{aligned} I_1' &= \varepsilon_1 + \varepsilon_2 + \varepsilon_3 \\ I_2' &= \varepsilon_1\varepsilon_2 + \varepsilon_2\varepsilon_3 + \varepsilon_3\varepsilon_1 \\ I_3' &= \varepsilon_1\varepsilon_2\varepsilon_3 \end{aligned} \tag{2.87}$$

The physical meaning of the first invariant of the strain tensor is quite clear, if we take a micrometric body with volume $\mathrm{d}x\mathrm{d}y\mathrm{d}z$ before deformation, and the volume after deformation as

$$(1 + \varepsilon_x)\left(1 + \varepsilon_y\right)(1 + \varepsilon_z)\mathrm{d}x\mathrm{d}y\mathrm{d}z = \left(1 + \varepsilon_x + \varepsilon_y + \varepsilon_z\right)\mathrm{d}x\mathrm{d}y\mathrm{d}z \tag{2.88}$$

The higher order microvolumes have been omitted from the above equation, and the increment of the volume of the microelement before and after the deformation is

$$\left(1+\varepsilon_x+\varepsilon_y+\varepsilon_z\right)\mathrm{d}x\mathrm{d}y\mathrm{d}z-\mathrm{d}x\mathrm{d}y\mathrm{d}z=\left(\varepsilon_x+\varepsilon_y+\varepsilon_z\right)\mathrm{d}x\mathrm{d}y\mathrm{d}z \tag{2.89}$$

So the ratio of the volume change to the original volume is defined as the volume strain $\varepsilon_v$, i.e.

$$\varepsilon_v=\varepsilon_x+\varepsilon_y+\varepsilon_z=\varepsilon_1+\varepsilon_2+\varepsilon_3 \tag{2.90}$$

#### 2.2.3.5 Principal Shear Strain

The value of the principal shear strain can be calculated in a similar way to that of the principal shear stress to obtain the maximum shear strain as

$$\gamma_{\max}=\varepsilon_1-\varepsilon_3 \tag{2.91}$$

It follows that the maximum shear strain is the difference between the maximum and minimum principal strains.

### Exercises

1. Describe how the state of stress in a rock mass is described by the stress components (normal and shear stresses) in a rock unit block and list these stress components in a stress matrix. What do the components that line up in the same row in the stress matrix have in common? What is common to the components that line up in the same column in the stress matrix?
2. What is the principal stress plane? What is the principal stress? What are the first, second and third invariants of stress tensor?
3. The two following two-dimensional stress states of the rock are summed to determine the principal stress and stress direction after synthesis (Fig. E1).
4. The stress state at a point in the rock mass is $\begin{bmatrix} 0 & 1 & 2 \\ 1 & 2 & 0 \\ 2 & 0 & 1 \end{bmatrix}$, Find the stress vector and the normal and tangential components acting on the plane $\overline{n}=\frac{1}{\sqrt{11}}e_1+\frac{3}{\sqrt{11}}e_2+\frac{1}{\sqrt{11}}e_3$.
5. Figure E2 shows a dam of triangular cross-section with a capacity of $\rho$, subjected to the action of a body of water with a specific gravity of $\gamma$ on the right-hand side, with a known stress component of $\sigma_x=\mathrm{a}x+\mathrm{b}y$, $\sigma_y=\mathrm{c}x+\mathrm{d}y-\rho gy$, $\tau_{xy}=-\mathrm{d}x-\mathrm{a}y$. Try to determine the constants a, b, c and d in the stress components from the boundary conditions.

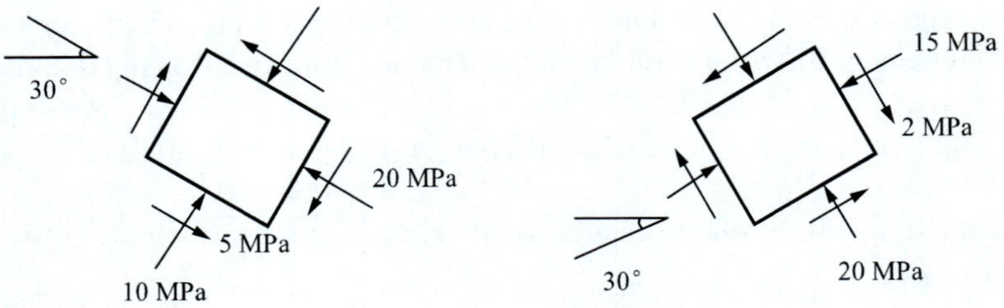


**Fig. E1** Stress distribution of microelements with different orientations

6. A resistance strain gauge is a device for measuring the relative elongation of a point on the surface of an object along a certain direction. It is usually used to measure the plane strain state of a point, as shown in the following Fig. E3. Strain gauges are respectively pasted in three directions of a point. If the relative elongation of these three strain gauges is measured as $\varepsilon_{0^\circ} = 0.0005$, $\varepsilon_{90^\circ} = 0.0008$, $\varepsilon_{45^\circ} = 0.0003$. Find the principal strain and the principal direction of the point.

**Fig. E2** A dam of triangular cross-section subjected to the action of water

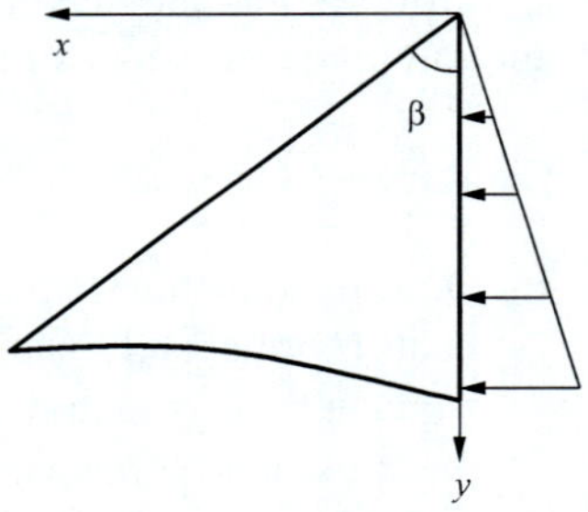


**Fig. E3** The distribution of strain gauges

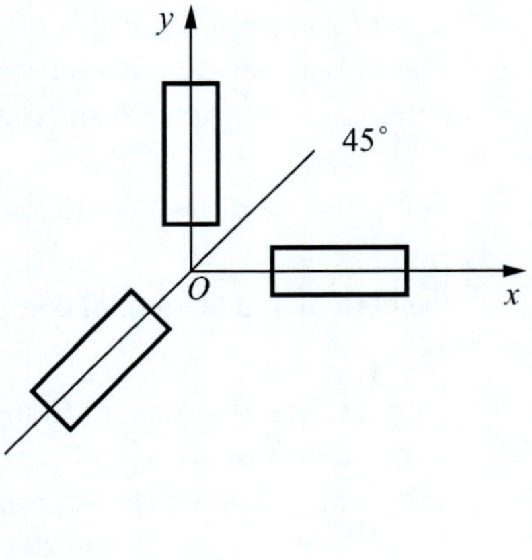

## References

Cai MF, He MC, Liu DY. Rock mechanics and engineering. Beijing: Science and Technology Press; 2002.
Goodman RE. International to rock mechanics. New York: Wiley; 1980.
Reismann H, Pawlik PS. Elasticity and application. New York: Wiley; 1980.
Sun G. Fundamentals of rock mechanics. Beijing: Science and Technology Press; 1983.
Yang G. Elasto-plastic mechanics. Beijing: People's Education Press; 1980.
Zhilun Xu. Elastic mechanics. Beijing: People's Education Press; 1980.

# Chapter 3
# Rock Composition and Physical Properties

## 3.1 Origin of Rock

The earth's crust is made up of rocks, which in turn are made up of mineral grains or crystals. Minerals are naturally occurring, relatively pure chemical compounds, such as quartz, feldspar, calcite, etc. Rocks and soils are aggregates of mineral particles. Rocks can be composed of several different mineral particles, such as granite, which is composed of quartz, feldspar, and mica particles. Rocks can also be composed of particles of the same mineral, such as limestone, which is composed of calcite mineral particles only.

There are three types of rocks that make up the earth's crust (Dai and Ji 1996): magmatic, sedimentary, and metamorphic rocks.

Magmatic rocks are formed when molten magma cools at the surface or underground. Intrusive rocks are magmatic rocks that crystallize underground. Rocks surrounding progressively cooler intrusive rocks are well adiabatic, and intrusive rocks often take thousands of years to solidify. The longer the magma cools, the larger the crystals that form. In general the mineral crystals of intrusive rocks are large and can be seen with the naked eye. Lava rocks are magmatic rocks that crystallize at the surface. When lava flows out of a crater, it immediately comes into contact with air or water and solidifies rapidly, crystallizing rapidly to form very small crystals that are difficult or impossible to distinguish with the naked eye.

Sedimentary rocks are formed by the deposition of sediments and have a complex history. Weathering decomposes the pre-formed rocks into mineral particles or dissolved salt deposits (from which sedimentary rocks are formed). These sediments were then transported by denudation (wind, waves, rivers, etc.) to another area, where they were deposited and buried in the ground. As the original sediments are covered by other sediments, the original sediments are buried deeper and deeper in the earth's crust and eventually become solid sedimentary rocks.

Metamorphic rocks are rocks formed by the recrystallization of other rocks at high temperatures and pressures. Metamorphic rocks have two types of structure, lamellar and non-lamellar structure. The lamellar structure is caused by recrystallization and



Y. Cheng et al., *Foundations of Rock Mechanics in Oil and Gas Engineering*,
https://doi.org/10.1007/978-981-99-1417-3_3

the separation of various minerals at high temperature and pressure resulting in inter-banding between light and dark minerals, and the dark band is formed by iron-rich and magnesium-rich minerals. Non-lamellar structures are formed by large mosaic crystals of the same composition and cannot form flakes when there are only a few common minerals in the original rock. For example, marble is a metamorphosed limestone and quartzite is a metamorphosed quartz sandstone. Since the temperature and pressure increase with depth of burial, metamorphism often occurs when rocks are buried deep underground. If the original rock is sedimentary, metamorphism usually destroys the fossil and sedimentary structures that help determine the genesis of that rock.

Sedimentary rocks are the main object of the search for oil and gas resources, i.e. they are the rocks that generate and store oil and gas. There are two types of sedimentary rock structures: clastic and crystalline sedimentary rocks, as shown in Fig. 3.1. Clastic rocks are composed of individual grains, such as sand grains, which are transported by exfoliation and deposited as intact grains. Because these particles do not stick together seamlessly, pore spaces exist between them, and oil, gas, and water are stored in the pore spaces of the subsurface clastic rocks. The structure of a crystalline sedimentary rock is determined by the crystals produced when that rock was deposited, and salt deposits are good examples of this. Since these crystals form a tightly arranged structure with no pore space, such rocks cannot store fluids in the subsurface, but can serve as a good cover for oil and gas reservoirs.

How do unconsolidated clastic grains like beach sand form a hard sandstone? When these original sand grains were buried underground, no matter how they were squeezed together, there were always pores between the grains, as shown in Fig. 3.2. Once the sand grains are buried into the ground, these pores are filled with water. Groundwater often contains salts dissolved in other subsurface rocks, and this water can flow from an area into the unconsolidated sandstone through the connecting pores. The salts often precipitate out of the groundwater and form arches between the unconsolidated particles, the process known as cementation, which bonds the unconsolidated particles into a solid sedimentary rock. There are two common types

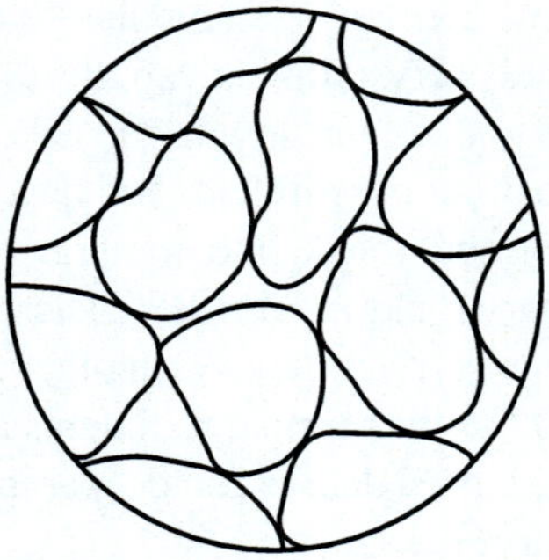

(a) Clastic structure

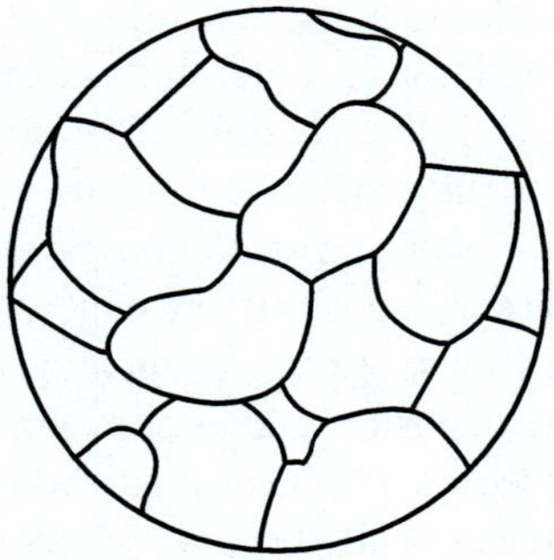

(b) Crystalline structure

**Fig. 3.1** Structure of sedimentary rocks

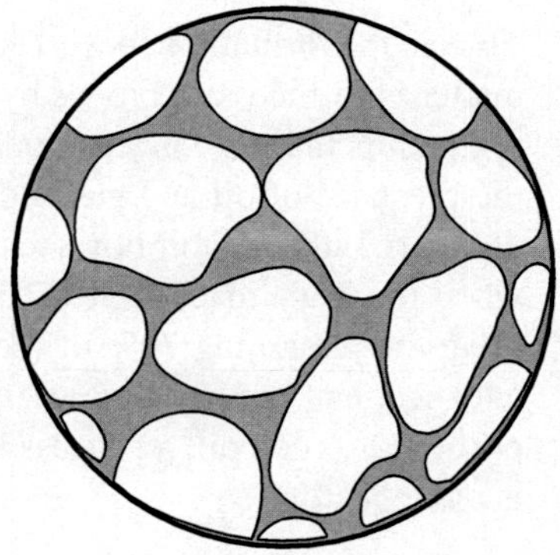

**Fig. 3.2** Pore space between clastic particles

of cementation and they are calcareous ($CaCO_3$) and siliceous ($SiO_2$). As sedimentary particles are buried deeper and deeper, the increasing weight of the overburden rocks exerts greater pressure on the particles, an effect that compacts the sediments and helps form solid sedimentary rocks. The physicochemical action that converts unconsolidated sediment into sedimentary rock is called diagenesis. Sedimentary rocks can be classified as conglomerates, sandstones, siltstones and mud shales depending on the size of the original clastic particles. The pore spaces in rocks differ depending on the size of the particles forming the rock. The finer the particles, the smaller the pore space in the rock and the more difficult it is for fluids to flow through the rock.

The age of the Earth is estimated to be about 4.5 billion years, and the rocks that generate the oil and gas, the rocks that store it, and even the oil and gas itself, are millions to hundreds of millions of years old. During this long geological period, sea levels have not remained constant; they have risen several times, spreading over continents in shallow seas, and then falling again to expose land. This phenomenon has happened many times on every continent. When the land became shallow, sediments were deposited in the shoreline areas along the beaches and in the shallow waters off the shoreline. These sediments were relatively simple materials, such as sand deposited along the shore and in the river, mud in the lagoons off the shore and behind the beaches, and ash that formed mesocosms and bioreefs. These ancient sediments accumulate layer by layer to form the sedimentary rocks that make up the uppermost part of the Earth's crust.

Oil and gas are derived from organic matter buried and preserved in ancient sedimentary rocks. In each of these environments of accumulated sediments, organic matter derived from dead animals and plants is deposited along with sand, mud and ash mineral grains and is buried deeper and deeper as the overburden sediments cover them. Dark mudstones are the most common organic-rich sedimentary rocks, and these are considered to be oil-bearing rocks. Shales were originally organic-rich soft mud deposited on the bottom of ancient oceans. Temperature is the main factor that drives the conversion of organic matter to oil, and as the shale buries at increasing depths, it is subjected to increasingly higher temperatures. The temperature at which oil is generally thought to form is between 100 and 200 °C. The reaction of organic matter to oil and gas conversion is complex and takes millions of years to complete.

Thus in the formation above 1500 m, where the ground temperature is lower, organic matter is formed into biogas by bacterial action. In formations deeper than 6000 m, high temperatures convert crude oil and other organic matter into natural gas. Once the presence of oil and gas is discovered and a borehole is drilled, the oil from the underground reservoir flows to the wellbore driven by formation pressure. Generally, when recovery reaches 30%, the formation pressure drops so much that it cannot drive the remaining 70% of the oil to the borehole. At this point, various recovery enhancement methods are used to rejuvenate the formation and discharge another portion of crude oil, a process known as secondary and tertiary oil recovery, which is quite costly.

## 3.2 Effect of Rock Structure on Strength

### *3.2.1 The Influence of Rock Composition on Rock Strength*

#### 3.2.1.1 Common Minerals and Common Rocks

Rocks are composed of minerals, and there are upwards of 1000 mineral species, with about 200 commonly listed in geology books, yet the most common rock-forming minerals are quite few, with only about 16, and only about 40 rocks relevant to engineering applications (Zhang et al. 1999). These 16 minerals are: quartz, feldspar (orthoclase, plagioclase), mica (black mica, white mica), chlorite, hornblende, pyroxene, olivine, calcite, dolomite, gypsum, hard gypsum, rock salt, pyrite, and graphite. Figure 3.3 shows a picture of common minerals, which are generally classified according to Mohs hardness, with talc minerals being the softest at grade 1 and diamond being the hardest at grade 10. Commonly seen objects used for hardness comparisons are fingernails (2.5), pocket knives (6), and glass (7).

The most common rocks in oil and gas engineering are sedimentary rocks, followed by metamorphic rocks, and occasionally granite, basalt, and gneiss. Granite is the most abundant intrusive rock, with coarse grains and rich quartz. The dark minerals make the granite show small speckled structure. If rich in iron, it is light red or pink. Basalt is the most abundant lava, which is fine-grained and dark in color. Gneisses are formed by strong metamorphism, with light and dark minerals interacting to form broad foliations and coarse grains. Conglomerate is a clastic rock that is itself highly variable in grain size and can range from rolled pebbles to clay grains. Sandstones consist mainly of sand-sized clastic grains. If the rock is loosely cemented, the sand grains can be separated and are visible to the naked eye when viewed at close range. Shale is made up of very fine clay particles and is generally well stratified and softer. When exposed to water, it is susceptible to hydration and decomposition and ranges in color from green to gray to black. Limestone is made up of calcite mineral grains. Calcite is soft enough to be carved with a knife and blisters when exposed to dilute hydrochloric acid. Biotite has the same characteristics as

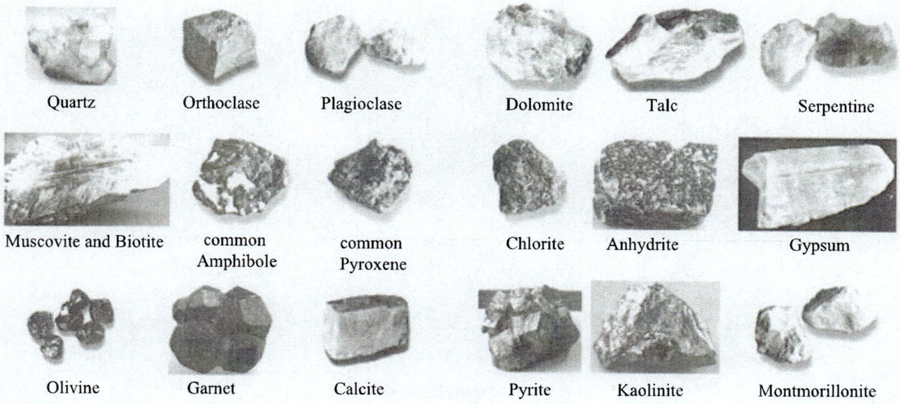


**Fig. 3.3** Pictures of common minerals

crystalline tuff except that it has fossil fragments. Dolomite is another carbonate rock, mainly composed of dolomite, often mixed with quartz, feldspar, calcite and clay minerals, and is stronger than limestone. Salt rocks are mainly chemically formed sedimentary rocks composed of halides and sulphates of potassium, sodium, calcium and magnesium, which are easily soluble in water. According to the main mineral composition, it can be divided into gypsum and hard gypsum rocks, rock salt, potash rocks, etc. It is a common evaporite sedimentary rock in the process of oil and gas field development. Coal is transformed from buried woody plant residues under pressure and heat, and it is brown to black in color, friable, and has cleat. Flint is amorphous quartz and cannot be incised by a knife. Because flint is amorphous, the rock breaks along curved surfaces. Color variants of flint include flint, jasper, chalcedony, and agate. Flint may be the result of direct precipitation by groundwater or recrystallization of fossil shell deposits with a silica composition. Figure 3.4 shows a picture of a typical rock.

Since rocks are composed of minerals, in order to study the composition, physical and mechanical properties of rocks, let us first examine the composition and physical and mechanical properties of minerals.

#### 3.2.1.2 Strength of Minerals

The most common rock-forming mineral in magmatic rocks is feldspar, followed by quartz, pyroxene, mica and olivine. The mineral composition in metamorphic rocks is similar to that of magmatic rocks, and in addition to the widely distributed quartz, feldspar, pyroxene and hornblende, there are typical metamorphic minerals: garnet, chlorite, sapphire, etc. The sedimentary rocks are dominated by quartz, feldspar and mica, also commonly seen are clay minerals, carbonatites, sulphate minerals and halites (Orlov et al. 1992).

**Fig. 3.4** Pictures of typical rocks

Almost all the minerals that make up rocks are crystalline structure, and crystalline minerals show a strict inherent regularity in their internal structure and external appearance. Only a very few minerals are amorphous, such as volcanic glass and silica. According to the knowledge of crystallography, the strength of a crystal depends mainly on the energy magnitude and directionality of the bonds between

**Table 3.1** Various chemical coupling bond energies

| Type of linkage | Coupling energy KJ/mol |
|---|---|
| Ionic bond | 300–1200 |
| Covalent bond | 200–1200 |
| Hydrogen bond | 4–40 |
| Van der Waals force | 0.4–12 |

the crystalline atoms (or ions), and this is also true for mineral crystals. The linkages between atoms in the crystal structures of major rock-forming minerals can be classified into four types: ionic, covalent, hydrogen, and van der Waals forces, which vary considerably in energy, with ionic and covalent linkages being the strongest. As shown in Table 3.1.

This makes the physico-mechanical properties of mineral structures very different. Minerals can be classified into three groups with different physico-mechanical properties according to their crystal structure and the way in which they are chemically bonded.

**Primary Silicate Minerals**

Their basic structural unit is the silicon-oxygen tetrahedron $(SiO_4)^{4-}$, with a core of $Si^{4-}$ and four vertices of $O^{2-}$. Silicon-oxygen tetrahedron can exist as individual groups or the tetrahedron can be linked to each other to form various combinations of shapes, thus forming various minerals, such as quartz, feldspar, pyroxene and hornblende. They are mainly ionic bonds—covalent bonds with higher strength and greater density.

**Simple Salt**

Such minerals are saltpeter (NaCl), potash (KCl), gypsum, barite, calcite, and dolomite. They are widely distributed in sedimentary rocks, and relatively few in metamorphic and magmatic rocks. They are all linked by ionic bonds, are readily soluble in water, and are poorly stable in water.

**Clay Minerals**

Clay minerals are a group of relatively stable water-bearing silicate minerals with a layered and layer-band structure, most of which are formed during chemical weathering of rocks and are distinguished from other silicate minerals by their high degree of dispersion and hydrophilicity and their ability to adsorb and exchange ions. The main ones are kaolinite, montmorillonite, illite, etc. Rocks composed of clay minerals are the most encountered and problematic formations in oil drilling engineering.

#### 3.2.1.3 Linkage Characteristics of Rocks

A rock is a complex system of minerals, pore fluids and its properties depend to a large extent on the relative content of the components and the characteristics and strength of their interactions. On the one hand, the strength of a rock depends on the strength of the minerals. The stronger the minerals are, the stronger the rock is, and vice versa. This law is more evident in igneous and metamorphic rocks. On the other hand, if we refer to the mineral grains that make up a rock as "structural units", then the rock is made up of "structural units" linked together by structure. This association determines to a large extent the basic properties of the rock. Studies have shown that the strength of mineral grains can be as high as several hundred MPa, while the strength of the rocks of which they are composed can be much lower. This depends largely on the linkage strength of the "structural unit". The structural linkages can be divided into three types according to their strength, as follows.

**Chemical Structure Linkage**

It is formed by covalent or ionic bonding, which arises in the case of mineral particles in direct contact with each other or in the case of high-strength cementing substances filling the intergranular pores. The latter case relies on chemical linkages between the cementing substance and the outer surface of the lattice network of the cemented particles to hold them firmly together, and in many cases the chemical structural linkages are no less strong than those within the lattice.

**Physical, Physicochemical Structural Linkages**

Physical and physicochemical structural linkage refers to structural linkage formed by forces such as molecular, electrostatic gravitational, magnetic, ion-electrostatic gravitational and capillary forces. It is mainly found in highly dispersed clay-based rocks (e.g. clays, sandstones, mud shales, etc.) and is characterized by the linkage strength being dependent on the type of clay, dispersion, water content and ion type and ion concentration.

**Mechanical Linkages**

It is the structural linkages produced by mechanical effects, i.e. caused by the intergranular embedding resulting from the microscopic inhomogeneity of the particle surface morphology.

In practice, these three structural linkages can co-exist in the same rock, but the degree of their role varies. For igneous, metamorphic and chemically sedimentary rocks, chemical structural linkages play a major role. The mechanism of their formation is that during rock formation, two contacting particles under higher temperature and pressure generate considerable normal and shear stresses at the contact site, which leads to plastic flow of the particles and the formation of coplanar chemical bonding associations along the recreated tiny shear surfaces. Thus for igneous, metamorphic, and chemically sedimentary rocks, the stronger the constituent minerals are, the stronger the rock will be. For clastic sedimentary rocks, the type of cementation is different and the structural linkage that plays a major role is still the chemical structural linkage, but the mechanism of formation is different from that of igneous rocks. During the process of diagenesis and post-formation, new crystalline or amorphous phases are precipitated from the circulating water solution, and the newly formed material is "linked" to the particle surface by chemical association, and a strong cementation "bridge" is created at the contact site. For clastic cementation, chemical and physicochemical structural linkages act simultaneously. For sedimentary rocks, the strength of the structural linkage is much lower than the strength of the minerals themselves. Thus the strength of sedimentary rocks depends mainly on the strength of the inter-mineral associations. In the case of sandstones, the strength of the associations of the different cementations varies, and their order of magnitude is: siliceous > ferrous > calcareous > muddy.

## 3.2.2 *Influence of the Structure and Tectonics of the Rock on Strength*

The strength of a rock is influenced by the structure and configuration of the rock, in addition to the strength of the minerals and the structural linkages between them.

### 3.2.2.1 The Structure of the Rock

The structure of intact rocks (Guo et al. 2010) refers to the size, shape, surface characteristics, quantitative relationships and interconnectedness characteristics of their constituent units (individual grains, aggregate grains, glassy). The structural characteristics of different kinds of rocks differ; for igneous and metamorphic rocks, the main features are the size of the structural units, the degree of crystallization, the degree of sorting and morphological characteristics. For example, basalt and granite are both igneous rocks, and the main rock-forming minerals are feldspar and quartz, but basalt minerals are fine-grained and granite grains are coarse, making basalt stronger and more elastic modulus than granite, as shown in Fig. 3.5.

**Fig. 3.5** Typical grain structure of basalt compared to granite

The structure of clastic sedimentary rocks mainly refers to the size and shape of the clastic fraction, the composition of the cement and its structure, and the interrelationship of the clastic particles and cement. According to the interrelationship between cement and debris can be distinguished as: (1) basal cementation, where debris particles are immersed in cement without contacting each other; (2) pore cementation, where cement fills the pores between particles in contact with each other; (3) contact cementation, where cement is distributed only on the contact points between particles and debris; (4) mosaic cementation, where under the consolidation effect during the diagenesis, especially when the pressure-solution effect is obvious, the clastic particles in sandy sediments come into closer contact, and the particles develop from point contact to line contact, concave-convex contact, or even form suture-like contact; as shown in Fig. 3.6. The strength magnitude of these four cementation types is: basal cementation > pore cementation > contact cementation mosaic cementation. In the actual sandstone these cementation types may exist at the same time, but the type of sandstone is different and the dominant cementation type is different. The space in a rock that is not filled with mineral particles, cement or other solid material is called the pore space of the rock. Pore spaces can be scattered uniformly throughout the rock, or they can be unevenly distributed in the rock to form pore clusters. The main components of rock pore space are pores and throats, and the larger space surrounded by rock particles is generally called a pore, while the narrow part connected only between two particles is called a throat, as shown in Fig. 3.7. The interconfiguration of pores and throat channels is very complex, resulting in a diversity of rock pore space variations. The pore space reflects the ability of the rock to store fluid, and the size and shape of the throat channel control the storage and permeability capacity of the pore space. The size and shape of pore spaces and throat channels depend mainly on the type of contact and cementation of sandstone particles, and the shape, size, roundness and sphericity of the particles themselves also have a direct effect on the geometry of pore spaces and throat channels. In the petrophysics, the geometry, size, distribution and interconnection of rock pores and roars are referred to as rock pore structure. Four basic pore structures exist in sandstones: intergranular pores, dissolved pores, microporosity and fractures.

Intergranular pores are the most dominant and common pore structure in sandstone reservoirs. It is the result of variations in the grain size, sorting, particle sphericity, roundness, particle orientation and filling factors of the sandstone. The distribution of

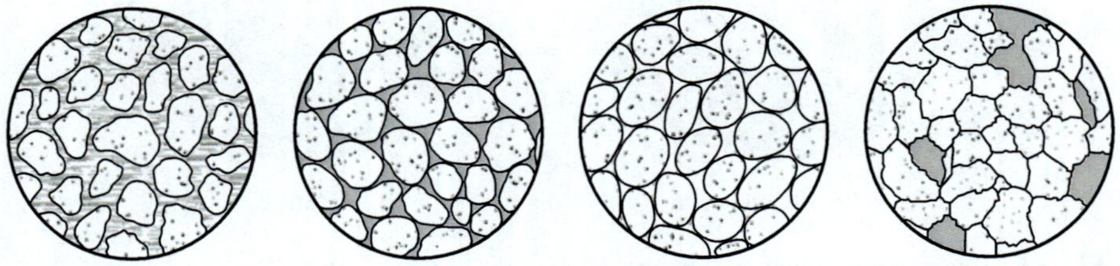

(a) Substrate cementation (b) Pore cementation (c) Contact cementation (d) Mosaic cementation

**Fig. 3.6** Types of cementation in clastic sedimentary rocks

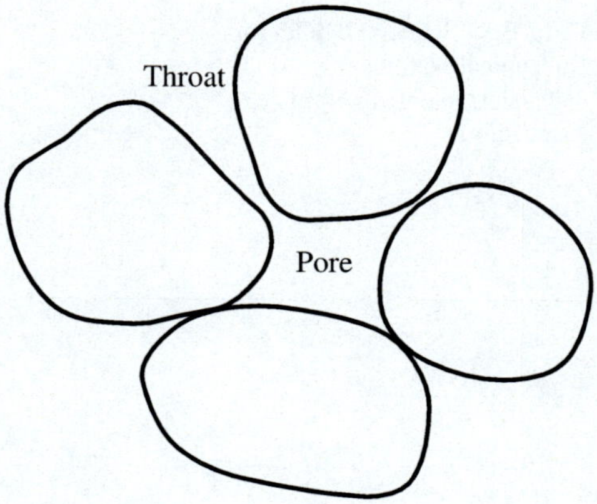


**Fig. 3.7** Pore and throat channels in sandstone

such pores is directly related to the depositional environment and diagenesis. Sandstone reservoirs dominated by intergranular pores with large pores, coarse throats and good connectivity are highly productive reservoirs. For example, the fine sandstone of the third series production layer in Daqing oilfield and the sandstone of the third series production layer in Shengli oilfield are dominated by intergranular pore space. A typical microscopic sketch of intergranular pore space is shown in Fig. 3.8.

Dissolved pores, formed by dissolution of carbonates, feldspars, sulfates, or other soluble components. For the less soluble silicate minerals, they can be replaced early by soluble minerals, which are then dissolved to produce secondary dissolution pores. Soluble components can be clastic particles, authigenic mineral cement, or substituted minerals, and Fig. 3.9 shows a schematic diagram of typical characteristics of dissolved pores within sandstones. If soluble minerals are scattered in the sandstone, then the dissolution pores formed are isolated; if soluble minerals are abundant enough to connect, then the dissolution pores are well permeable and can be good reservoir sandstones.

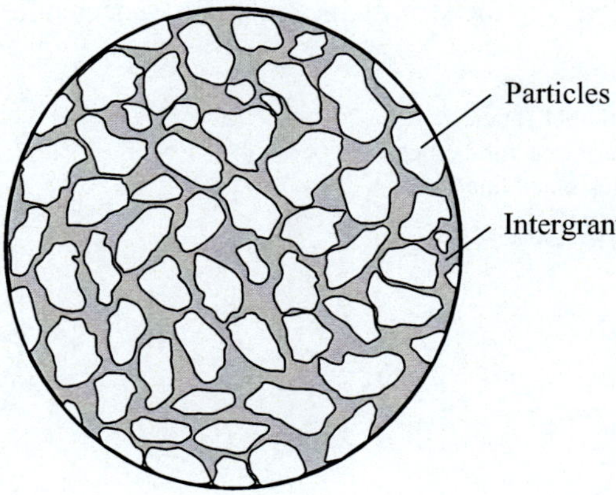


**Fig. 3.8** Microscopic sketch of a typical sandstone intergrain pore

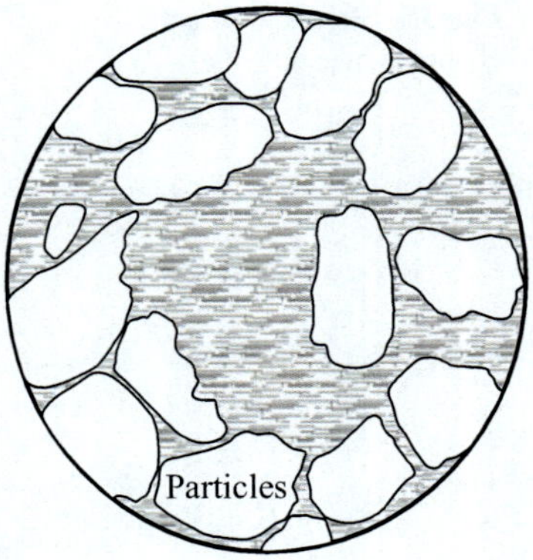


**Fig. 3.9** Schematic diagram of typical features of dissolution pores within sandstone

Microporosity, which is pore diameter less than 0.5 μm, has more causes, such as intergranular pores produced during recrystallization of clay minerals, mineral deconstruction joints, laminae joints, etc. The permeability of microporosity is very poor.

Fracture pores, in sandstone reservoirs, microfractures formed due to the action of tectonic forces can sometimes be very developed. The microfractures are in the form of fine sheets, with curved seam surfaces, bypassing the grain boundaries, and their alignment direction is controlled by tectonic forces. The fracture width is controlled by the tectonic horizontal stress field. In sandstone reservoirs, fracture widths are typically a few micrometers to tens of micrometers, as shown in Fig. 3.10. Fractures provide at most a few percent of porosity but will increase the overall permeability of the sandstone reservoir. Porosity caused by fractures alone is small, usually less than 1%, and such reservoirs are characterized by high initial production followed by a sharp decline.

In addition to the consistency of the pore structure of carbonate rocks with that of sandstones, their pore structure has special characteristics due to the development of secondary pore spaces such as caves and fractures. The pore structure of carbonate rocks refers to the size, shape and interconnectedness of the pores, holes and seams

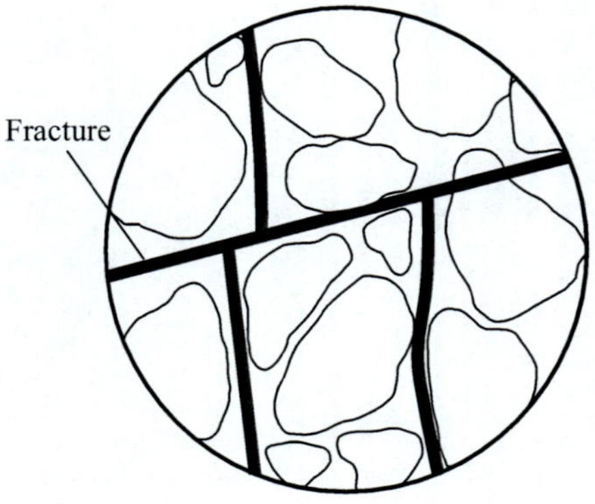


**Fig. 3.10** Schematic diagram of structural cracks (The thick black line shows the slit void)

that the rocks have. The law of the influence of pore structure on the strength of rocks is complex, and it is closely related to the rock type. At present, there is only a correlation between the strength and porosity of a certain type of rock; the greater the porosity, the lower the strength of the rock.

In summary, the diversity of rock microstructures makes the mechanical properties of rocks exhibit non-homogeneity and anisotropy, a feature that will become apparent in subsequent experimental chapters.

#### 3.2.2.2 Rock Tectonics

The scale of tectonics in the crust and lithosphere varies greatly, roughly globally and down to the micrometer scale. The tectonics of rocks here refers to structural surfaces such as beddings, cleavages, joints, and faults (Dai and Ji 1996; Guo et al. 2010). Their influence on the mechanical properties of the formation or rock masses is significant.

Planar structure is widely developed tectonic phenomena in the Earth's crust, and are the most fundamental objects of study and tectonic landmarks in tectonic research. There are many types of foliations and various causes. From the analysis of the formation and development process of foliations, they can be divided into two major categories: primary and secondary. Primary foliation includes bedding and rhythmic layers in sedimentary rocks and compositional differentiation layers in magmatic rocks, etc.; the secondary foliation refer to various foliations formed during the deformation and metamorphism, and cleavage is one of them. Bedding is the primary structure of sedimentary rocks, and is the most basic reference surface for studying tectonics. Sedimentary rock bedding is identified by the variation in composition, structure, and tone of the rock. A cleavage is a secondary foliation formation consisting of a potential splitting surface that divides the rock into parallel dense sheets or plates in a certain direction, as shown in Fig. 3.11. It develops in strongly deformed and lightly metamorphosed rocks with obvious anisotropic characteristics. And the development is often closely related to the amount of flake minerals contained in the rock, the grain size of the rock, and the degree of its orientation. The microscopic feature of cleavage is a domain structure, which is manifested by the parallel arrangement of cleavage domains and micro split stone between each other in the rock. Cleavage domains are usually parallel or interlaced thin strips or films enriched by layered silicates or insoluble residual materials. Split stones are narrow flat or lenticular sheets of rock sandwiched between cleavage domains. It is the directional arrangement of the layer silicate minerals within the cleavage domains that makes the rock potentially cleavable.

Joints are cracks in rocks, which are fractures without obvious displacement. They are the most widely developed structures in the upper crust. Joints can provide channels for oil and gas uplift and infiltration, and provide space and sites for mineral precipitation. A large number of developed joints often pose a potential hazard for projects such as reservoirs and dams. The nature, production and distribution patterns of joints are genetically linked to folds, faults and regional tectonics. According to

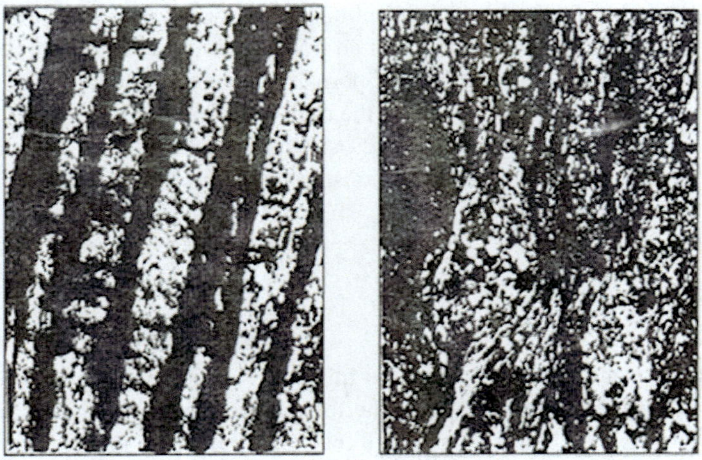

**Fig. 3.11** Cleavage domain and micro split stone in cleavage

the mechanical properties of joints, they are divided into two categories: shear joints and tension joints, as shown in Figs. 3.12 and 3.13.

Shear joints are fracture surfaces produced by shear stress, and have the following characteristics: the production is more stable, extending farther along the strike and tendency; the shear joints are relatively straight and smooth, sometimes with scratches; Shear joints developed in glutenite generally cut through gravel and cement. The typical shear joints often form a conjugate "X" type joint system; both sides of the main shear plane are often accompanied by feathery microcracks. Tension joints are rupture surfaces produced by tensile stress and have the following characteristics: they are not very stable, do not extend far, single joints are short and curved, and joints are often produced laterally; they are rough and uneven, without abrasion; they often pass around gravels or coarse sand grains in less solidly cemented gravels;

**Fig. 3.12** Two sets of common shear joints developed in Cretaceous sandstones in Zhucheng, Shandong Province

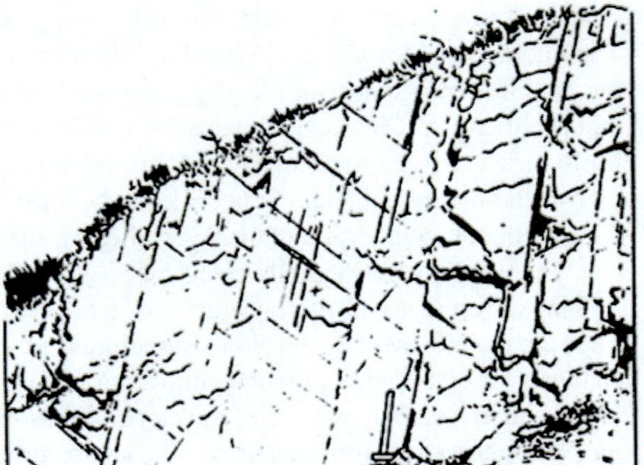

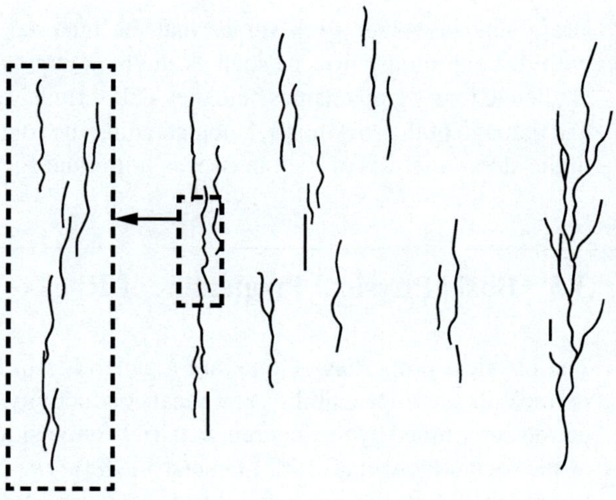

**Fig. 3.13** Lateral cracking of tensile joints in Cretaceous tertiary sandstones in Hubei Province

they are open and generally filled by veins; they are sometimes irregularly dendritic in shape; tension joints are formed under the action of tensile stresses perpendicular to the junction surface.

A fault is a type of rupture tectonics in the earth's crust in which a body of rock is significantly displaced along a rupture plane. Faults are widely developed and are the most important type of tectonics in the Earth's crust. Large faults often control the regional geological framework, not only controlling the structure and evolution of regional geology, but also controlling and influencing regional mineralization. Some small and medium-sized faults often directly determine the production of certain deposits and ore bodies. Active faults directly affect the stability of hydraulic structures and oil well casing. Based on the relative motion of the two sides of the fault, the faults can be classified as positive, reverse, and walking-slip faults, as shown in Fig. 3.14.

Rock tectonics (beddings, cleavages, joints, faults, etc.) are formed during the process of rock formation and the dramatic changes in load throughout its geological history. The multi-period nature of the tectonic movements makes the structural surfaces superimposed and the degree of development highly variable. The usual

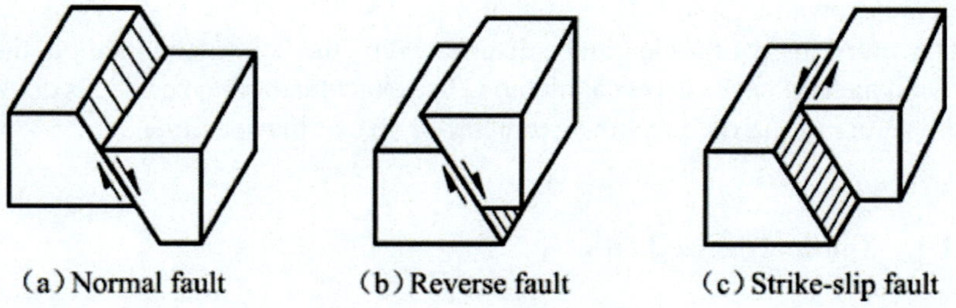


**Fig. 3.14** Faults classified according to the relative movement of the two sides of the fault

stress analysis generally assumes that the material is continuous, so how to deal with discontinuous surfaces such as faults, joints, beddings and fissures is a major challenge for rock mechanics, and they will seriously affect the deformation, strength and damage of the rock mass. Understanding the role of such discontinuous surfaces in the stress analysis of rock masses is important for engineering rock mechanics.

## 3.3 Basic Physical Properties of Rocks

The physical properties of rock are a general term for the bulk capacity, porosity, permeability, compressibility, electrical conductivity, heat transfer, and sound propagation determined by the inherent material composition and structural characteristics of the rock (Orlov et al. 1992; Umetaro 1982). The physical properties of rocks are inseparable from the mechanical properties, and understanding the physical properties of rocks is important for understanding the mechanical properties of rocks, and it is also the basis for interpreting the engineering mechanical parameters of the formation from logging data.

### *3.3.1 Bulk Density of Rocks*

The mass per unit volume of rock (including the volume of pores within the rock) is called the bulk density of the rock. Its expression is

$$\rho = \frac{m}{V} \tag{3.1}$$

where $\rho$ is the bulk density of the rock, $m$ is the mass of the sample being tested, and $V$ is the volume of the sample being tested.

The bulk density of rocks depends on the mineral composition of the constituent rocks, the degree of pore development and their water content. For sedimentary rocks, the magnitude of rock bulk density reflects to some extent the level of porosity of the rock and is therefore one of the basic logging items. The bulk densities of common rocks are shown in Table 3.2.

The bulk density of rocks can be determined by the volumetric method, the water intrusion method or the wax seal method. The exact method to be used is determined by the nature of the rock and the geometry or size of the specimen.

#### 3.3.1.1 Volumetric Method

Any rock that can be prepared as a regular specimen can be used to determine its bulk density by the volumetric method. To determine the bulk density of a rock by

**Table 3.2** Common rock bulk densities

| Rock name | Bulk density (g/cm$^3$) | Rock name | Bulk density (g/cm$^3$) | Rock name | Bulk density (g/cm$^3$) |
|---|---|---|---|---|---|
| Granite | 2.30–2.80 | Conglomerate | 2.40–2.66 | Granite gneiss | 2.90–3.30 |
| Diorite | 2.52–2.96 | Quartz sandstone | 2.61–2.70 | Hornblende gneiss | 2.76–3.05 |
| Gabbro | 2.55–2.98 | Siliceous cemented sandstone | 2.50 | Mixed gneiss | 2.40–2.63 |
| Porphyry | 2.70–2.74 | Siltstone | 2.20–2.71 | Gneiss | 2.30–3.00 |
| Porphyrite | 2.40–2.86 | Hard shale | 2.80 | Schist | 2.90–2.92 |
| Dolerite | 2.53–2.97 | Sandstone | 2.60 | Hard quartzite | 3.00–3.30 |
| Rough-face rock | 2.30–2.67 | Shale | 2.30–2.62 | Schistose quartzite | 2.80–2.90 |
| Andesite | 2.30–2.70 | Siliceous limestone | 2.81–2.90 | Marble | 2.60–2.70 |
| Basalt | 2.50–3.10 | Dolomitic chert | 2.80 | Dolomite | 2.10–2.70 |
| Tuff | 2.29–2.50 | Marl | 2.30 | Slate | 2.31–2.75 |
| Tuffaceous conglomerate | 2.20–2.90 | Limestone | 2.30–2.77 | Serpentine | 2.60 |

*Data source* Mei-Feng Cai, Rock Mechanics and Engineering

the volumetric method, the average cross-sectional area *A*, the average height *h* and the mass *m* of the regular specimen are determined and the bulk density of the rock is calculated by the following formula

$$\rho = \frac{m}{A \cdot h} \tag{3.2}$$

#### 3.3.1.2 Water Intrusion Method

The mass *m* of the irregular rock sample is first weighed, the volume *V* of the irregular rock sample is determined according to Archimedes' principle, and then the bulk density of the rock is found from Eq. (3.2). This method cannot be used for rocks that dissolve in water or absorb water very rapidly.

#### 3.3.1.3 Wax Seal Method

The wax seal method is suitable for rocks whose bulk density cannot be determined by the volumetric method or by water intrusion method. First of all, the specimen will be tied to a thin line, weigh the rock sample $g_s$, hold the line to slowly immerse the rock sample into the wax solution just past the melting point, pull out immediately after the dip, check the wax film around the specimen, if there is an bubble should be punctured with a needle, and then use the wax solution to fill the flat. After cooling the wax seal rock sample mass is $g_1$, and then submerge the wax seal rock sample in pure water and weigh its mass $g_2$, then the bulk density of the rock is

$$\rho = \frac{g_s}{g_1 - g_2 - \frac{g_1 - g_2}{\rho_n}} \tag{3.3}$$

where, $\rho_n$ is the density of the wax.

### 3.3.2 *Porosity of the Rock*

Natural rocks contain a variable number of pores and fissures of varying genesis and are one of the important structural features of rocks. The porosity of a rock is the ratio of the total volume of rock pores to the total volume of the rock and is often expressed as a percentage. The total volume of a rock $V_{\text{total}}$ consists of two parts, the volume of the matrix $V_{\text{matrix}}$ and the volume of the pores $V_{\text{pore}}$, then the porosity of a rock $\varphi$ is expressed as

$$\varphi = \frac{V_{\text{pore}}}{V_{\text{total}}} = \frac{V_{\text{pore}}}{V_{\text{matrix}} + V_{\text{pore}}} = 1 - \frac{V_{\text{matrix}}}{V_{\text{total}}} \tag{3.4}$$

From formula (3.4), the value $\varphi$ is a dimensionless quantity called absolute porosity.

For clastic sedimentary rocks, the cements or fillings in the rock may partially or completely block and close part of the pore space, making part of the pore throat very small, with very low permeability, so that the pores of different sizes in the rock play a very different role in fluid storage and seepage. According to the connectivity of pore space can be divided into two kinds of connected pore space and closed pore space. In oil field development, the connected pores involved in seepage are the effective ones, for those closed pores, although there is oil and gas storage in them, but it is difficult to extract. Hence they are called ineffective pore spaces. In this way, for the convenience of analysis, porosity is further divided into two types of connected porosity and flow porosity.

The effective porosity of a rock is the ratio of the volume of effective pore space in the rock to the total volume of the rock, expressed as a percentage. The effective pore volume is the volume of connected pore space involved in seepage. The connected porosity and connected pore volume of the rock can be expressed as $\varphi_{\text{connect}}$ and $V_{\text{connect}}$, respectively, then

$$\varphi_{\text{connect}} = \frac{V_{\text{connect}}}{V_{\text{total}}} \tag{3.5}$$

The flow porosity of a rock is the ratio of the volume of pore space to the total volume of the rock with a volume of liquid flowing in the rock containing oil, expressed as a percentage, and expressed as $\varphi_{\text{flow}}$ and $V_{\text{flow}}$ the flow porosity of the rock and the volume of liquid flowing in the pore space, respectively, then

$$\varphi_{\text{flow}} = \frac{V_{\text{flow}}}{V_{\text{total}}} \tag{3.6}$$

Flow porosity is conceptually different from connected porosity. It does not take into account not only the ineffective pores, but also the capillary pores occupied by those fluids bound by capillary forces, as well as the volume of the liquid film on the surface of the rock particles. In addition, the flow porosity varies with the pressure gradient in the formation and the physicochemical properties of the fluid.

In summary, it is not difficult to understand that: absolute porosity > connected porosity > flow porosity.

The range of actual reservoir rock porosity values is as follows. The porosity of sandstone ranges from 10 to 40%, which depends on the nature of the sandstone and its cementation state. Limestone and dolomite porosity ranges from 5 to 25%. The porosity of clay or shale is 20–45%, which depends on the genesis of the clay (or shale) and the depth of burial.

### 3.3.3 Permeability of the Rock

The property of a rock to permit the passage of fluids under pressure is called the permeability of the rock. It is customary to refer to those rocks, such as sandstones and fractured carbonates, in which fluids flow more easily along connected pores, throats, fissures, and cavities under formation pressure conditions as permeable rocks; and to those rocks, clays or shales, gypsum, rock salt, etc., in which fluids have difficulty in flowing, as non-permeable rocks. But this formulation is not strict, because on the one hand permeability is conditional, and on the other hand the permeability of the same rock is constantly changing.

The permeability of rock is a measure of the permeable ability of fluid through rock. French engineer Darcy first studied this phenomenon. He studied the percolation of water using artificial sand bodies. Darcy's experiments showed that the rate of change $Q/A$ in volume of water flow per unit area of the artificial sand body, is proportional to the difference in head $\Delta H$ between the inlet and outlet ends and inversely proportional to the length of the sand body. That is

$$\frac{Q}{A} = k\frac{\Delta H}{L} \tag{3.7}$$

Later researchers found that Darcy's law can be applied to other fluids and the proportionality constant $k$ can be rewritten as $k/\mu$, $\mu$ is the viscosity of the fluid and $k$ is a coefficient characterizing the properties of the rock, i.e. the permeability. Later, with further modification, the general formula of Darcy's law was obtained as

$$\frac{\mathrm{d}p}{\mathrm{d}x} = \frac{\mu}{k} \cdot \frac{Q}{A} \tag{3.8}$$

where, $Q$ is the flow rate through the pore medium. $A$ is the cross-sectional area through which the fluid passes. $\mu$ is the viscosity of the fluid. $k$ is the permeability of the rock. $\mathrm{d}p/\mathrm{d}x$ is the pressure drop per unit length.

Darcy's law is premised on (1) the assumption that no physicochemical reaction occurs between the fluid and the rock and (2) the presence of only one fluid in the permeability medium.

The unit of permeability is D, and its physical meaning is that the pore medium allows a fluid with a viscosity of 1 cps to flow through a cross-sectional area of 1 $\mathrm{cm}^2$ at a pressure gradient of 1 atm/cm at a rate of 1 $\mathrm{cm}^3$/s, at which point the permeability of the pore medium is called 1D (1D $= 0.9869 \times 10^{-8}$ $\mathrm{cm}^2$).

### 3.3.4 Water Content of the Rock

The ratio of the mass $m_{\mathrm{w}}$ of water in the rock in its natural state, to the drying mass $m_{\mathrm{rd}}$ of the rock, is called the natural water content of the rock and is expressed as a percentage, i.e.

$$w = \frac{m_{\mathrm{w}}}{m_{\mathrm{rd}}} \times 100\% \tag{3.9}$$

The water content of a rock under certain conditions varies and depends on the number, size, opening and closing of the pores of the rock and their distribution. The maximum water content of a rock under certain conditions is called saturated water content. Many physical and mechanical properties of rocks are closely related to the water content of rocks.

### 3.3.5 Particle Size Composition and Specific Surface Area of Rocks

The particle size composition of a rock is the content of various particles of different sizes that make up the rock, expressed as a percentage. The common method of analysis of particle size composition is sieve analysis. That is, a series of sieves of different apertures are used to divide the core particles into groups by size, and then the mass of each group is weighed to obtain the percentage of mass in each size range. Sieve holes are usually expressed in terms of the number of holes per inch of length, called mesh. For fines with particle sizes less than 0.074–0.053 mm, it is generally determined by the sedimentation analysis method.

The particle size composition of a rock can determine many physical properties of the rock. For example, the porosity, bulk density, permeability, capillary force, etc. of a rock can be determined based on the ratio of the number of various particles. Clastic rocks are classified into four main categories based on particle size as follows

Sandstone, main particle size composition 1–0.1 mm.
Fine sandstone, main particle size composition 0.1–0.01 mm.
Siltstone, main particle size composition 0.01–0.0039 mm.
Mudstone, major particle size composition less than 0.0039 mm.

In this case, the main particle size composition is the particles whose mass fraction must be 50–80%.

The specific surface area of a rock is the total surface area of particles within a unit volume of rock. The units are expressed in $cm^2/cm^3$. The more fine particles in a rock, the larger its specific surface, and vice versa, the smaller it is. When classified according to the aforementioned particle size composition, their specific surface areas are:

- Sandstone, less than 950 $cm^2/cm^3$
- Fine sandstone, 950–2300 $cm^2/cm^3$
- Siltstone, more than 2300 $cm^2/cm^3$
- Mudstone, much larger than siltstone.

The size of the specific surface area of reservoir rocks has an obvious influence on the flow of fluids in the pores of the rocks. The specific surface area of reservoir rocks is one of the important parameters to characterize the pore structure of rocks, and the surface phenomena and surface properties in the rock-oil-gas-water system are closely related to the specific surface area of rocks. The specific surface area is generally determined by the adsorption method.

### *3.3.6 Acoustic Properties of Rocks*

The propagation of vibration in rock is called sound wave, which has all the characteristics of fluctuation and can produce reflection, refraction, interference and diffraction. Sound waves can travel not only in gases, but also in liquids and solids. Sound waves with frequencies below 20 Hz are called infrasound, those with frequencies between 20 and 20,000 Hz are called sound waves, and those with frequencies above 20,000 Hz are called ultrasonic waves. The propagation law of acoustic wave in rock is called the acoustic characteristics of rock. It can be used to judge the structural characteristics, density characteristics and mechanical characteristics of rock. Therefore, the acoustic wave of rock is widely used.

According to the characteristics of the vibration direction and propagation direction of acoustic wave, it can be divided into longitudinal wave and transverse wave. If the direction of sound wave vibration and the direction of propagation is consistent, called longitudinal wave; if the direction of sound wave vibration and the direction of propagation is perpendicular, it is called shear wave. Since the velocity of the longitudinal wave is faster than that of the shear wave, the longitudinal wave is also called the p-wave and the shear wave is also called the s-wave. According to the knowledge of wave mechanics, the relationship between rock wave velocity, rock bulk density and elastic parameters (Young's modulus and Poisson's ratio) is as follows.

The relationship between longitudinal wave velocity and elastic parameters is as follows:

$$V_p = \sqrt{\frac{\lambda + 2G}{\rho}} = \sqrt{\frac{E(1-\mu)}{\rho(1+\mu)(1-2\mu)}} \tag{3.10}$$

where: $\lambda$ is Lame constant, G is shear modulus, is $\rho$ bulk density of rock, E is Young's modulus, and $\mu$ is Poisson's ratio.

The relationship between shear wave velocity and elastic parameters is as follows:

$$V_s = \sqrt{\frac{G}{\rho}} = \sqrt{\frac{E}{2\rho(1+\mu)}} \tag{3.11}$$

The elastic modulus and Poisson's ratio obtained from Eqs. (3.10) and (3.11) are called dynamic elastic modulus and dynamic Poisson's ratio.

In logging engineering, sonic differential time is commonly used to represent the wave velocity of rock, that is, the reciprocal of the longitudinal wave velocity is called the longitudinal wave differential time, with the unit of μs/m; The reciprocal of shear wave velocity is called shear wave differential time, in μs/m.

## Exercises

1. What are the types of rocks that make up the Earth's crust? And how are they created?
2. What are the two types of sedimentary rocks according to their structure? What are the characteristics of each of these two types of sedimentary rocks?
3. What types of clastic rock structures can be distinguished based on the interrelationship between the cement and the clasts? Please list them in order of strength.
4. What is the permeability of a rock and what is its significance?
5. Describe the sonic feature of the rocks? How to determine the dynamic Yaung's modulus and dynamic Poisson's ratio using primary wave, shear wave and bulk density?

# References

Dai QD, Ji YL. Geology of oil and gas reservoirs. Dongying: Petroleum University Press; 1996.
Guo JX, Wang JX, Zhang LQ. Geology of oil and gas field development. Dongying: Petroleum University Press; 2010.
Orlov et al. Petrophysical studies of oil and gas reservoirs. Beijing: Petroleum Industry Press; 1992.
Umetaro Y. Fundamentals of rock mechanics. Beijing: Metallurgical Industry Press; 1982.
Zhang HF, Fang CL, Gao XZ. Petroleum geology. Beijing: Petroleum Industry Press; 1999.

# Chapter 4
# Strength and Deformation Characteristics of Rocks

## 4.1 Mechanical Properties of Rocks at Ambient Temperature and Pressure

Rock mechanical properties mainly refer to the deformation characteristics of rocks and the strength of rocks. Since wellbore stability, sand production, hydraulic fracturing, and reservoir physical properties changes are closely related to rock mechanical properties in petroleum engineering, it is necessary to study the mechanical properties of rocks and their reflection under physical environment and stress conditions. There are many factors that affect the mechanical properties of rocks, such as rock type, composition, confining pressure, temperature, strain rate, water content, loading time, and nature of load, among others. To study the effect of these complex factors on the mechanical properties of rocks can only be carried out in the laboratory under strict control of certain factors. For any engineering phenomenon, it is only by studying the mechanical properties of rocks under the influence of certain factors one by one that it is possible to recognize which are the main influencing factors and which are the secondary ones. Thus, certain parameters can be derived to establish the intrinsic equations and damage criterion of the rock, which provide certain models and bases for the analysis of theoretical studies. Therefore, theoretical research is based on experimental studies of the mechanical properties of rocks under various physical environments and influencing factors. Of course, the results obtained from the theory need to be verified and corrected back in practice until they match with the reality.

The deformation properties of rocks are most intuitively expressed by means of stress–strain relationship curves and curves of strain variation with time. The mechanical properties of rocks at room temperature and pressure (i.e., room temperature conditions) are usually studied first, and then other influences are taken into account. In this way, the influence of a combination of factors on the mechanical properties of rocks under in situ conditions of the formation can be gradually clarified.



Y. Cheng et al., *Foundations of Rock Mechanics in Oil and Gas Engineering*,
https://doi.org/10.1007/978-981-99-1417-3_4

Laboratory experiments on the mechanical properties of rocks should begin with the collection of rock specimens from the formation under study. In order to maintain the original physical and mechanical properties of the rock sample (e.g., mineral composition, grain size, structure, tectonics, fractures, degree of joint development, etc.) it is best to perform a pressure coring in drilling. The drill core is then processed into ($\phi$50 mm × 100 mm or ϕ 25 mm × 50 mm) cylinders, but sometimes rectangular columns of (50 mm × mm 50 × 100 mm) can also be used. The ratio of length to width (or diameter) of the specimen is between 2.5 and 3.0 as recommended by the International Society for Rock Mechanics and Engineering (between 2.0 and 2.5 is mostly used in China) (ISRM 2007). The rock specimen is placed on a conventional press for loading, and its strain can be determined by a deformation transducer or by strain gauges attached to the specimen by a resistance strain gauge. As the load is incremented (as can be seen by the press readings), the compressive stress applied to the specimen $\sigma = P/A$ (where $P$ is the load and A is the cross-sectional area of the specimen) and the corresponding strain ($\varepsilon = \Delta h/h$) are obtained. In continuous loading (generally experiments are loaded at 0.5–0.8 MPa/s), the curve of stress and strain plotted in a right-angle coordinate system is called the stress–strain curve. According to Deere and Miller (1966), who analyzed the experimental results of the mechanical properties of 28 types of rocks, the stress–strain curves under uniaxial compression are broadly classified into six types as shown in Fig. 4.1.

The first type is elastic deformation, from loading until damage is produced, with an approximately linear stress–strain curve characteristic of, for example, basalt, quartzite, gabbro, dolomite and hard limestone.

The second type is elastic–plastic deformation, where the stress–strain curve shows continuous inelastic deformation as it approaches the failure load. Examples include soft limestones, siltstones and tuffs.

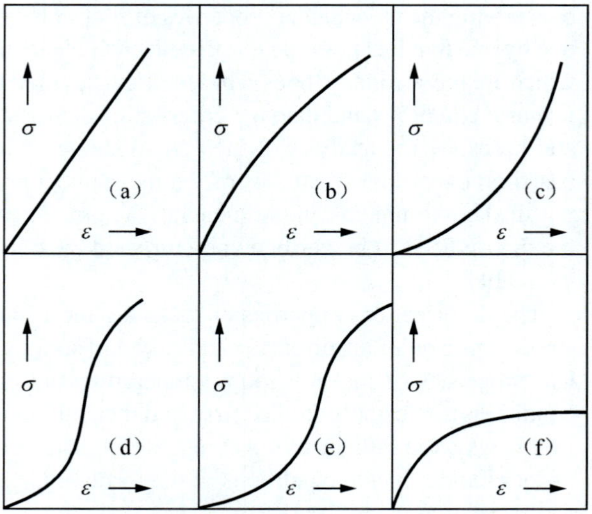


**Fig. 4.1** Typical stress–strain curves for uniaxial compression experiments on rocks

The third type is plastic-elastic deformation, where the stress–strain curve exhibits an upward bending at low stresses, followed by an approximately linear relationship until destruction. Examples include sandstone and granite.

The fourth type and the fifth type are plastic-elastic–plastic deformation, and both stress–strain curves show S-shaped curves. These two curves differ in that the former is steeper in the near-linear part and less compressible in the initial stage. Examples include marble and gneiss in metamorphic rocks. The latter has a slower linear part, which indicates a higher deformation under the same stress and a high compressibility in the initial stage. The common feature between them is that they all show varying degrees of inelastic deformation as they approach damage.

The sixth type is elastic–plastic-creep deformation with a short linear portion of the curve followed by inelastic deformation and continuous creep, e.g., sparse sandstone, soft mudstone, etc.

The downward concavity in the initial stage of the stress–strain curve is a result of the compression of the microfracture or foliation surface, which occurs more or less frequently in all rocks with microfractures.

In uniaxial compression experiments on conventional testing machine, the rock specimen is deformed and a certain amount of strain energy is stored as the pressure gradually increases. At the same time, the testing machine, due to its frame stiffness, also deforms accordingly and stores part of the strain energy. When the stress exceeds the strength limit of the rock, with the rapid development of micro-rupture of the rock specimen, the ability to resist the load decreases and the strain energy on the frame of the testing machine is rapidly released to the rock specimen, resulting in sudden destruction of the rock. In order to get a more accurate and detailed picture of the full process of the rock mechanical stress–strain curve, a rigid testing machine with a much greater stiffness than the rock specimen is used, it is guaranteed to still produce stable damage in the post-peak region. Experiments can be performed by placing the rock specimen in a rigid testing machine, yielding a typical whole stress–strain curve as shown in Fig. 4.2. The curve (Jager and Cook 1981) can be divided into four stages:

(1) The first stage *OA* curve, the load gradually increases from zero to point *A*, and the curve shows a slightly upward curved shape. This is the result of the existence of certain micro-fractures inside the rock specimen, and the specimen micro-fractures gradually close when the load increases. The degree of concave curvature of this section of the curve depends on the degree of fracture in the rock that can be easily compacted, and this phenomenon is less obvious for dense rocks or under high circumferential pressure.
(2) The second stage *AB* curve, generally *AB* line segment is approximately straight line, and its slope is called elastic modulus *E*. In the *OB* interval below point B, if the load is removed, the deformation is completely recoverable and there is no permanent deformation, so the *OB* interval is called the elastic deformation stage. The point *B* on the curve is the stress limit value that produces elastic deformation and is called the elastic limit. In fact most rocks, even when

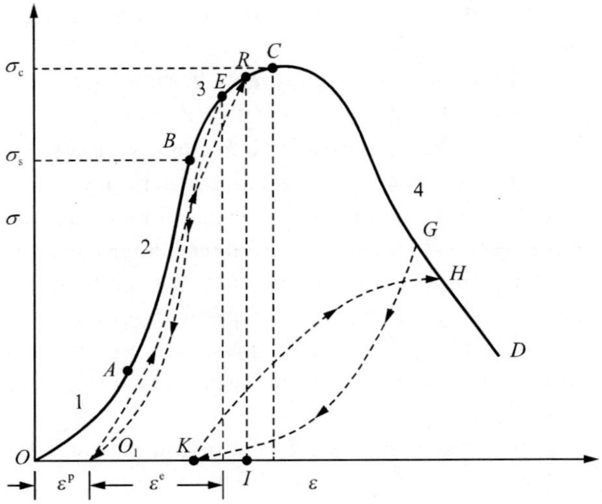


**Fig. 4.2** Total stress–strain curve of the rock

producing very small strains, will always retain more or less partial permanent strains when the load is removed, which is a result of the impossibility of complete recovery of the compressed and dense microfractures.

(3) The third stage *BC* curve, which has a downward curved shape as the load continues to increase beyond point B, indicates that there is little increase in stress and much increase in strain. Unloading at any point on the curve beyond point *B* (for example, point *E*), the stress–strain curve will fall along the $EO_1$ path until it is completely unloaded down to the point of intersection with the horizontal axis $O_1$, which indicates that the stress within the rock specimen completely disappears, but the strain does not fully recover, and the part of the strain still retained $OO_1$, is called plastic strain or permanent deformation. The stress at point *B* is referred to in rock mechanics as the yield stress $\sigma_s$. After unloading and then reloading, it rises along the curve $O_1R$ to join the original curve *BC*, thus forming a hysteresis loop that continues to increase after point R as the load continues to rise still along the curve *BC* to the highest point of that curve *C*. If unloading occurs again after point R, a new plastic strain appears, which appears to raise the elastic limit from point B to point *R*. This phenomenon is called strain hardening.

The stress value at the highest point *C* of the stress–strain curve is called the compressive strength, and it indicates the maximum compressive stress that the rock can withstand under these conditions. For normal rocks, the compressive strength is about 1.5–3 times the elastic limit. Starting from point B, within the *BC* line segment, the rock specimen continuously produces micro-rupture as well as slip within or between grains, which is the apparent inelastic deformation that the rock has before it is damaged, and this phenomenon is called **dilation**. As the number of microruptures and the length of expansion increase sharply when point *C* is reached, the rock has a significant inelastic volume expansion until a distinct rupture surface is formed at point *C*. If experimented by rigid testing machine, a fourth stage curve will also appear.

(4) The fourth stage *CD* curve, the rock specimen under the action of the rigid testing machine, the stress–strain curve reaches point *C*, there has been a macroscopic rupture surface formation, but has not yet completely broken into several pieces, the rock is still partially connected internally, still able to withstand part of the load, but its load-bearing capacity is getting smaller and smaller, reflecting the gradual decline of the curve. If any point *G* on the *CD* curve is unloaded in time it falls along the *GK* curve until it is completely unloaded, reaching the point *K*, indicating that the rock produces a large permanent strain OK. If it is loaded again, the curve rises again along the *KH* line until the point H is coupled to the *CD* curve, but the stress at the point *H* is lower than the stress at the point *G* at the beginning of unloading. This is quite different from the case of unloading followed by reloading in the curve *BC* line, where the stress value rises after unloading and falls in the latter, which indicates that the strength of the rock in the *CD* line keeps decreasing until a certain point on the *CD* line, when the rock specimen breaks into several pieces due to complete loss of cohesion on the rupture surface.

The characteristics of rocks after reaching the peak stress can be divided into two types: one is called the stable rupture propagation type and the other is the unstable rupture propagation type. The former is characterized by the fact that when the load exceeds the peak of the bearing capacity of the rock specimen, the strain energy stored in the specimen is not sufficient to cause the rupture to continue to expand, unless the work done by the external force is increased further to cause further damage to the specimen, otherwise the specimen will not be damaged, but its bearing capacity has been reduced accordingly. The latter is characterized by the fact that when the load exceeds the peak of the bearing capacity of the rock specimen, although the testing machine no longer does work on the rock specimen, the strain energy stored in the rock specimen is sufficient to cause the rupture to continue to expand, finally leading to the destruction of the specimen and the rapid loss of the bearing capacity.

In summary, under the action of load, the rock specimen first produces microfracture compaction deformation inside the specimen, and when the load gradually increases and reaches the yield limit, it starts to produce new microruptures, and then the microruptures gradually expand. When the damage strength is reached, the macroscopic rupture surface has been formed, and with the further increase of deformation, it finally leads to the complete rupture of the specimen. Therefore, deformation and rupture are two different development processes that are interdependent, and the deformation reaches a certain stage that contains both the rupture factor and the arrival of the damage stage is also the result of the continuous development of deformation. Therefore, destruction is essentially a process of rupture from quantitative to qualitative change.

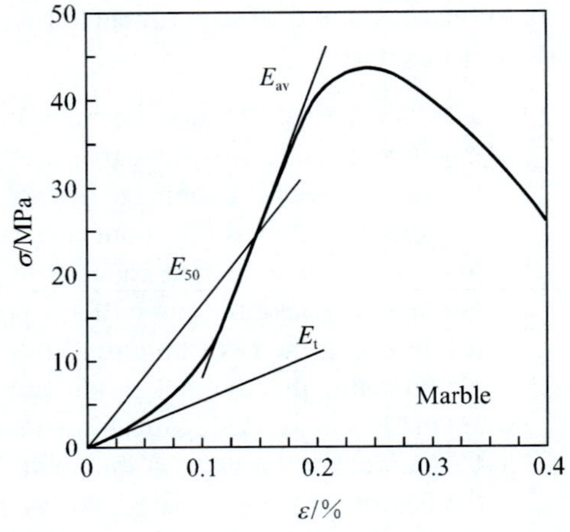


**Fig. 4.3** Characterization of the three moduli on the full stress–strain curve

As can be seen from the whole stress–strain curve in Fig. 4.2, the stress–strain relationship for the rock is not a simple linear relationship. How can the elastic parameters of the rock be determined at this point? There are three ways to characterize the elastic modulus of a rock: tangent modulus, secant modulus, and mean modulus, as shown in Fig. 4.3.

The tangent modulus $E_t$ is generally chosen as the slope of the tangent line of the stress–strain curve at the origin or at the point where the stress is half the strength of the rock sample. The secant modulus $E_s$ is mostly used as the ratio of stress to strain when the stress is half the strength of the rock sample, i.e., $E_{50}$, also known as the deformation modulus of the rock. The mean modulus $E_{av}$ is the slope of the approximately straight line portion of the stress–strain curve.

The presence of a significant depression in most of the full stress–strain curves of the rock makes the three moduli significantly different. The magnitude of $E_{50}$ depends on the strain of the stress at 50% of the strength, and this value is significantly influenced by the deformation of the OA section at the beginning of the loading, and the dispersion is large in all cases. The mean modulus, on the other hand, is the slope of the approximately straight line portion of the stress–strain curve, expresses the proportional relationship between the amount of change in stress and strain, is less affected by experimental conditions, and has a clear mechanical meaning. Hard brittle rocks still have good linear deformation characteristics when they are close to damage, and it is also more adequate to use the average modulus than the secant modulus. On the other hand, in the structural analysis of actual rock masses, the initial deformation of the rock mass is not determined, and it is the response of the material when the load or deformation changes that needs to be studied. Therefore, the mean modulus is mostly used to express the deformation characteristics of rocks when studying the mechanical properties of rocks.

## 4.2 Effect of Confining Pressure and Intermediate Principal Stress on Mechanical Properties of Rocks

Rocks generally produce brittle damage at room temperature and pressure, but rocks buried deep underground exhibit significant ductility. Changes in this property of rocks are caused by changes in the physical environment to which they are subjected.

The so-called brittleness and ductility have not been very clearly defined so far, and generally the so-called brittle damage refers to the rapid destruction of rocks by elastic deformation, with less plastic deformation after the destruction. Ductility refers to the destruction of a rock by a large plastic deformation after elastic deformation, or the direct development of ductile flow. By ductile flow is meant the property of having a large amount of permanent deformation without damage. For rocks, a pre-damage strain of less than 3% can be considered as brittle damage, more than 5% as ductile damage, and 3–5% as a transitional stage.

In order to simulate the mechanical properties of rocks in this state, triaxial stress experiments are generally performed indoors. They can be divided into two types: conventional triaxial stress experiments ($\sigma_1 \neq \sigma_2 = \sigma_3$) and true triaxial stress experiments ($\sigma_1 \neq \sigma_2 \neq \sigma_3$). Most of the current experiments belong to the former.

For conventional triaxial stress experiments, a certain size cylindrical core specimen is usually wrapped in a rubber sleeve or metal foil and placed in the cell of a triaxial testing machine surrounded by liquid or gas loading, and the axial load experiment is applied by a piston, using the differential stress $\sigma_1 - \sigma_2$ as the vertical axis of the right angle coordinate system and the axial strain $\varepsilon$ as the horizontal axis to plot the stress–strain curve.

Experiments on the mechanical properties of rocks under confining pressure were first carried out by von Karman (1912) in Germany. The experiments were carried out on white cylindrical marble specimens, which were very fine-grained and very homogeneous. It was found that at zero or low circumferential pressure the marble specimens broke in a brittle manner, producing an inclined rupture surface; as the circumferential pressure increased, the ductile deformation and strength of the specimens increased until full ductile or plastic flow occurred. The conventional triaxial stress test method he used, which is still widely used today, is shown in Fig. 4.4.

From Figs. 4.5, 4.6, 4.7 and 4.8 shows the stress–strain curves of Carrara marble, Cranport limestone, dolomite, and siltstone at different envelope pressures.

The experimental results show that the rocks gradually change from brittle to ductile as the confining pressure increases. the Carrara marble shows a brittle state at zero or lower confining pressure; when the confining pressure increases to 50 MPa, the marble shows a transitional state from brittle to ductile; when the confining pressure increases to 68.5 MPa, the marble shows ductile flow; the confining pressure of 165–320 MPa all show strain hardening. This is a good indication that the increase in confining pressure is one of the conditions for the conversion of brittle to ductile flow. However, the values of the transformed confining pressure vary with the rock type. For example, Carrara marble reaches ductile flow at a confining pressure of 68.5 MPa (Fig. 4.5), while dolomite is about 145 MPa (Fig. 4.7).

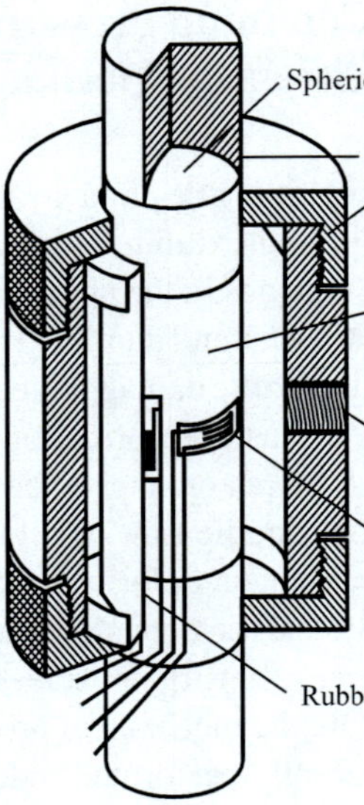


**Fig. 4.4** Conventional triaxial rock mechanics experiment system designed by V. Karman, Germany

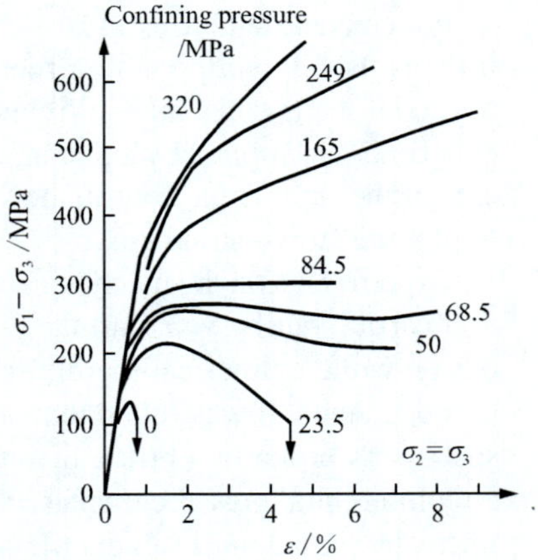


**Fig. 4.5** Stress–strain curves for Carrara marble at different confining pressures (Karman 1912)

Both the rock strength and the pre-damage strain increase with the increase in the confining pressure. For example, dolomite, when the confining pressure increases from zero to 145 MPa (Fig. 4.7), its strength $(\sigma_1 - \sigma_3)_{\max}$ almost doubles, while the confining pressure is 200 MPa strength increases further, but the strength increase is not proportional to the confining pressure. A similar situation is observed for marble and limestone. The pre-damage strain of marble rocks also increases with the increase of the confining pressure, when the confining pressure is zero the pre-damage strain is about 0.3%, when the confining pressure increases to 50 MPa, the strain is about 7%; when the confining pressure increases to 165 MPa, the pre-damage strain reaches 9%. The pre-destructive strain of most rocks can reach more than 10% as the confining

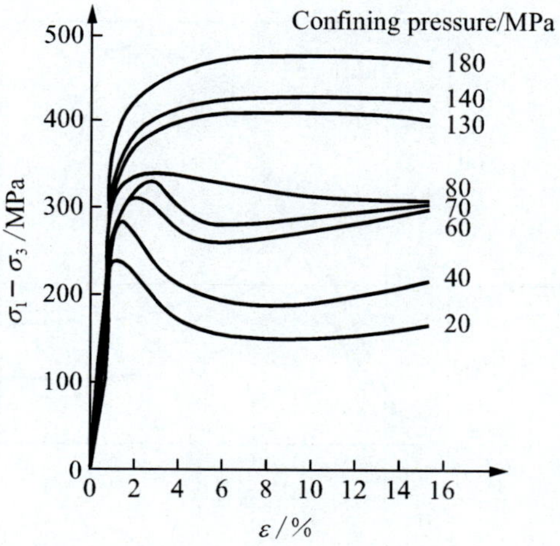


**Fig. 4.6** Stress–strain curves for the Clonport limestone under confining pressure (Spencer 1981)

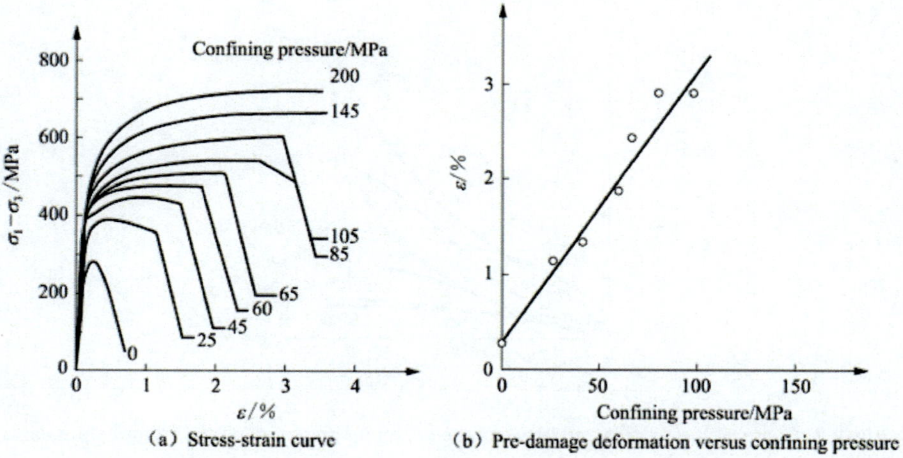


**Fig. 4.7** Stress–strain curve of dolomite under confining pressure (Mogi 1966)

pressure increases. There is an almost linear relationship between the confining pressure and pre-destructive strain for dolomite, but not all rocks have a linear relationship between the confining pressure and pre-destructive strain. Figure 4.9 shows the relationship between confining pressure and pre-destructive strain for several rocks. It can be seen from the figure that the pre-destructive strain varies with the rock type, even at the same confining pressure.

Confining pressure also affects the residual strength of the rock. As can be seen from Figs. 4.5, 4.6, 4.7, and 4.8, if the confining pressure is zero or very low, the stress value reaches a peak and then its curve rapidly decreases to zero, indicating that there is no residual strength in the rock under such conditions. However, as the

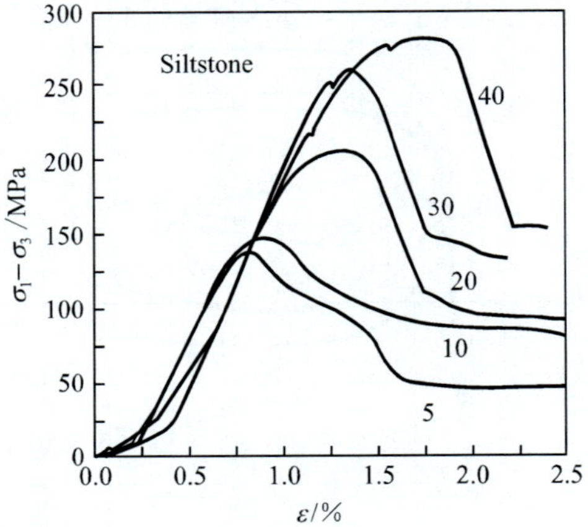


**Fig. 4.8** Stress–strain curve of siltstone under confining pressure (You 2003)

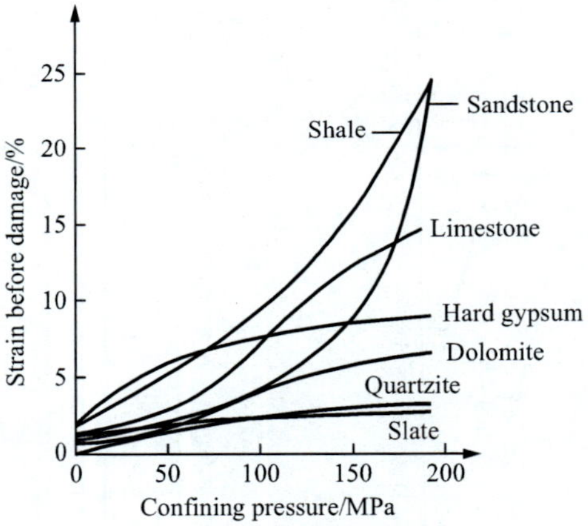


**Fig. 4.9** Curve of the relationship between the confining pressure and the strain before damage (Spencer and NUr 1977)

confining pressure increases, the residual strength of the rock gradually increases until ductile flow or strain hardening occurs.

The Japanese scholar Kiyoh Mogi divided the rocks into two categories according to the mechanical properties of rocks under confining pressure: Class A rocks (mainly carbonate rocks) and Class B rocks (mainly silicate rocks) as shown in Fig. 4.10.

Class A rocks have relatively little effect of their confining pressure on yield stress, i.e., the yield stress increases relatively little as the confining pressure increases, but their pre-damage strain increases monotonically as the confining pressure increases. In other words, carbonate type rocks are prone to transition from brittleness to ductility at room temperature.

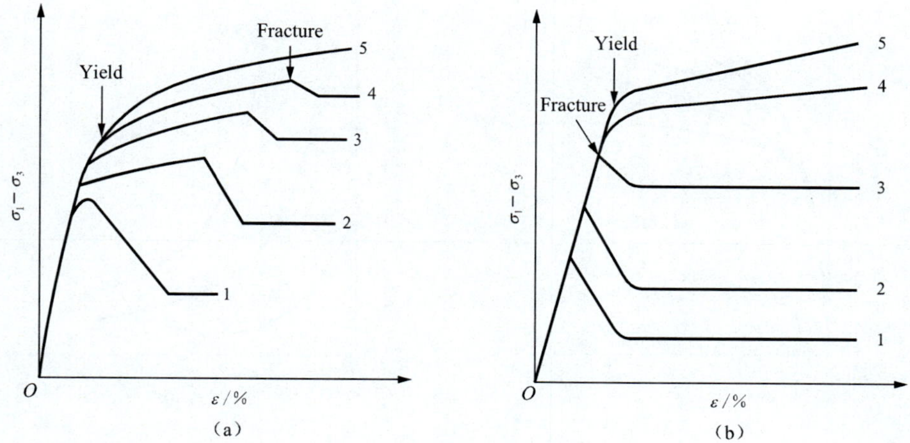


**Fig. 4.10** Stress–strain curves for rock type A (**a**) and rock type B (**b**) (Mogi 1966)

Class B rocks have a large effect of confining pressure on strength, which increases with increasing confining pressure, but the transition from brittle to ductile at room temperature often requires an increase in higher confining pressure unless the temperature is increased. Some silicate rocks remain in a brittle state when the confining pressure is increased to several hundred MPa. For example, basalt and granite do not convert from brittle to ductile until they reach about 1000 MPa at room temperature; general rocks do not convert to ductile until roughly 1200 MPa; quartzite is still brittle even at 2000 MPa. The strain before failure is supposed to be small, about a few percent, but sometimes some strain is caused by friction on the damage surface of the rock and by debris flow on the damage surface.

The effect of confining pressure on the elastic modulus of rocks can generally be divided into two cases: less for hard, low porosity rocks and more for soft, highly porous rocks. Hoffmann (1958) experiments on sandstones indicate that the elastic modulus increases by 20% with increasing confining pressure and decreases by 20–40% near destruction. However, in general, the elastic modulus E and Poisson's ratio ν of the rock increase to some extent with increasing confining pressure.

Mogi (1966) conducted true triaxial ($\sigma_1 \neq \sigma_2 \neq \sigma_3$) rock mechanics experiments to study the effect of intermediate principal stresses $\sigma_2$ on the mechanical properties of rocks, as shown in Fig. 4.11.

When $\sigma_3$ is a fixed value ($\sigma_3 = 125$ MPa), the rock strength ($\sigma_1 - \sigma_3$) Max increases with the increase of the intermediate principal stress $\sigma_2$ and when the value of $\sigma_2$ is larger than $\sigma_3$, the rock tends to show a brittle state. When $\sigma_2$ decreases gradually, then the rock gradually transitions from a brittle to a ductile state. When $\sigma_2 = \sigma_3$, which corresponds to the confining pressure case, the rock exhibits a ductile flow state. It is clearer from Fig. 4.11b that if $\sigma_3$ is constant, then the strain before damage decreases exponentially with increasing intermediate principal stress $\sigma_2$; but for the same value of $\sigma_2$, the strain decreases with decreasing of $\sigma_3$. In other words, the mechanical properties of rocks in general are influenced by the intermediate principal

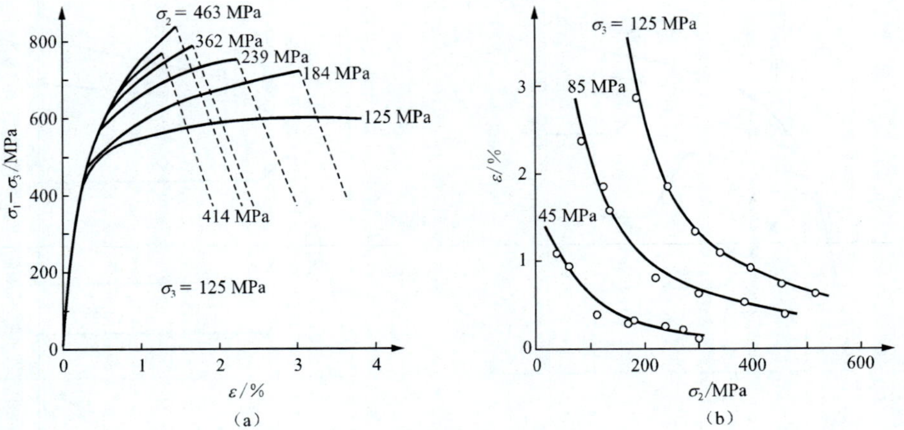


**Fig. 4.11** True triaxial stress–strain curves for dolomite (Mogi 1972)

stress, and as the intermediate principal stress $\sigma_2$ increases, the rock becomes more brittle and less ductile. If the intermediate principal stress $\sigma_2$ is kept as a constant and the effect of the change in $\sigma_3$ on the stress–strain is studied, as shown in Fig. 4.12, the barite gradually transitions from brittle to ductile as the minimum principal stress $\sigma_3$ is increased, but its yield stress remains constant.

The results of the above experiments are summarized as follows.

- The strength, yield stress and ductility increase as the confining pressure increases.
- In the triaxial unequal compression case, as the intermediate principal stress $\sigma_2$ increases, its strength, yield stress increases, but its ductility decreases.
- In the triaxial unequal compression case, the strength and ductility increase as the minimum principal stress $\sigma_3$ increases, but the yield stress remains constant.

**Fig. 4.12** Experimental curves for true triaxial marble (Mogi 1966)

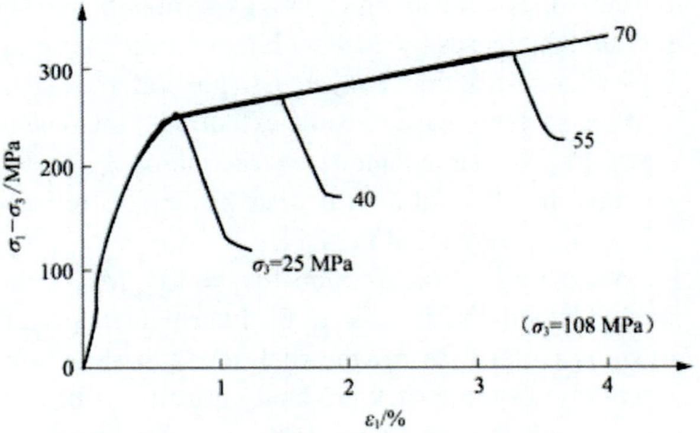

## 4.3 Effect of Temperature on Mechanical Properties of Rocks

The temperature of the Earth's interior gradually increases with depth due to the metamorphosis of radioactive material. Underground mineral development and drilling engineering practice show that the temperature gradient below the surface varies with different regions. In general, the temperature gradient is about 20–30 °C/km, and the average temperature gradient is about 25 °C/km in the Asian continent, about 50 °C/km in Daqing Oilfield, and up to 40–80 °C/km in regional metamorphic areas. If these figures are estimated, the temperature can reach more than 100 °C at a depth of a few kilometers to a dozen kilometers, making the mechanical properties of rocks significantly different compared to those at room temperature and pressure. Even typical brittle rocks may be converted to ductile damage in this physical environment. Therefore, the study of rock mechanical properties under the influence of temperature and confining pressure factors is of great importance for understanding the mechanical properties of subsurface rocks in petroleum engineering.

Experiments show that the yield stress and strength of rocks under a certain confining pressure are reduced with increasing temperature, either in tension or compression, accelerating the transformation from brittleness to ductility. The degree of influence varies with different rock types and stress states.

Figure 4.13 shows the conventional triaxial stress–strain curves of marble, granite, and gabbro at 500 MPa confining pressure with temperature variation. From the experimental results, it can be seen that the yield stress and strength are high at room temperature (25 °C); as the temperature increases, the yield stress and strength decrease and are gradually converted to ductility. Granite at an confining pressure of 500 MPa and a temperature of 25 °C, the strength $(\sigma_1 - \sigma_3)_{\max}$ can reach 2000 MPa and exhibits brittle damage; however, when the temperature increases by 800 °C, the strength decreases to about 600 MPa, which is about 1/3 of the strength at room temperature (25 °C), and ductile flow occurs. Therefore, the temperature under certain confining pressure is the main factor in the transformation of rock deformation from brittle to ductile. The reason for ductility at elevated temperature is the enhanced thermal movement of molecules within the rock, which weakens the cohesive forces between them and makes the grain faces susceptible to slip. Figure 4.14 shows the stress–strain curves for conventional triaxial compression of the Solnhofen limestone at an confining pressure of 300 MPa with a change in temperature. These results illustrate not only the effect of temperature on strength, yield stress, and conversion of brittle to ductile, but also the varying degree of effect for different types of rocks. For example, at an confining pressure of 300 MPa, the Solnhofen tuff becomes ductile at about 400 °C and has a strength of about 450 MPa, whereas the dolomite needs to reach 800 °C before it becomes ductile and has a strength of about 550 MPa.

Heard (1960) carried out triaxial compression tests on the Solenhofen limestone under the influence of confining pressure and temperature factors and derived the confining pressure and temperature limits for the conversion of limestone from brittle

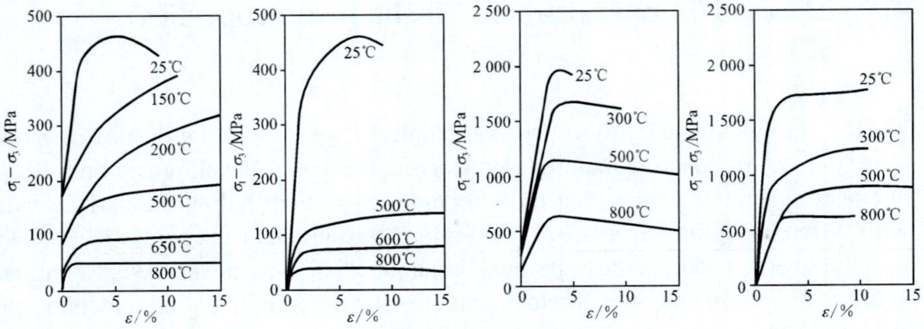


(a) Marble compression with a strain rate of 0.03/s (b) Marble extension with a strain rate of 0.02/s (c) Granite compression (d) Gabbro compression

**Fig. 4.13** Stress–strain curves for rock at 500 MPa perimeter pressure with temperature change

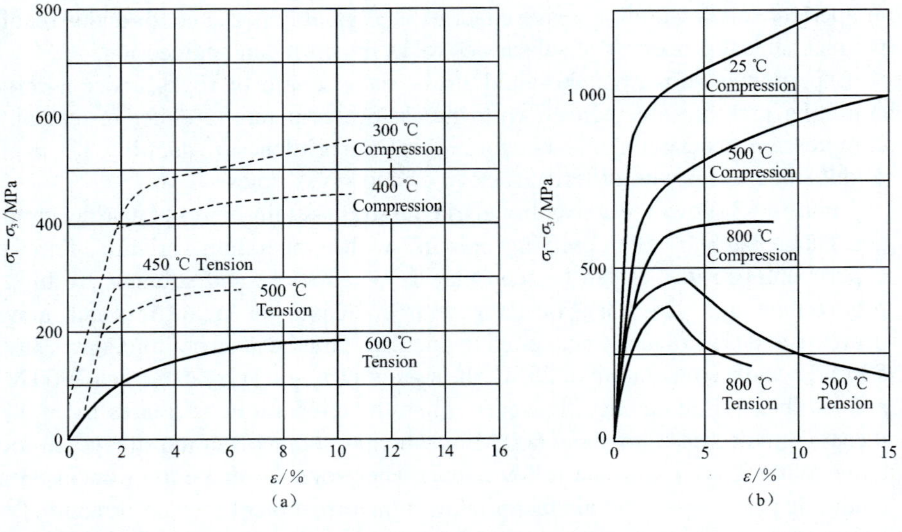


**Fig. 4.14** Stress–strain curves of **a** Solenhofen limestone, **b** dolomite with temperature for an confining pressure of 300 MPa (Spencer 1981)

to ductile, as shown in Fig. 4.15. Since the loading properties in tension and compression are different, the brittle to ductile conversion limits also vary. The temperature and the confining pressure required to convert brittleness to ductility in tension are much greater than in compression. For example, at a temperature of 400 °C and an confining pressure of 200 MPa, the tuff is converted from brittle to ductile in compression; however, at the same temperature and confining pressure, it is still in a brittle state in tension. Only by increasing the temperature or the confining pressure can the limestone be converted to a ductile state. For example, at a temperature of 400 °C, the confining pressure increases to more than 400 MPa, or the confining pressure remains at 200 MPa, while the temperature increases to more than 600 °C,

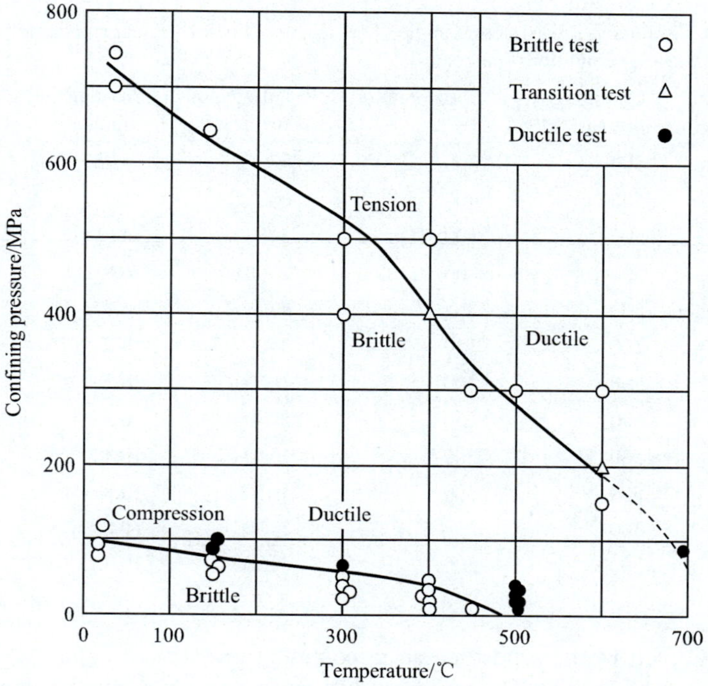


**Fig. 4.15** Temperature and pressure conditions for the conversion of limestone from brittle to ductile (Heard 1960)

before it is possible to convert the stretched limestone from brittle to ductile. This indicates that when the strata are subjected to compression at a certain temperature and confining pressure, they are prone to ductile deformation and produce folds, lamellae and other tectonic types; when subjected to tension, they tend to show a brittle state and are prone to normal faults, joints and other tectonic forms.

Rock salt is representative of minerals that produce ductility, other minerals such as barite and azurite (a type of plagioclase) can also produce ductility at an confining pressure of $2 \times 10^3$ MPa and a temperature of 400 °C. However, quartz crystals show a brittle state at an confining pressure of $5 \times 10^3$ MPa and a temperature of 400 °C or an confining pressure of $2 \times 10^3$ MPa and a temperature of 800 °C. The strain and shear fracture angle of several rocks before rupture is produced at temperature and confining pressure are listed below for reference (Table 4.1).

The effect of temperature on the modulus of elasticity of rocks also depends on the type of rock. The results of experiments conducted by Handin and Hager (1958) on Barns sandstone at temperatures increasing from room temperature to 300 °C show that the value of the modulus of elasticity decrease gradually with increasing temperature. This decrease can also be seen in Figs. 4.13 and 4.14, among others. However, the extent of the effect varies with the type of rock and with the tensile or compressive nature of the rock. experiments conducted by Hughes and Jones (1950)

**Table 4.1** Strain and shear fracture angle of several rocks before damage under temperature and confining pressure conditions (Paterson 1978)

| Rock type | Confining pressure/MPa | Temperature/°C | Differential stress/MPa | Pre-damage strain/% | Shearing angle/(°) |
|---|---|---|---|---|---|
| Dolomite | 500 | 300 | 470 | 1.25 | 31 |
| | 500 | 400 | 600 | 1.6 | 37 |
| Quartzite | 500 | 500 | 3050 | 4.8 | – |
| | 500 | 800 | 2500 | 6.8 | 45 |
| Limestone | 500 | 25 | 470 | 6.8 | 25 |
| | 500 | 150 | 450 | 4.9 | 30 |
| Granite | 500 | 300 | 1660 | 8.1 | 33 |
| | 500 | 25 | 2080 | 6.4 | 32 |
| Olivine | 500 | 300 | 1540 | 10.4 | 34 |
| | 500 | 500 | 1050 | 10.8 | 32 |
| Dolerite | 500 | 25 | 1740 | 9.8 | – |
| | 500 | 300 | 1250 | 10.3 | – |

on the Caplen Dorne sandstone at an confining pressure of about 50 MPa and a temperature of 25–200 °C showed a reduction in elastic modulus of about 20%.

## 4.4 Effect of Pore Pressure on Mechanical Properties of Rocks

Rocks in the earth's crust contain a wide variety of pores and fissures. There are many causes of these pores and fissures, ranging from primary pores created during rock formation to secondary fissures created by internal and external dynamics. As far as their primary pores are concerned, their causes are also various. For example, during the process of magmatic rock formation, the volatile components in the magma precipitate out in a gaseous state due to the decrease in pressure, forming pores, which become pores containing solutions when the temperature drops and are dispersed in the magmatic rock. Due to different generation depths, different conditions of magma solidification, and different gas exhalations contained, magmatic rocks have different pore volumes. Primary pore space in sedimentary rocks depends mainly on factors such as sediment shape, sorting and filling. During rock-forming compaction, there are often two types of pore space, one type of pore space belongs to the joint pore space, under the rock-forming compaction, the liquid then expels from the pore space, and the pore space then decreases. The other type is a closed pore space, which is filled with water retained during deposition. Due to compaction, the pore water between the clay-like rock particles can not be freely discharged. In the overburden pressure, in the rock particles and particles contact edge, the inevitable formation

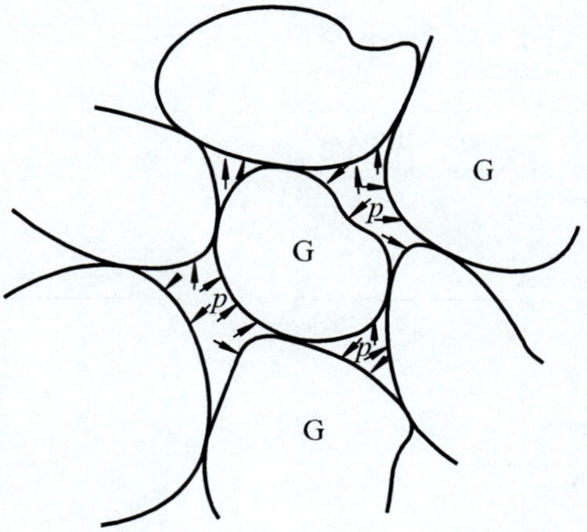


**Fig. 4.16** Schematic diagram of pore pressure formation (Fyfe et al. 1978)

of stress concentration, resulting in the deformation of the particles near the contact point, this deformation is bound to cause the volume of the intergranular pore space reduction. If the pore contains a closed static liquid, due to the incompressibility of the liquid, it is bound to resist the reduction potential of the pore. Therefore, the intergranular liquid will produce a pressure on the particles, this pressure is perpendicular to the surface of the particles (Fig. 4.16), called pore pressure.

Rocks in nature contain more or less water or other liquids (e.g., oil, natural gas, etc.), and under normal pressure systems, pore pressure increases linearly with depth:

$$p = \rho_w g h \tag{4.1}$$

where $\rho_w$ is the density of the liquid, g is the acceleration of gravity, and $h$ is the depth of rock burial.

During the development of oil and gas, actual measurements of pore pressure have been obtained. Pore pressure is not always normal pressure, and Fig. 4.17 shows the variation curve of pore pressure with well depth in the Monte Cristo field, while the variation of overburden pressure with well depth is given, which shows the relationship between the ratio of pore pressure and the confining pressure.

The effect of the liquid in the pores of the rock on the properties of the rock can be divided into two effects, one of which is the decrease of the internal surface free energy due to the adsorption of the liquid on the surface of the pores, which increases the possibility of dislocation at the boundaries of the particles. It also produces the effects of pressure solubility, diffusion, dissolution, lubrication, etc., which is conducive to the generation of new minerals. The other effect mainly shows the effect of pore pressure on the mechanical properties of the rock.

The concept of pore pressure and effective stress was first introduced by Terzaghi (1933) in his analysis of saturated soils: when a load is applied, the compressive stress $\sigma$ in the soil is carried by two components, the effective compressive stress $\sigma'$ at the point of contact of the particles and the pore pressure Pp generated by the

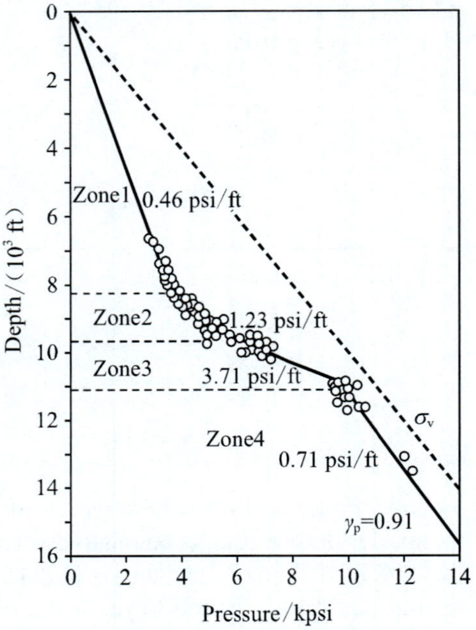


**Fig. 4.17** Shows the variation curve of pore pressure with well depth in the Monte Cristo field (Engelder and Leftwich 1997)

saturated water in the pores (assuming that the pore water cannot drain freely). So the stress at any point in the saturated soil is

$$\sigma = \sigma' + P_p$$

The effective stress is

$$\sigma' = \sigma - P_p \tag{4.2}$$

Griggs (1936) performed the first triaxial stress experiments with rock specimens without jackets. At this time the pore pressure was nearly equal to the confining pressure and it was observed that the Solnhofen tuff was brittle failure at a confining pressure of 405 MPa. Later Griggs (1940), Turner et al. (1950) and Handin et al. (1963) added impermeable jackets to most of the rock specimens so that the pore pressure was zero, or pore water could not drain freely, and compared the results of the jacketed and unjacketed experiments. It was found that when the rock was stressed, the pore pressure of the rock specimen counteracted the effect of the jacket pressure. Handin et al. (1963) performed triaxial stress experiments on Berea sandstone, Marianna limestone, Hosmark dolomite, Reptto siltstone, and Muddy shale. In the experiments pore pressures up to 200 MPa were applied. The experimental results showed that the strength of porous rocks depends on the difference between the confining pressure and the pore pressure, i.e. the effective confining pressure. Thus the concept of effective stress becomes widely used in rocks.

Skempton (1961), inspired by experiments, modified Terzaghi's theory of effective stress, namely

$$\sigma' = \sigma - P(1 - \beta) \tag{4.3}$$

where $\beta$ is the material constant, which is approximately zero for some rocks, but can be 3–5% for other types of rocks. And $\beta = K/Ks$, where $K$ is the bulk modulus of the whole rock and Ks is the bulk modulus of the constituent minerals. This shows that the effective stress is not only related to the confining pressure and pore pressure, but also to the nature of the material.

Confining and pore pressure experiments were performed for dry and saturated rocks. The volumetric strain $\varepsilon_V$ is used as the vertical axis and the effective stress $\sigma'$ is used as the horizontal axis. Three different cases are considered, as shown in Fig. 4.18. (a) no pore pressure is considered and the horizontal coordinate is the total stress $\sigma$; (b) the pore pressure P is considered but not the material properties, and the horizontal coordinate $\sigma' = \sigma - p$ (Terzaghi formula); (c) the pore pressure $p$ and the material properties β are considered, i.e., the horizontal coordinate $\sigma' = \sigma - p(1 - \beta)$ (Skempton) formula. As can be seen from Fig. 4.18, the total stress versus volume strain relationship in (a) is scattered; the Terzaghi effective stress versus volume strain relationship in (b) is more densely distributed over a very narrow strip; and the Skempton effective stress versus volume strain relationship in (c) overlaps on a single curve.

Generally, for the rocks containing pores, $K_s > K$, for high permeability sandstone β is approximately equal to 0, so $\sigma' \approx \sigma - p$, so the simpler Terzaghi formula is still used in general rock mechanics. However, in the study of hydraulic fracturing, the significance of β should not be neglected.

Porosity and pore pressure have a large effect on the deformation, strength, compressibility and elastic modulus of the rock.

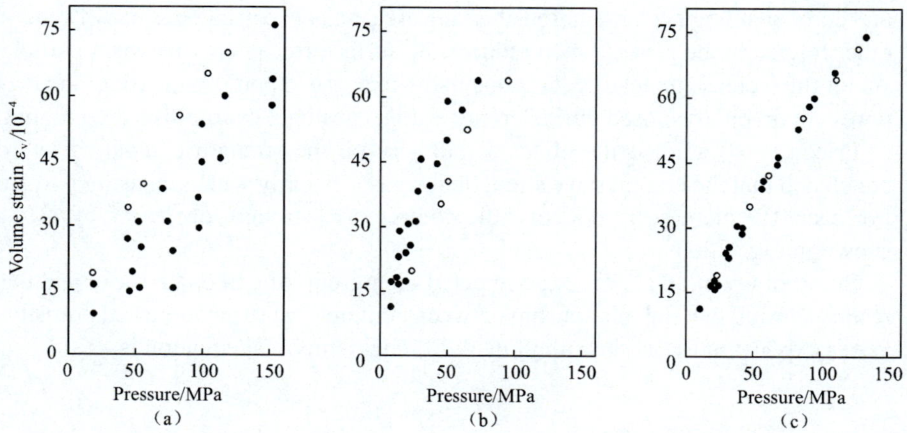


**Fig. 4.18** Relationship between effective stress and volume strain (Wang et al. 1980)

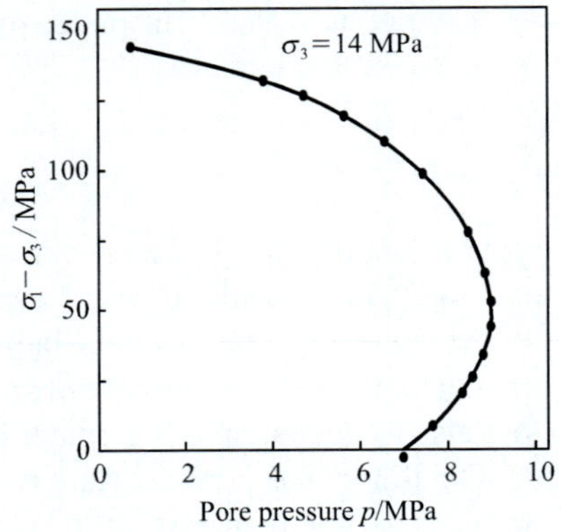


**Fig. 4.19** Pore pressure versus differential stress for undrained experiments in Berea sandstone (Paterson 1978)

In the undrained test (i.e., when the rock is saturated and hydraulically isolated from the outside world so that the pore fluid content remains constant), a curve can be obtained for the change in differential stress versus pore pressure. Figure 4.19 shows the experimental results for the Berea sandstone at 18% porosity, 14 MPa confining pressure, and 7 MPa initial pore pressure. It can be seen from the Fig. that under the triaxial stress experiment, the initial stage causes the pore pressure to increase with the increase in differential stress because the pores are compacted. After the pore pressure reaches 9 MPa, the pore pressure gradually decreases with the increase in differential stress. This is a result of the gradual expansion and development of microfractures inside the specimen after the specimen reaches the yield stress, and the rock produces dilation (expansion). The change in pore volume in the rock specimen can also be seen in such experiments.

Price (1960), Smorodinov et al. (1970) and others conducted experiments on coal rocks, carbonate rocks and quartzites and concluded that the increase in porosity in rocks decreases the strength but increases the ductility. The reasons for the decrease in strength are (1) stress concentration due to pore boundaries, (2) increase in porosity and corresponding decrease in rock bearing area, and (3) partial presence of water or other liquids in the pores, which reduces the surface free energy between particles. As for the increase in ductility, it is due to the increase in pores in the rock, and these pores are gradually closed during compression, causing a ductile-like deformation.

Price (1960) studied the effect of porosity on the strength of coal rocks and concluded that the compressive strength decreases linearly with increasing porosity. For each 1% increase in porosity, the compressive strength decreases by 4%, as shown in Fig. 4.20.

Smorodinov et al. (1970) experimented on a group of carbonate rocks and their results showed that the relationship between compressive strength σc and porosity ϕ varies exponentially, as shown in Fig. 4.21. Their empirical equation is

$$\sigma_c = 2590\exp(-0.09\varphi) \tag{4.4}$$

where $\phi$ is the rock porosity.

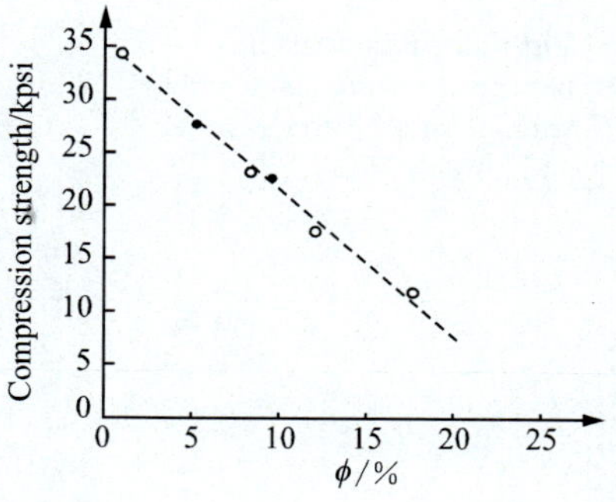


**Fig. 4.20** Relationship between porosity and compressive strength of coal-derived rocks (Price 1960)

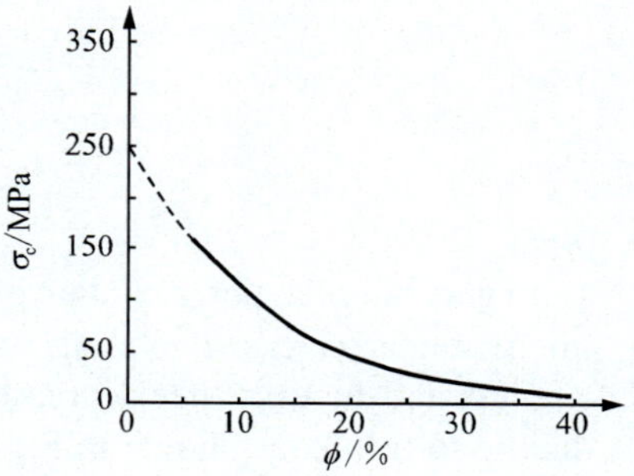


**Fig. 4.21** Relationship between porosity and compressive strength of carbonate rocks (Smorodinov et al. 1970)

The compressive strength versus porosity curve for quartzite is shown in Fig. 4.22. The empirical equation is

$$\sigma_c = 3500\exp(-0.108\varphi) \tag{4.5}$$

Dube and Singh (1972) experimentally studied the relationship between tensile strength and porosity of sandstones and compared dry with wet sandstones. As shown in Fig. 4.23, the tensile strength of wet sandstone is somewhat lower than that of dry sandstone for the same porosity. However, the tensile strength of sandstones, whether wet or dry, decreases with increasing porosity.

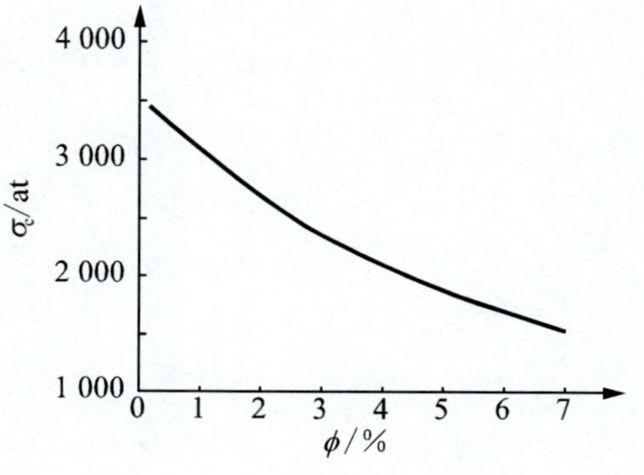


**Fig. 4.22** Relationship between porosity and compressive strength of quartzite (Smorodinov et al. 1970)

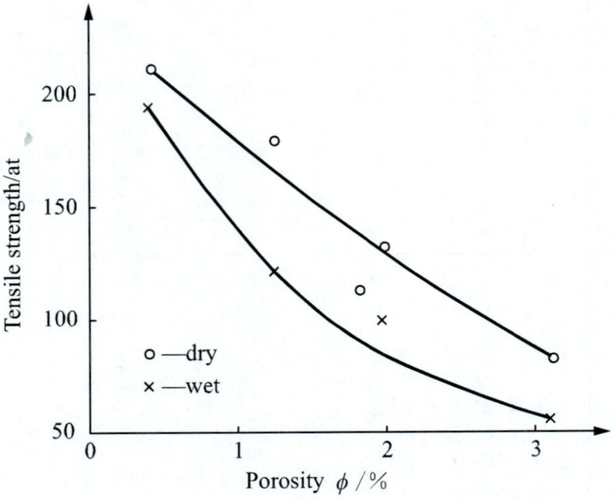


**Fig. 4.23** Relationship between porosity and tensile strength of sandstone (Dube and Singh 1972)

In general, rock pores contain pore pressure. For rock triaxial compression experiments under certain confining pressure, as the pore pressure increases, the rock strength and ductility then decreases, and the deformation gradually transforms from ductile to brittle, as shown in Fig. 4.24. If the pore pressure is zero (i.e., the rock has pores but no pore pressure), the limestone not only increases in strength but also appears in the strain-hardening stage at higher confining pressures. When the pore pressure is equal to the confining pressure, the stress–strain curve and strength are equivalent to uniaxial compression. In general, the rock contains pores, but the pore pressure is less than the confining pressure, and its stress–strain curve tends to be between the two curves mentioned above, and its strength is also in between.

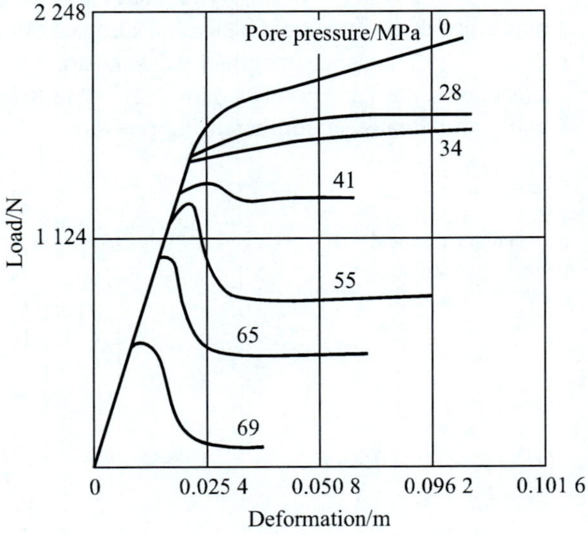


**Fig. 4.24** The pattern of influence of limestone pore pressure on rock strength (Spencer 1981)

Figures 4.25 and 4.26 show the relationship between strength, permanent strain before failure, and depth for dry and saturated rocks with pore pressure, respectively. Comparing the two plots shows that the strength and ductility of both depend on the effective confining pressure. For example, dry limestone has a strength of about 200 MPa at a depth of 2 km and a permanent strain before failure of about 8–12% at a depth of 3 km, whereas with normal pore pressure, its strength decreases to 120 MPa at 2 km and the permanent strain before failure is about 2–7% at 3 km, which is entirely due to the presence of pore pressure that reduces the effective confining pressure. The influence of pore pressure on porous sandstone is particularly significant; as the pore pressure increases, its internal friction angle and cohesion also gradually weaken, and it becomes brittle, and its rupture tends to occur between the grains of the rock. If the effective confining pressure increases, the particles are compressed with each other and tend to cut through the particles to produce rupture. Similar situations exist for other types of rocks, but the strength and permanent strain before failure also vary with the type of rock.

Figure 4.27 shows a series of experiments conducted by Rutter (1974) on the Solnhofen limestone considering the effects of differential stress, confining pressure, pore pressure, temperature, and depth of burial separately. Separate experiments were conducted on dried and saturated rock samples to derive a limit for the conversion of limestone from brittle to ductile. Where $\lambda$ is the ratio of the pore pressure P to the confining pressure $\sigma_3$ (i.e., $\lambda = P/\sigma_3$), the weakening effect of pore pressure can be seen in experiments varying in the range $\lambda = 0$ (drying) to $\lambda = 1$ (pore pressure equal to the confining pressure). $\lambda = 1.0$, the rock enters the brittle domain completely, and the limestone remains in the brittle domain even when the limestone is buried at a depth of 15 km, a temperature of 450 °C, and an confining pressure of 300 MPa.

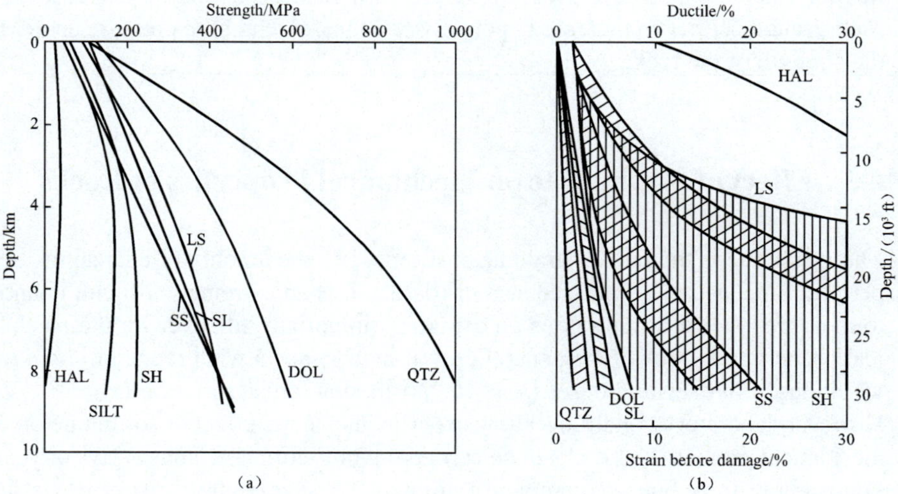


**Fig. 4.25** Relationship between dry rock strength, permanent strain before fracture, and depth (Handin et al. 1963) from left to right: HAL—marlstone, SITL—siltstone, SH—shale, Ls—limestone, SS—sandstone, SL—slate, DOL—dolomite, QTZ—quartzite

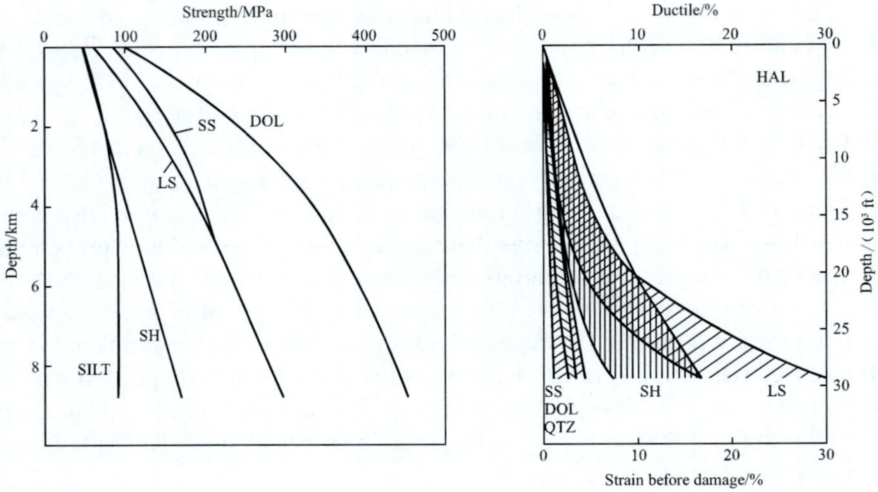


**Fig. 4.26** Relationship between strength, permanent strain before fracture, and depth for saturated rocks with pore pressure (Handin et al. 1963) from left to right: SITL—siltstone, SH—shale, Ls—limestone, SS—sandstone, DOL—dolomite

While for $\lambda = 0.9$, the limestone is transformed from brittle to ductile state under the same conditions. For $\lambda = 0.1$, the rock is in the ductile state even with an confining pressure of 100 MPa, a rock burial depth of 5 km, and a temperature of 150 °C.

It is clear from the above that certain effects of rock pore pressure on the mechanical properties of rocks are precisely the opposite of the effects of the presence of only pores in the rock. For example, for rock ductility, when porosity increases, ductility increases; however, when pore pressure increases, ductility decreases. For rock strength, either an increase in porosity or an increase in pore pressure decreases the strength of the rock.

## 4.5 Effect of Strain Rate on Mechanical Properties of Rocks

The strain rate is the rate of strain against time, i.e., the amount of change in strain per unit time, usually expressed as $\dot{\varepsilon}$ or ($\mathrm{d}\varepsilon/\mathrm{d}t$). It is an extremely difficult quantity to measure, yet the strain rate is an extremely important influence on the mechanical properties of rocks. The general crustal motion has a wide range of strain rate variations, from seismic onset ($\varepsilon = 10^{-4}$/s) to slow crustal motion ($\varepsilon = 10^{-15}$/s). However, laboratory conditions allow experiments on the effect of strain rate on the mechanical properties of rocks to be carried out only for a few hours, days or years. For example, three hours to complete a strain of 10% gives a strain rate of $10^{-5}$/s. The slowest experimental strain rate can be up to $10^{-8}$/s. At a certain confining pressure and temperature, the strain rate has a large effect on the deformation characteristics, elastic modulus and strength of the rock. In general, the peak stress (strength) and

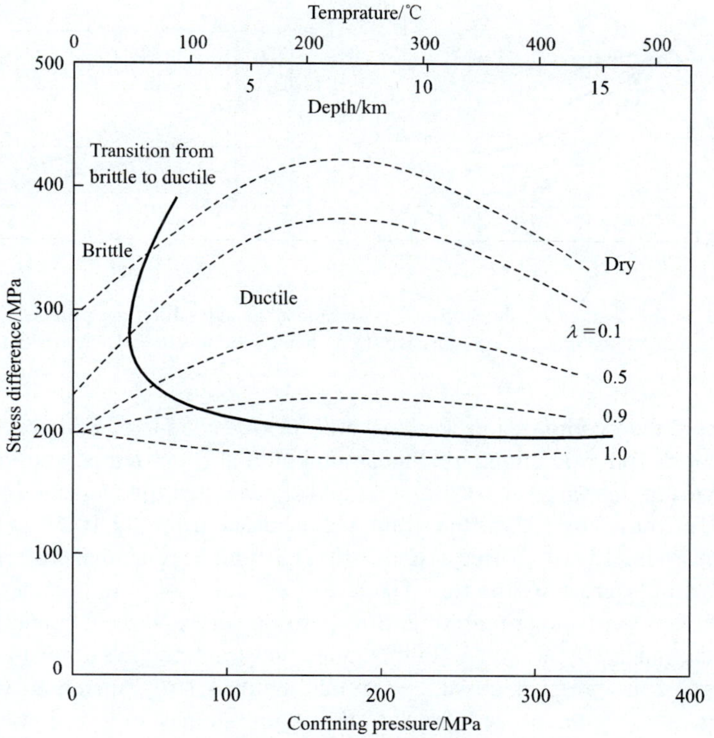


**Fig. 4.27** Effect of limestone pore pressure on strength and mode of destruction (Rutter 1972, 1974)

elastic modulus of the rock decrease with decreasing strain rate, while the pre-damage strain increases with decreasing strain rate.

Bieniawski (1970) conducted experiments on sandstone cylindrical specimens ($\phi$21.6 mm × 10.8 mm) on a rigid test machine using different strain rates, and obtained stress–strain relationship curves and the effect of strain rate on uniaxial compressive strength and modulus of elasticity, as shown in Fig. 4.28. When the strain rate increases, the peak stress (compressive strength) also increases, for example, the compressive strength exceeds 120 MPa at a strain rate of $10^{-4}$/s, while the compressive strength decreases to about 93 MPa at a strain rate of $10^{-10}$/s, with a difference of about 30% between the two. The modulus of elasticity increases with increasing strain rate. The relationship between the modulus of elasticity of sandstone and strain rate is approximately linear (Fig. 4.28b). The strain before damage, however, increases as the strain rate decreases; the curve after the peak stress decreases sharply as the strain rate increases, and gradually transforming into brittle damage. Horibe and Kobayashi (1965) subjected a group of rocks to compression and tension experiments using a stress rate of $10^{-2}$–$10^{-5}$ MPa/s. Their results indicate that the elastic modulus increases with increasing stress rate and that the tensile experiments are

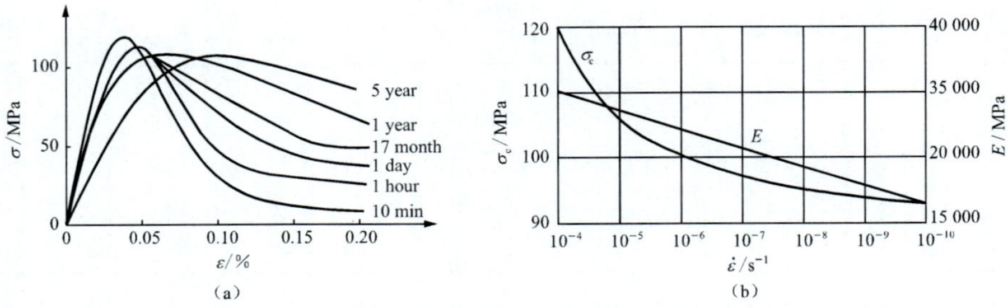


**Fig. 4.28** **a** Stress–strain relationship curves for sandstone under different strain rate conditions, **b** effect of strain rate on uniaxial compressive strength and modulus of elasticity (Bieniawski 1970)

more rapid than compression. Perkins et al. (1970) conducted compression experiments on porphyritic clinopyroxene amphibolite at room temperature (25 °C), at strain rates in the range of $10^{-4}$ to $10^{3}$/s and showed that the tangential modulus of the rock increased by 15% as the strain rate increased from $3 \times 10^{-4}$/s to $6 \times 10^{-1}$/s. Peng and Podnieks (1972) used a rigid testing machine for tuffs that performed experiments with different strain rates. The test piece was a 31.5 mm diameter cylinder with a height to diameter ratio of 2. Stress–strain curves were obtained for a strain rate in the range $10^{-2}$ to $10^{-7}$/s. In this range the strain rate has almost no effect on the shape of the pre-peak curve, except that the peak stress increases as the strain rate increases; for the post-peak curve, the strain rate has a more significant effect (Fig. 4.29).

Heard (1972) carried out a series of experimental studies on polycrystalline salt rocks, at different temperatures and strain rates, as shown in Figs. 4.30 and 4.31. A comparison of the two figures shows that: the effect of strain rate on the stress–strain curve is more pronounced as the temperature increases, with almost the same initial modulus of elasticity at higher temperatures; and a small difference at lower temperatures. At the same strain rate, the yield stress also decreases relatively due to the decrease in strain hardening coefficient at higher temperature.

Based on Rutter's (1972) experimental data on the effect of different strain rates on the strength of Solnhofen limestone at temperatures of 20 and 400 °C, Fyte et al. (1978) state that the change in strength with strain rate is related to the surrounding environment and especially to temperature. At low temperatures the strength decreases less as the strain rate decreases. For example, when comparing stresses at 10% strain, a reduction in strain rate by one order of magnitude (e.g., $10^{-4} \rightarrow 10^{-7}$/s) results in a reduction in strength of about 3%. However, at high temperatures (400 °C), strain rates below $10^{-8}$/s result in a rapid reduction in strength (about 40%).

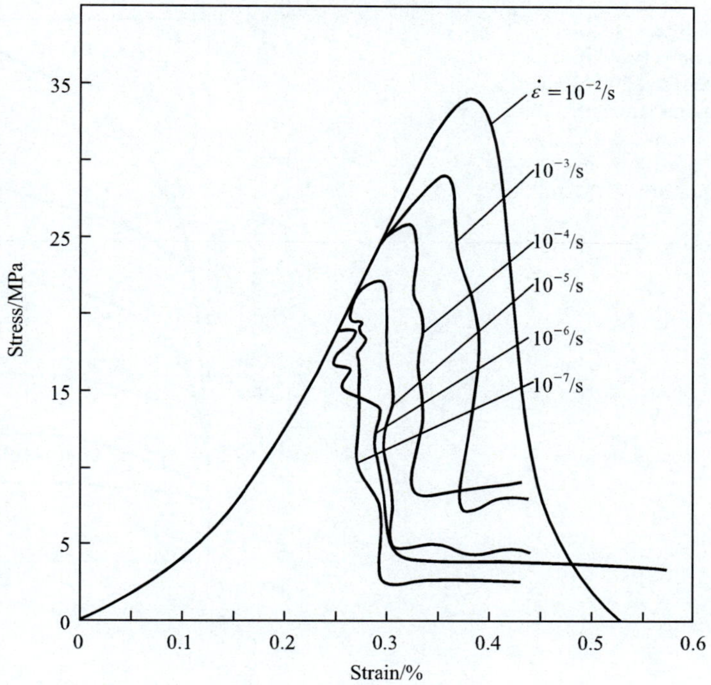


**Fig. 4.29** Stress–strain relationship curves for tuffs under different strain rate conditions (Peng and Podnieks 1972)

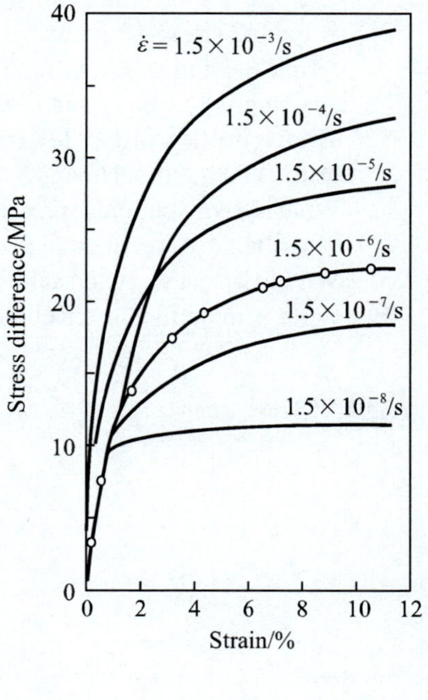


**Fig. 4.30** Stress–strain curve for polycrystalline salt rock stretched at 200 MPa confining pressure with strain rate from $1.5 \times 10^{-3}$/s–$1.5 \times 10^{-8}$/s, and temperature 100 °C

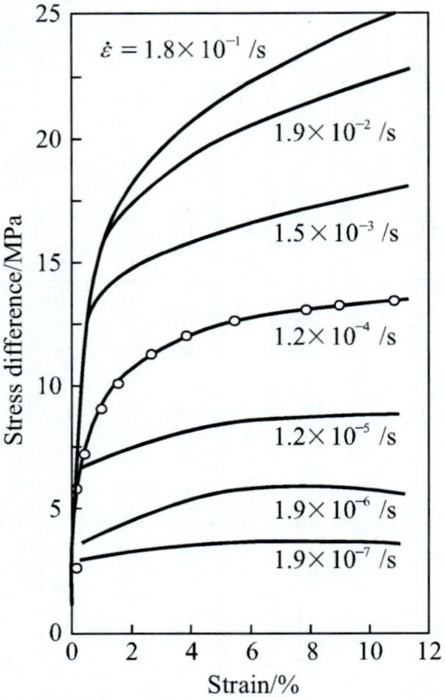


**Fig. 4.31** Stress–strain curve for polycrystalline salt rock stretched at 200 MPa confining pressure with strain rate from $1.8 \times 10^{-1}$/s–$1.2 \times 10^{-7}$/s and temperature 248 °C (Heard 1972)

## Exercises

1. What are the geometrical forms of stress–strain curves for rocks under uniaxial compression, and describe them briefly.
2. A typical stress–strain curve can be obtained by placing the rock specimen in a rigid testing machine for experiments as shown in Fig. E1, with the basic characteristics illustrated in test segments.
3. What is brittle and ductile damage of rocks? What are the criteria for determining brittle and ductile damage?
4. What is a conventional triaxial stress experiment, explain the concept and briefly describe the experimental procedure.
5. Why is it necessary to seal wrap the experimental core in a triaxial press?
6. What is the effect of envelope pressure on rock properties?

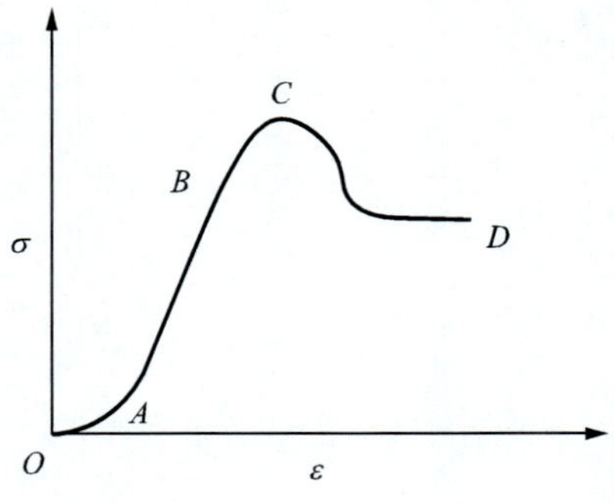


**Fig. E1** Stress–strain curve

7. What effect does temperature have on the properties of rocks?
8. What is the pore pressure of a rock and how does it affect the properties of the rock?
9. What is strain rate and how does it affect the properties of the rock?
10. In the laboratory, what is the strain rate for a rock specimen with 10% strain completed in 3 h? What is it again if converted to one year?

## References


Bieniawski ZT. Time-dependent behaviour of fractured rock. Rock Mech. 1970;2(3):123–37.

Deere D, Miller R. Engineering classification and index properties for intact rock. Tech. Report No AFWL-TR-65-116, Air Force Weapons Lab., Kirtland Air Base, New Mexico; 1966.

Dube AK, Singh B. Effect of humidity on tensile strength of sandstone. J Min, Metals Fuels. 1972;20(1):8–10.

Engelder T, Leftwich JT. A pore-pressure limit in overpressured south Texas oil and gas fields. Seals, Traps and the Petroleum System: AAPG Memoir 67 R.C. Surdam. Tulsa, OK; 1997, p. 255–67.

Fyte WS, Price NJ, Thompson AB. Fluids in the Earth's crust. Amsterdam: Elsevier; 1978.

Griggs DT. Experimental flow of rocks under conditions favoring recrystallization. Geol Soc Am Bull. 1940;51(7):1001–22.

Griggs DT. Deformation of rocks under high confining pressures: I. Experiments at room temperature. J Geol. 1936, p. 541–77.

Handin J, Hager Jr RV, Friedman M, et al. Experimental deformation of sedimentary rocks under confining pressure: pore pressure tests. AAPG Bull. 1963;47(5):717–55.

Handin J, Hager RV Jr. Experimental deformation of sedimentary rocks under confining pressure: tests at high temperature. AAPG Bull. 1958;42(12):2892–934.

Heard HC. Transition from brittle fracture to ductile flow in Solenhofen limestone as a function of temperature, confining pressure, and interstitial fluid pressure. Geol Soc Am Mem. 1960;79:193–226.

Heard HC. Steady-state flow in polycrystalline halite at pressure of 2 kilobars. In: Flow and fracture of rocks; 1972, p. 191–209.

Hoffmann H. Investigations into carbonic rocks under tiaxial pressure for the purpose of rock stress computation. In: Proceedings of the international of Strata control conference Leipzig, 1958.

Horibe T, Kobayashi R. On mechanical behavior of rock under various loading rates. J Soc Mater Sci (Jpn). 1965;14:498–506.

Hughes DS, Jones HJ. Variation of elastic moduli of igneous rocks with pressure and temperature. Geol Soc Am Bull. 1950;61(8):843–56.

ISRM. The complete ISRM suggested methods for rock characterization, testing and monitoring: 1974–2006. International Society for Rock Mechanics, Commission on Testing Methods. 2007.

Jager JC, Cook NGW. Fundamentals of rock mechanics. Beijing: Science Press; 1981.

Mogi K. Fracture and flow of rocks. Tectonophysics. 1972;13(1):541–68.

Mogi K. Pressure dependence of rock strength and transition from brittle fracture to ductile flow. Bull Earthquake Res Int Japan.1966;44:215–232.

Paterson MS. Experimental rock deformation: the brittle field. New York: Springer; 1978.

Peng S, Podnieks ER. Relaxation and the behavior of failed rock. Int J Rock Mech Min Sci Geomech Abstracts. Pergamon 1972;9(6):699–700.

Perkins RD, Green SJ, Friedman M. Uniaxial stress behavior of porphyritic tonalite at strain rates to $10^3$/s. Int J Rock Mech Min Sci Geomech Abstracts. Pergamon 1970;7(5):527–35.

Price NJ. The compressive strength of coal measure rocks. Colliery Eng. 1960;37(437):283–92.

Rutter EH. The effects of strain-rate changes on the strength and ductility of Solenhofen limestone at low temperatures and confining pressures. Int J Rock Mech Min Sci Geomech Abstracts. Pergamon 1972;9(2):183–89.

Rutter EH. The influence of temperature, strain rate and interstitial water in the experimental deformation of calcite rocks. Tectonophysics 1974;22(3):311–34.

Skempton AW. Horizontal stresses in an overconsolidated Eocene clay. In: Proceedings of the 5th international conference on soil mechanics and foundation engineering 1961;1:351–57.

Smorodinov MI, Motovilov EA, Volkov VA. Determinations of correlation relationships between strength and some physical characteristics of rocks. In: International society of rock mechanics, proceedings, 1970;1(1–19).

Spencer JW, NUr AM. The effect of pressure temperature and pore water on velocity in Westerly granite. J Geophys Res. 1977;v81N5:pp8990904.

Spencer JW. STRess relaxation at low frequencies in fluid saturated rocks: attenuation and modulus dispersion. J Geophys Res. 1981;86:1803–1812.

Terzaghi K. Structural mechanics of soil based on Its physical properties. Moscow-Leningrad: Gosstroiizdat;1933.

Turner FJ, Griggs DT, CHIH CS. Fabric of Yule marble deformed experimentally at ordinary temperatures in absence of water, Geological Society of America Bulletin. 1950;61(12):1512.

von Karman T. Festigkeits versuche unter allseitigem druck. Mittelungen Uber Forschungsarbeite Des Ver Deutscher Ing. 1912;55:1749–57.

Wang CY, Mao NH, Wu FT. Mechanical properties of clays at high pressure. J Geophys Res: Solid Earth. 1980;85(B3):1462–8.

You MQ. Young's modulus of rock specimens as a function of peritectic pressure. J Rock Mech Eng. 2003;22(1):53–60.

# Chapter 5
# Characterization and Indoor Determination of the Strength of Rocks

## 5.1 Types of Tock Damage and Destruction

The study of rock damage and damage types is essential for engineering rock mechanics, and only if the types of rock damage are known can the damage criterion be correctly selected to provide a good basis for stress analysis of underground structures.

### *5.1.1 Preparation of Rock Samples*

The study of the strength characteristics of rocks necessitates indoor experiments with cores, so the first question to be answered is the question of how to select rock samples and how to prepare them.

We know that the mechanical properties of a rock depend on the interaction between the constituent crystals, particles and cement, and on the presence of, for example, fractures, joints, layers and smaller faults. On the one hand, it is difficult to state the mechanical properties of the rock, especially its strength, on the basis of the nature of its constituent grains; on the other hand, the distribution of fractures, joints, layers and faults is so variable that the mechanical properties of a large mass of rock affected by such separation have few common associations for any other large mass of rock. In determining the most basic mechanical properties of rocks, therefore, a sufficient number of constituent particles should be included, but sufficient to exclude large structural discontinuities and to give the specimens a roughly homogeneous character, and rock samples of a few to several tens of centimeters in size are generally suitable for this requirement and can be easily experimented on in the laboratory. It can be seen that the rock must be sampled in such a way as to avoid structural facets of the rock, but also to be representative and extensive. The sample obtained must be marked with its provenance (origin, well depth, stratigraphy and lithology). Depending on the need, experimental samples can be divided into three



Y. Cheng et al., *Foundations of Rock Mechanics in Oil and Gas Engineering*,
https://doi.org/10.1007/978-981-99-1417-3_5

main categories: regular rock samples (cylindrical, prismatic or cubic), irregular rock samples and specially shaped rock samples.

#### 5.1.1.1 Regular Rock Sample

Most of the experimental samples are cylindrical, and the diameter of the processed samples is mostly used from 2.5 to 5.0 cm. According to the International Society of Rock Mechanics (ISRM) and the national standard (GB/T23561-2010), the ratios of height and diameter used in common experiments are as follows.

- Compressive strength test 2.5 to 3.0/(2.0 to 2.5)
- Bending test 3.0 to 7.0
- Brazilian experiment 0.5 to 1.0
- Stamping test 0.2 to 0.25.

To prepare these samples, cores obtained from underground excavations or boreholes are mostly used. In preparing the samples, the cores are first cut to the desired length with a saw. When samples are prepared from large pieces of rock, they are first cut into smaller pieces by machine or by hand, with the equipment shown in Fig. 5.1a. A modified bench drill or small quarry drill is then used to drill a core sample through the block, as shown in Fig. 5.1b.

When a prismatic sample is prepared using a saw, the rock is first cut into slabs, then into strips, and finally processed into a prism. If the rock is very brittle or of low strength, in which case, when the sample is about to be cut, processing is often found difficult because of the shearing of the material. At this point the rock sample can be buried in paraffin and then cut.

Prior to the experiment, the sample end faces should be made to meet certain standards. This is because stress concentration points can form at grooves or holes in the end face of the sample, causing the sample to break at fairly low loads.

(a) Slicing machine (b) Coring machine (c) Grinding machine

**Fig. 5.1** Diagram of core preparation equipment

When preparing the sample by lathe or surface grinding, the perimeter of the sample must be trimmed smooth if it is rough, as shown in Fig. 5.1c.

For cylindrical rock samples for compressive strength tests, the ISRM specifies the following criteria.

- The sample end face should be ground flat to 0.02 mm.
- (b) The sample end face should be perpendicular to the sample axis to within 0.001 radians.
- (c) The perimeter of the sample should be smooth and free of irregular projections and have a diameter difference of not more than 0.3 mm over the entire length of the sample.

#### 5.1.1.2 Irregular Samples

Samples for irregular experiments were taken to remove sharp corners by tapping with a small hammer, and the size was chosen according to the weight of the sample.

#### 5.1.1.3 Specially Shaped Rock Samples

Specially shaped rock samples need special processing according to the requirements of special applications, such as circular, spherical, etc.

Figure 5.1 shows a schematic diagram of the equipment required to perform rock sample preparation, including (a) a slicer for cutting sections, (b) a coring machine for drilling specimens, and (c) a grinder for grinding end faces.

Due to the non-homogeneous nature of rocks, experimental data from a single rock cannot be representative of the properties of a particular layer and a sufficient number of rock samples must be made. The number of experimental samples depends on the deviation factor of the results and the accuracy and reliability of the average values. In general, the recommended number of samples for homogeneous rocks is as follows: 2–3 pieces for marble, 5 pieces for shale, and 5–10 pieces for sandstone.

### *5.1.2 Types of Rock Damage*

The damage of rocks depends on the type of rock and the physical environment to which it is subjected (confining pressure, temperature, strain rate, intermediate principal stresses and pores, pore pressure, etc.). At low temperatures, low confining pressures and high strain rates, rocks tend to exhibit brittle damage, while at high temperatures, high confining pressures and low strain rates, rocks exhibit ductile damage or ductile flow.

Rock damage can be classified into five types based on the percentage of strain before rock failure (Fig. 5.2). The figure provides a general illustration of the effects

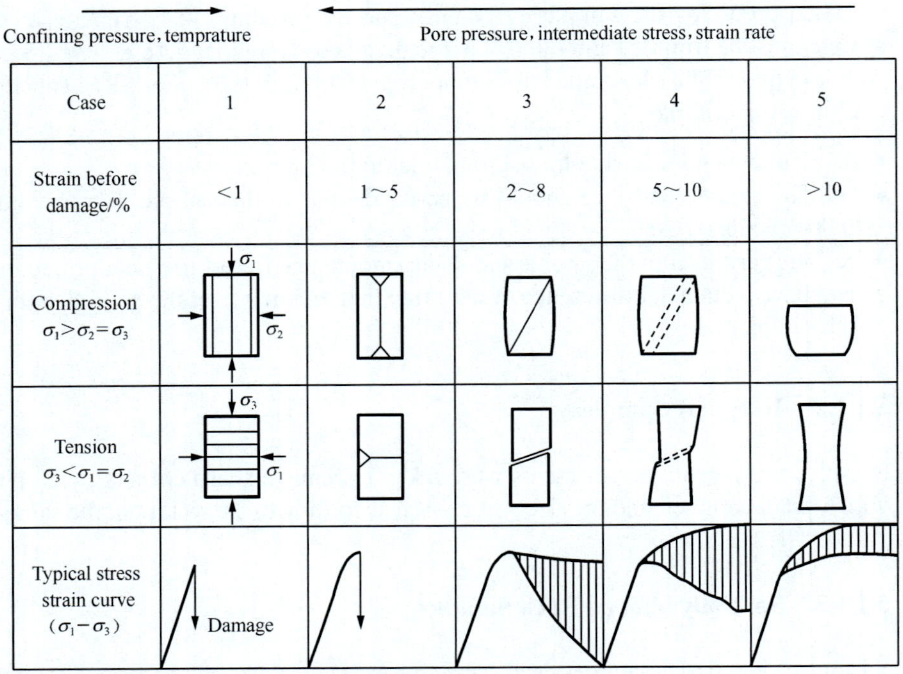


**Fig. 5.2** Schematic diagram of the change from brittle damage to ductile flow (Griggs and Handin 1960)

of the confining pressure, temperature, pore pressure, intermediate principal stress, and strain rate conditions on the rock damage pattern (Chen 1986).

The first type corresponds to the case where the rock is at or near the surface. In a normal temperature and pressure environment or when there is a slight confining pressure and the stress difference $\sigma_1$–$\sigma_3$ is very large, the permanent strain before damage is < 1% and the rock behaves in a brittle state with an approximately linear relationship between stress and strain. Under the above premise, when the stress reaches a certain value, the rock ruptures tensely perpendicular to the direction of the minimum principal stress, as a result of the lower confining pressure, temperature or larger pore pressure and strain rate, allowing some lateral expansion of the rock specimen.

The second type corresponds to the case where the rock is at a certain depth from the surface and its confining pressure and temperature have been slightly higher than in the first case, but the strain rate and pore pressure are still very high. The rock exhibits a small amount of ductility, with the ductility increases the permanent strain before damage increases to 1–5%, the rock rupture appears at the edge of the specimen with the maximum principal stress $\sigma_1$ direction angle of less than 45° local shear rupture surface, while the large area is still along the maximum principal stress $\sigma_1$ direction into tension rupture, its stress–strain curve initially shows an approximately linear relationship, beyond the elastic limit is slightly downward bending part, still can be classified as brittle rupture range.

The third type corresponds to the rock being at a greater depth from the surface (about 2–5 km). Damage in this case can be produced by higher confining pressure, higher temperature or higher confining pressure than before. The damage surface is a single shear surface, and the angle between the rupture surface and the direction of the maximum principal stress $\sigma_1$ is less than 45° The strain before rupture is 2%–8%, a brittle to ductile transition state. This depth is the main range involved in oil and gas engineering. Schwartz (1964) derived the condition of brittle to ductile damage from experimental results: for limestone and marble, when the confining pressure increases from zero to 69 MPa, then it can be converted from brittle to ductile damage; for granite and sandstone, when the confining pressure reaches 69 MPa, the ductile damage phenomenon is still not seen. The angle of shear damage surface varies with the confining pressure, and the Conjugate shear failure angle increases with the increase of the confining pressure.

The fourth type corresponds to rocks at 10–20 km from the surface. Its confining pressure and temperature is higher, or the strain rate is very low, or the rock itself has a certain ductility (such as carbonate rocks), and its total strain before damage is 5–10%, the former is the limit of brittle into ductile, and the latter is already in a ductile state. The shear fragmentation zone is wider and has a certain relative misalignment when the damage occurs, and the angle between the geological fault surface and the direction of maximum compressive stress is slightly less than 45° or close to 45° at this time, producing a wider fault zone.

The fifth type corresponds to the case where the rock is deeper to the surface, the confining pressure is greater than 500 MPa, the temperature exceeds 500 °C or the strain rate is less than $10^{-12}$/s, the rock shows a fully ductile state and the plastic deformation may tend to soften or to harden. Finally, increased ductile deformation can result in permanent deformation greater than 10%.

Similar phenomena were obtained from damage experiments on limestone, marble specimens, as shown in Fig. 5.3. For marble, compression produced brittle damage at room temperature and 25 MPa confining pressure; compression produced brittle to ductile transition at room temperature and 28 MPa confining pressure. For Solenhofen limestone, compression produced ductile damage at room temperature and 100 MPa confining pressure; tensile testing produced ductile damage at 150 °C and 650 MPa confining pressure.

From Fig. 5.2, it can be summarized that rock damage types are divided into rupture (including tensile rupture and shear rupture) and flow, without considering time and other factors, under the premise that the confining pressure and temperature are low or the strain rate is high, it tends to produce tensile rupture, whose rupture direction is parallel to the direction of the maximum compressive stress or perpendicular to the direction of the maximum tensile stress. As the confining pressure and temperature increase, the rock is in the transition stage from brittle to ductile, which tends to produce a single shear rupture surface or conjugate shear rupture surface, and the angle between the initial direction of the rupture surface and the maximum compressive stress is less than 45°. The angle of the shear rupture surface becomes

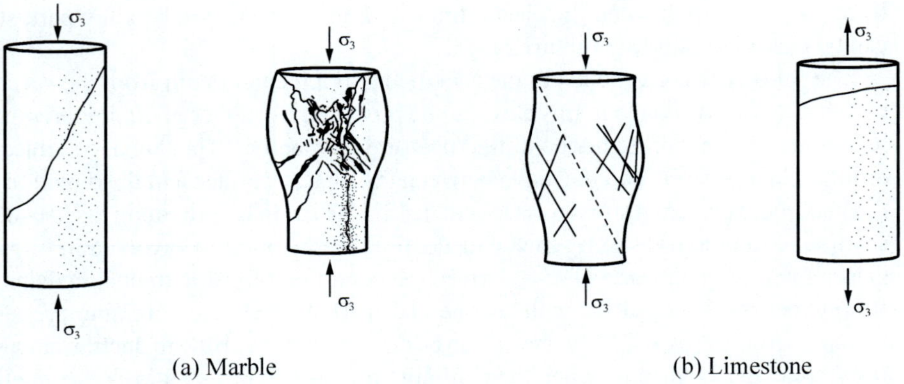


**Fig. 5.3** Limestone and marble destruction patterns (Heard 1960)

larger as the ductility of the rock grows. And at high temperature and high confining pressure the rock shows a slow flow state. The question now is how to determine the failure stress of the rock, that is, to determine the degree of stress at which the rock produces damage, and this is the problem of rock strength that will be described below.

## 5.2 Compressive Strength of Rocks and Its Influencing Factors

### *5.2.1 Compressive Strength of Rocks*

Rock strength is defined as the maximum stress that a rock can resist without producing damage, and the breaking stress is often defined in rock mechanics as the rock strength.

Compression of rock specimens of a certain scale and shape using a testing machine to obtain the relationship between load and deformation, i.e., the stress–strain curve, is an important method for revealing the mechanical properties of rock materials. In terms of experimental principles and conditions, uniaxial compression of cylindrical specimens is the simplest and the first method used to determine the strength and deformation properties of rocks. To this day, the uniaxial compression test of rocks is still one of the most used experimental methods, but the mechanical phenomena exhibited by the uniaxial compression test are exceptionally complex and research on them is still being conducted.

Uniaxial compressive strength is referred to as compressive strength, usually a cylindrical ($\phi 5.4 \times 11\ \text{cm}^3$) or cubic column 5 cm × 5 cm × 11 cm of rock specimen is placed on the test machine for uniaxial compression test, when the pressure reaches damage, the specimen damage stress is called the compressive strength of rock.

Namely

$$\sigma_c = P_c/A \tag{5.1}$$

where $P_c$ is the breaking load and $A$ is the original cross-sectional area of the specimen.

The uniaxial compressive strength of rock, which is the most widely used parameter for the mechanical properties of rock in underground engineering today, determines the strength and modulus of elasticity, Poisson's ratio and other parameters of the rock. This index is relatively simple, easy to calculate, and it has a certain proportional relationship with the tensile and shear strength, such as the tensile strength is 3–30% of it, thus it can be roughly estimated with the help of it. Therefore, the compressive strength of rocks is one of the most basic indices in rock mechanics experiments.

### 5.2.2 *The Influencing Factors of Compressive Strength*

Compressive strength is influenced and constrained by many factors (Cai 2002). These influencing factors can be broadly divided into two aspects: on the one hand, the rock itself, such as rock-forming mineral composition, particle size, cementation, generation conditions, laminar structure, bulk weight, porosity and water content; on the other hand, they belong to the experimental methods and physical environment, such as the size of the specimen, shape, specimen processing, the friction between the press head and the specimen, the loading rate and the surrounding physical environment (such as temperature and other factors).

#### 5.2.2.1 Intrinsic Factor of Rock

Rocks composed of different minerals have different compressive strengths. Generally, if a rock contains higher amounts of quartz, feldspar, pyroxene and other minerals, the compressive strength of the rock is relatively higher. Conversely, rocks containing more mica, kaolin, chlorite, talc, chlorite, etc., which have lower strength, will have relatively lower compressive strength. However, even if the rocks are composed of the same minerals, the compressive strength varies greatly depending on the size of the particles, the cementation condition and the generation conditions.

Quartz is the stronger of the rock-forming minerals, and if quartz grains are interconnected in the rock to form a skeleton, increasing the quartz content can increase the strength. Another example is Price's (1934) study of the relationship between quartz content and uniaxial compressive strength for a series of sandstones containing calcite inclusions and mudstones containing clay mineral inclusions, which concluded that the strength of the rock increased with increasing quartz content in both rocks.

However, it is not always the case that the strength increases with the increase in quartz composition. For example, in granites where the grains of quartz are dispersed (not forming a skeleton), an increase in quartz content does not increase the strength of the rock. However, granites containing mica-like flake minerals and feldspars with very developed jointing surfaces in both directions cause hidden weak surfaces in granites, thus decreasing the strength. Therefore, when the granite contains more of these minerals and the particles are larger, it has an adverse effect on the strength of the granite and becomes the main factor in determining the strength of the granite.

The size of the particles also affects the strength of the rock, and the strength of fine-grained rocks is often higher than that of coarse-grained rocks. For example, coarse-grained granite has a strength of 120 MPa, while fine-grained granite can reach 260 MPa. dense limestone has a compressive strength of 140 MPa when dry, while oolitic limestone with a diameter of 0.5 mm has a compressive strength of 118 MPa when dry and 97 MPa when full of water.

For sedimentary rocks, the cementation situation has a strong influence on strength. The cement in the rock is strongest with siliceous cement. This is followed by iron and calcareous cement, while mud cement is the least strong. Siliceous cement has high strength, such as siliceous cemented sandstone compressive strength of 200 MPa or more; limy cement has lower strength, such as ash cemented sandstone its compressive strength of 20–100 MPa; mud cement has the lowest strength, soft rocks often belong to this category. In terms of clay particles, siliceous cemented mud slate can have a strength of 200 MPa or more, while mud shale, cemented by mud, will not exceed 100 MPa at most.

The bulk weight of the rock also often affects the strength, for example, the bulk weight of white limestone increases from 1.5 to 2.7 g/cm$^3$ and its compressive strength increases from 4.9 to 176.6 MPa; the bulk weight of sandstone changes from 1.87 to 2.57 g/cm$^3$ and its strength increases from 14.7 to 88.3 MPa.

Rock porosity has a significant effect on compressive strength, as shown in Fig. 5.4, where compressive strength decreases significantly as rock porosity increases. If there is water infiltration to promote cementation and softening, this can significantly reduce the strength of the rock, as shown in Fig. 5.5. An increase in water to 4% in sandstone reduces the strength value by about 50% of that in dryness, while an increase in water to 2.5% in muddy slate reduces the strength to 70%.

It can be seen that even for rocks of the same name, the strength varies greatly due to different origins, internal rock structure, particle size, cement, bulk weight, porosity and water content, etc. The data in Table 5.1 can only provide a reference.

Sedimentary rocks have laminae, and various schist have laminae, and the minerals are mostly in directional arrangement, all forming anisotropic features, and the compressive strength of such anisotropic rocks differs greatly in the direction of parallel laminae and in the direction of perpendicular laminae. Table 5.2 lists the compressive strengths of several major sedimentary rocks perpendicular and parallel to the laminae, and Table 5.3 shows the compressive strengths of several coals perpendicular and parallel to the laminae. As can be seen, except for fine-grained sandstones, which are approximately equal, the compressive strengths of rocks and coals perpendicular to the laminae are generally greater than those parallel to the laminae, the

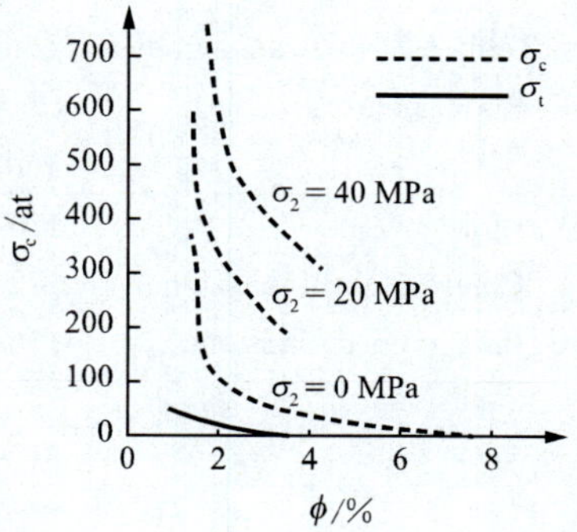


**Fig. 5.4** Rock strength versus porosity (Gao 1979)

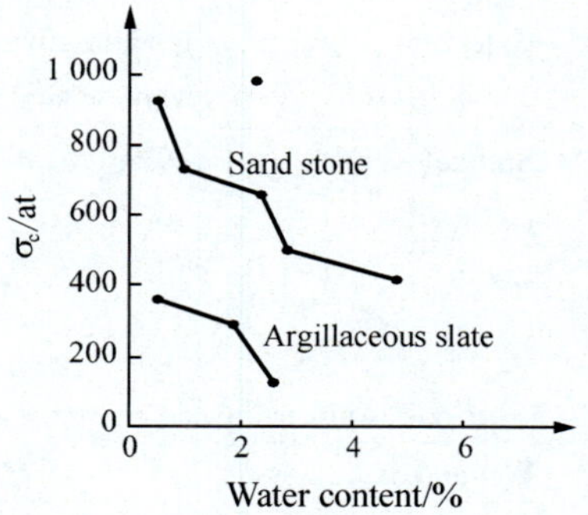


**Fig. 5.5** Effect of rock moisture content on compressive strength (Burshtein 1969)

**Table 5.1** Changes in strength of rocks under water immersion

| Rock name | Compressive strength/MPa | | Rock name | Compressive strength/MPa | |
|---|---|---|---|---|---|
| | Dry | Immersion | | Dry | Immersion |
| Granite | 40–220 | 25–205 | Sandstone | 17.5–250 | 5.7~241.5 |
| Diorite | 97.7–232 | 68.8–159.7 | Claystone | 20.7–59 | 2.4–31.8 |
| Gabbro | 118.1–272.5 | 58–145.8 | Shale | 57–136 | 13.7–75.1 |
| Basalt | 102.7–200.5 | 32.5–153.7 | Slate | 123–199.6 | 72–149.6 |
| Tuff | 61.7–178.5 | 7.8–189.2 | Phyllite | 30.1–49.4 | 28.1–33.3 |
| Limestone | 13.4–250.1 | – | Schist | 59.6–218.9 | 29.5–174.1 |
| Quartzite | 145.1–700 | 50–176.8 | | | |

ratio of the two being about 1.3 for rocks and 1.5 for coals, and the harder the rock, the smaller the ratio of vertical to parallel strength, and the softer the rock, the larger the ratio (Li 1983). The ratio of vertical laminar compressive strength to parallel laminar compressive strength is greatest for coal. This is because the interlaminar bond is poor and when pressure is applied parallel to the direction of the laminae, it tends to crack along the inside of the laminae.

**Table 5.2** Rock anisotropy and compressive strength

| Rock name | Compressive strength/MPa | | $\sigma_c^{\perp}/\sigma_c^{//}$ |
|---|---|---|---|
| | Vertical stratification | Parallel stratification | |
| Limestone | 180 | 151 | 1.19 |
| Coarse-grained sandstone | 142.3 | 118.5 | 1.20 |
| Fine-grained sandstone | 156.8 | 159.7 | 0.98 |
| Sandstone | 78.9 | 51.8 | 1.52 |
| Shale | 51.7 | 36.7 | 1.41 |

**Table 5.3** Coal rock anisotropy and compressive strength

| Categories | Compressive strength/MPa | | $\sigma_c^{\perp}/\sigma_c^{//}$ |
|---|---|---|---|
| | Vertical stratification | Parallel stratification | |
| Smokeless coal | 19.60 | 13.19 | 1.49 |
| Anthracite | 10.76 | 7.41 | 1.45 |
| Lignite | 13.50 | 8.00 | 1.69 |

**Table 5.4** Anisotropy and compressive strength of dry and wet rocks

| Wet and dry | Dry/MPa | | | Wet/MPa | | |
|---|---|---|---|---|---|---|
| Direction of pressurization | $\sigma_c^{\perp}$ | $\sigma_c^{//}$ | $\sigma_c^{\perp}/\sigma_c^{//}$ | $\sigma_c^{\perp}$ | $\sigma_c^{//}$ | $\sigma_c^{\perp}/\sigma_c^{//}$ |
| Siltstone | 80.0 | 77.0 | 1.04 | | | |
| | 95.0 | 75.0 | 1.27 | 40.0 | 30.0 | 1.33 |
| | 134.0 | 128.0 | 1.05 | 85.0 | 60.0 | 1.42 |
| Shale | 40.0 | 24.0 | 1.67 | | | |
| | 116.0 | 100.0 | 1.16 | 55.0 | 40.0 | 1.38 |

As can be seen from Table 5.4, the effect of laminations in the wet state of the rock (mean value of the ratio of vertical to parallel strength 1.38) is greater than the effect of laminations in the dry state (mean value of the ratio of vertical to parallel strength 1.24).

#### 5.2.2.2 Experimental Methods and Physical Environmental Factors

The experimental method also has a significant effect on the compressive strength of the rock, and the two main influencing factors are the ratio of height to diameter of the specimen and the non-uniform distribution of pressure and friction on the specimen end surfaces (Hudson and Harrison 2008; Jaeger and Cook 1979). If the prepared specimens are loaded directly under the test machine, the specimen end surfaces will have non-uniformly distributed stresses, as shown in Figs. 5.6.

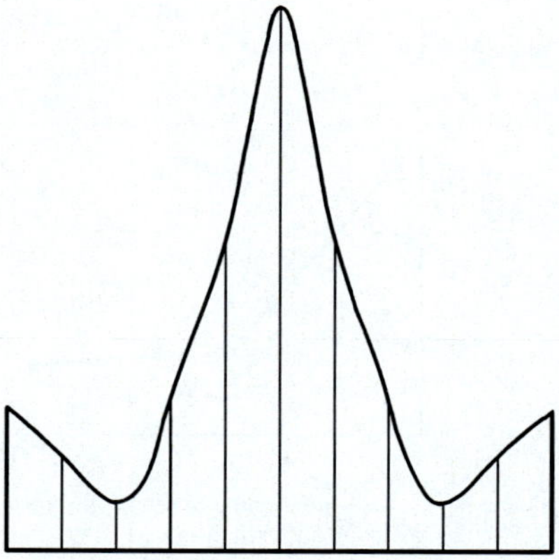

**Fig. 5.6** Compressive stress distribution on the end face of the rock specimen

The complexity of the stress distribution in the specimen under uniaxial compression determines the three types of rock damage during the damage experiments: tensile rupture, shear rupture and tapered rupture against the top. This type of damage is caused by the end face effect. Due to the different size of friction between the indenter of the testing machine and the end face of the specimen, causing the different types of rock specimen rupture. If the contact surface is coated with lubricant such as paraffin or padded with Teflon to reduce its friction, the rock specimen is damaged by the tensor rupture in the direction of parallel pressure, and the strength is reduced; if the test is loaded directly on the test machine, the rock specimen is shear rupture or conical rupture against the top.

Both the shape and size of the rock specimen have an effect on the strength of the rock. If the specimen is cylindrical in shape, the stress distribution is more uniform than that of cubic rock specimens due to axisymmetry. In general, the compressive strength of a cylinder with a height-to-diameter ratio of 1 is higher than that of a cube with the same ratio. The compressive strength of a cylinder with a height-to-diameter ratio of 1.5 is lower than that of a rectangular body with a height-to-diameter ratio of 1.5:1, and is approximately the same as that of a rectangular body with a height-to-diameter ratio of 2:1. This shows that the shape of the specimen and the height-to-diameter ratio also affect the strength of the rock. In general, if the height of a cylindrical specimen is shorter than its diameter, the compressive strength increases, and vice versa, the strength decreases. The relationship between height-to-diameter ratio and compressive strength of rock specimens is shown in Figs. 5.7, and 5.8.

As can be seen from Fig. 5.7, the compressive strength tends to be constant at a height-to-diameter ratio of 2.5. The International Society of Rock Mechanics (2007) recommends that rock specimens with a height-to-diameter ratio of 2.5–3 are appropriate. However, considering the difficulty in obtaining test samples, the sample height-to-diameter ratio is sometimes less than 2, when the measured uniaxial compressive strength needs to be corrected by the following equation (You 2007a, b).

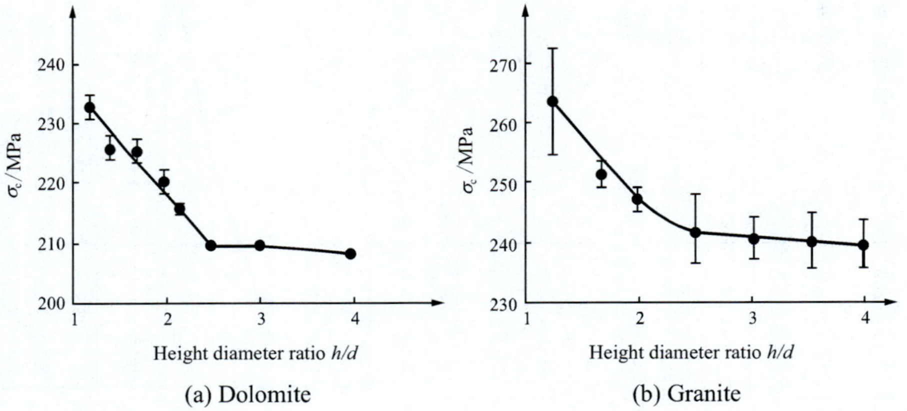


**Fig. 5.7** Relationship between compressive strength of rocks and height-to-diameter ratio (Mogi 1971)

**Fig. 5.8** Effect of specimen size on rock strength

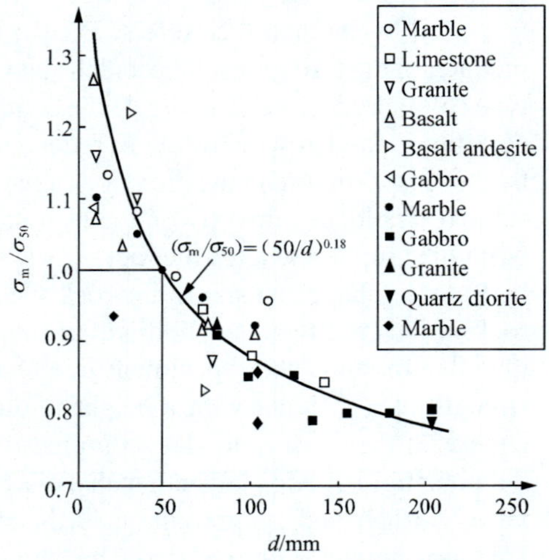


$$\sigma_s = \frac{8\sigma_L}{7 + 2D/L} \tag{5.2}$$

where $\sigma_{\mathrm{L}}$ is the uniaxial compressive strength of a specimen with diameter $D$ and height $L$, and $\sigma_{\mathrm{s}}$ is the uniaxial compressive strength at a height-to-diameter ratio of 2.

The diameter of the specimen also has an effect on the uniaxial compressive strength of the rock due to the variation in the size of the mineral grains and the fracture scale that make up the rock. Kramadibrata and Jones (1993) summarized the relationship between strength and diameter for some typical rocks, using the strength of a rock sample with a diameter of 50 mm for dimensionless processing, and the results are shown in Fig. 5.8. It is suggested that Eq. (5.3) be used to characterize the

**Table 5.5** Effect of loading rate on compressive strength

| Rock name | Uniaxial compressive strength/MPa | | |
|---|---|---|---|
| | Time to destruction 30 s | Time to destruction 0.03 s | Increase in strength/% |
| Beryl sandstone | 56.25 | 84.37 | 50 |
| Gabbro (geology) | 217.9 | 282.15 | 29 |

law of influence of the diameter of rock specimens.

$$\sigma_m = \sigma_{50}(50/D)^{0.18} \tag{5.3}$$

where $D$ is the diameter of the rock sample in mm;$\sigma_{50}$ is the uniaxial compressive strength of a 50 mm diameter rock sample;$\sigma_{\mathrm{m}}$ is the uniaxial compressive strength of a $D$ diameter specimen.

The loading rate has a significant effect on the compressive strength of the rock. Usually the compressive strength of rock specimens increases with the rate of loading, and at high rates of loading, such as impact or acoustic loading, the compressive strength is several times that of low rate loading, as shown in Table 5.5.

Generally, the loading rate is controlled within the range of 0.1–1 MPa per second when rock strength experiments are conducted, so the effect of loading rate on rock strength is negligible.

Temperature has a great influence on the tensile and compressive strength of rocks, especially at a temperature of several hundred degrees. General crystalline rocks (such as granite) both tensile and compressive strengths decrease as the temperature rises; however, the compressive strength of amorphous rocks (such as tuff) increases as the temperature rises. The rate of increase and decrease of compressive and tensile strength of various rocks with the increase of temperature is shown in Fig. 5.9. When the temperature of crystalline rocks rises to 800 °C, the rate of decrease of compressive strength is 20–40% of normal temperature; when the temperature of andesite rises to 600 °C, the rate of increase of compressive strength is 170% of normal temperature, and the rate of decrease of tensile strength is about 100% of normal temperature.

## 5.3 Tensile Strength of Rocks and Its Influencing Factors

The tensile strength of a rock is the ultimate stress at which the specimen reaches damage under uniaxial tensile conditions. Fractures and weak surfaces within rocks can carry compressive stresses and carry shear stresses by friction, but not tensile stresses. The rock is a non-homogeneous material with great variation in internal strength, which is gradually destroyed when carrying tensile stresses, i.e., the internal

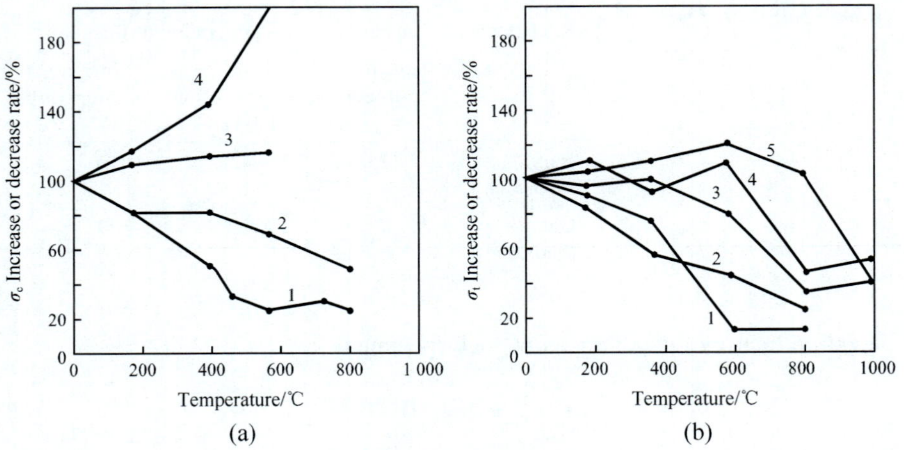


**Fig. 5.9** The pattern of temperature effect on compressive strength (**a**) and tensile strength (**b**) of rocks (Vutukuri 1974)

micro-elements do not reach their respective load-bearing limits simultaneously, and thus the tensile strength of the macroscopic specimens is lower than the average of the micro-elements' strengths. Due to the difficulty of clamping specimens, direct tensile tests are generally not used to determine the tensile strength of rocks, and various indirect methods are mostly used. The fact that the tensile strength of rocks is much lower than the compressive strength is an inherent factor in the implementation of indirect methods.

## *5.3.1 Direct Stretching Method*

The direct tension method is similar in principle to the tensile damage test for metals. Using a special fixture and adhesive, the specimen is clamped to the testing machine and stretched. The difficulty, however, lies in the clamping of the specimen and in keeping the load parallel to the specimen axis. It is important that the specimen is mounted in the tensile jig and that the surface of the specimen is not damaged but can be clamped; it is also important to consider the simplicity of operation, mainly the hemispherical steel die method and the rope sleeve cap method. In addition, if the load is not parallel to the specimen axis in a very close range, bending will be caused and abnormal stress concentrations will arise.

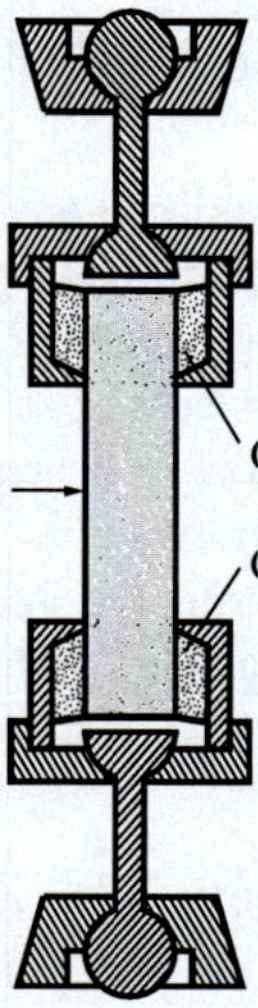


**Fig. 5.10** Axial tension steel die

#### 5.3.1.1 Hemispherical Steel Die Method

The fixture for the hemispherical steel die method is composed of two parts, the joint and the connecting rod. Figure 5.10 shows its structural dimensions. One end of the joint is connected to the specimen and glued with epoxy resin in the flat contact part; the ball seat of the connecting rod is set into the other end of the joint for adjusting the eccentricity, and the pressure head of the testing machine is directly clamped to the connecting rod and the experiment is carried out afterwards.

#### 5.3.1.2 Rope Capping Method

The apparatus for the rope-cap method is shown in Fig. 5.11; this method uses a cylindrical specimen and glues it to a steel cap with epoxy resin, which is loaded with a flexible rope tied to a single arm. There is a 0.008 cm high flange at the end of the sleeve cap to ensure good alignment at the intersection of rock and steel when glued, so the specimen can be accurately centered on the plate and the cross section of the end steel plate should have the same shape and dimensions as the specimen.

For a cylindrical specimen, if the tensile force at damage is $P_t$, the tensile strength of the specimen $\sigma_t$ can be expressed by the following equation.

$$\sigma_t = \frac{P_t}{A} \tag{5.4}$$

Due to the difficulties in performing tensile damage experiments by the direct method, they are generally performed by the indirect method.

### 5.3.2 Indirect Stretching Method

The indirect tensile method is generally used in the splitting method, also known as the Brazilian test. The rock specimen is cut into a cylinder and a uniform wiring load is applied along the diameter of the cylinder, as shown in Fig. 5.12, the specimen is placed across the indenter of the press, a steel bar is placed on each of the upper and lower bearing plates of the specimen, and then pressure is applied until the specimen splits along the diameter direction, if it is assumed that the material is a uniform isotropic elastomer, the compressive strength can be derived from the theory of elasticity as follows.

$$\sigma_t = \frac{2P}{\pi dl} \tag{5.5}$$

where d is the diameter of the specimen, l is the thickness of the specimen, and P is the breaking load.

According to the theory of elasticity, the distribution of tensile stresses on the diameter of a cylindrical specimen under a uniform load is shown in Fig. 5.13. However, in practice, the splitting experiment with the press does not achieve the ideal linear distribution of load, only in the central part of the specimen, the tensile stress is more uniformly distributed; at 0.8r (radius) above and below from the center, the stress value is zero; greater than 0.8r becomes compressive stress, and the compressive stress at both ends of the loading is the largest, and its value is 12 times the tensile stress at the center of the specimen. However, since the tensile strength of the rock is much lower than the compressive strength (about 1/10–1/50), the specimen can still be split under the tensile stress. The size of the cylindrical specimen (diameter and thickness) also has an effect on the tensile strength of the rock. As the diameter and thickness of the specimen increase, the tensile strength also increases. Generally,

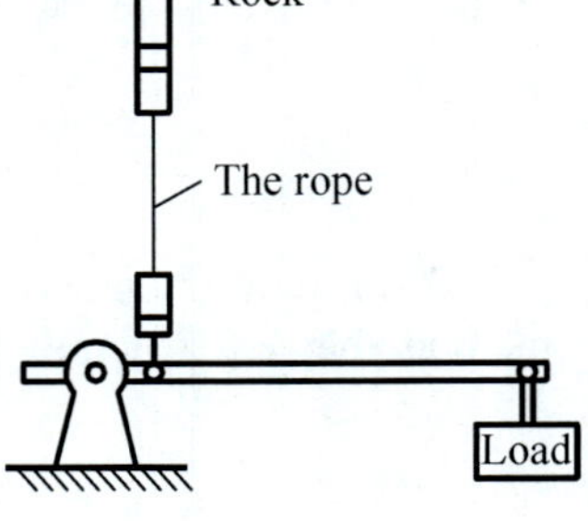


**Fig. 5.11** Rope cap method device

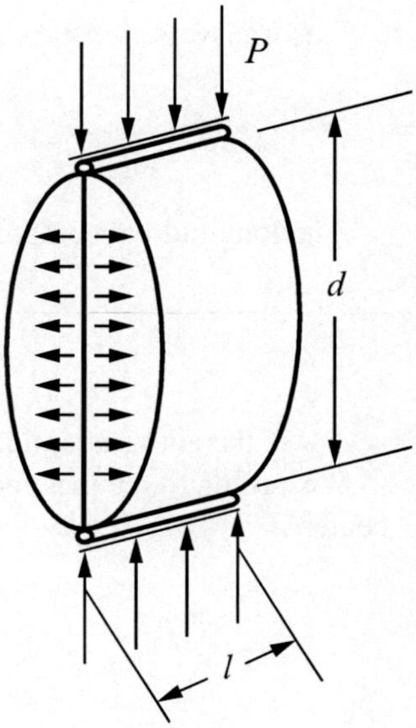


**Fig. 5.12** Tensile experiment of the indirect method

with a diameter of 5.1 cm and a thickness of 2.5 cm, the strength basically tends to be stable.

Incidentally, the elastic parameters E, $\upsilon$ of the rock can also be obtained by observing the longitudinal and transverse strains of the resistive strain gauges at the centre of the specimen using a resistive strain gauge when performing rock tensile strength tests using the splitting method with cylindrical rocks.

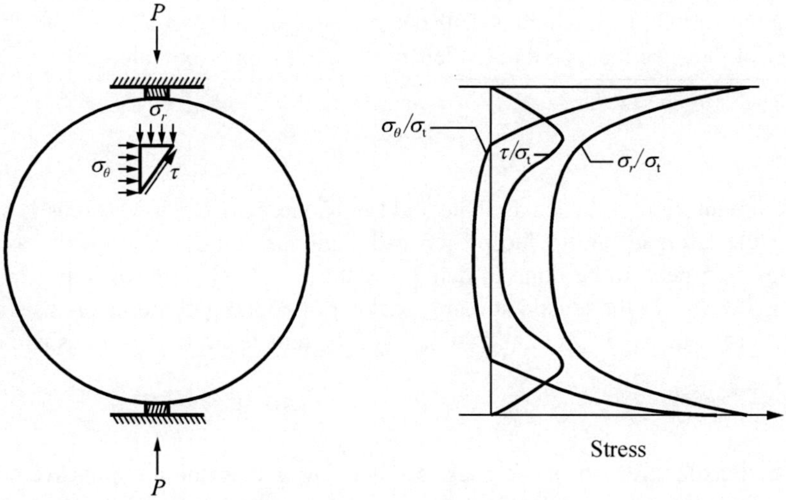


**Fig. 5.13** Stress distribution along the vertical section in the splitting method experiment

The transverse tensile stress at the center of the cylindrical specimen is

$$\sigma_1 = \sigma_t = \frac{2P_0}{\pi d}$$

The longitudinal compressive stresses are.

$$\sigma_2 = \sigma_c = \frac{-6P_0}{\pi d} \tag{5.6}$$

$P_0$ in the above equation is the load per unit thickness, i.e., $^P/_l$.

According to the generalized Hooke's law for plane stress states, the strain at the center is

$$\varepsilon_1 = \frac{2P_0}{\pi dE}(1 + 3\nu)$$

$$\varepsilon_2 = -\frac{2P_0}{\pi dE}(3 + \nu)$$

Further rewritten as.

$$\begin{aligned} E &= \frac{16P_0}{\pi d(3\varepsilon_2 + \varepsilon_1)} \\ \upsilon &= -\frac{3\varepsilon_1 + \varepsilon_2}{3\varepsilon_2 + \varepsilon_1} \end{aligned} \tag{5.7}$$

If a cubic specimen is used, as shown in Fig. 5.14a, the horizontal tensile stress along the vertical line, which is distributed as shown in Fig. 5.14b, is 0.98 $\frac{2P}{\pi dl}$ at the center of the specimen, so that the tensile strength is approximated as

$$\sigma_t = \frac{2P}{\pi dl} \tag{5.8}$$

Sedimentary rocks, such as shale and sandstone, have the lowest tensile strength along the laminae due to the presence of laminae. Three experiments as shown in Fig. 5.15 need to be done to determine the tensile strength of laminated rocks (Chen 1986). Let the tensile strength between the load and the layers as shown in Fig. 5.15a–c be $\sigma_t$, $\sigma_t^{'}$, $\sigma_t^{''}$, and the ratio of the tensile strength of the shale is

$$\sigma_t : \sigma_t' : \sigma_t'' = 100 : 83 : 68$$

The tensile strength of the rock is much lower than the compressive strength, generally the former is 1/10–1/20, or even 1/50 of the latter. the reason for its low tensile strength is mainly due to the influence of the internal pores of the rock, generally due to the internal microfractures, pores are more developed, this defect is

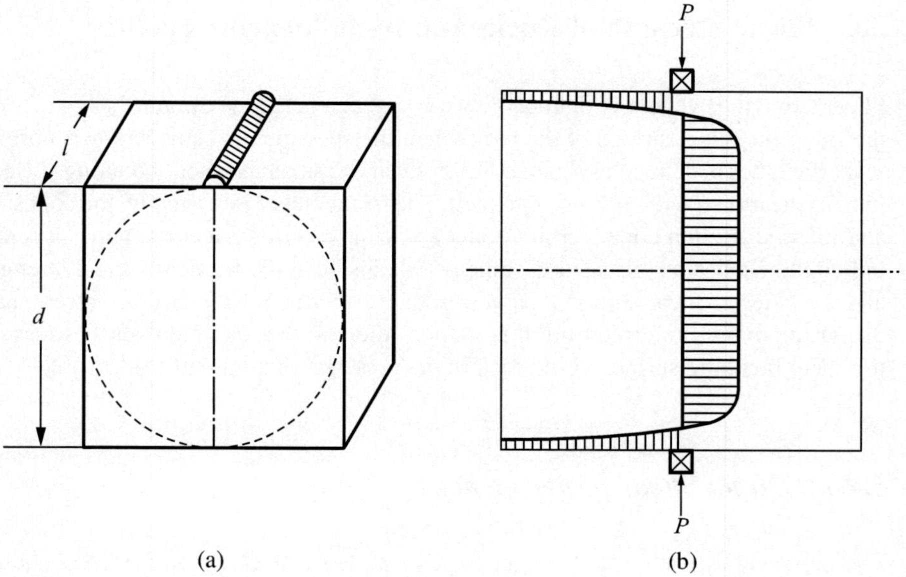


**Fig. 5.14** Distribution of horizontal stresses in the vertical centerline of a cubic specimen by the splitting method

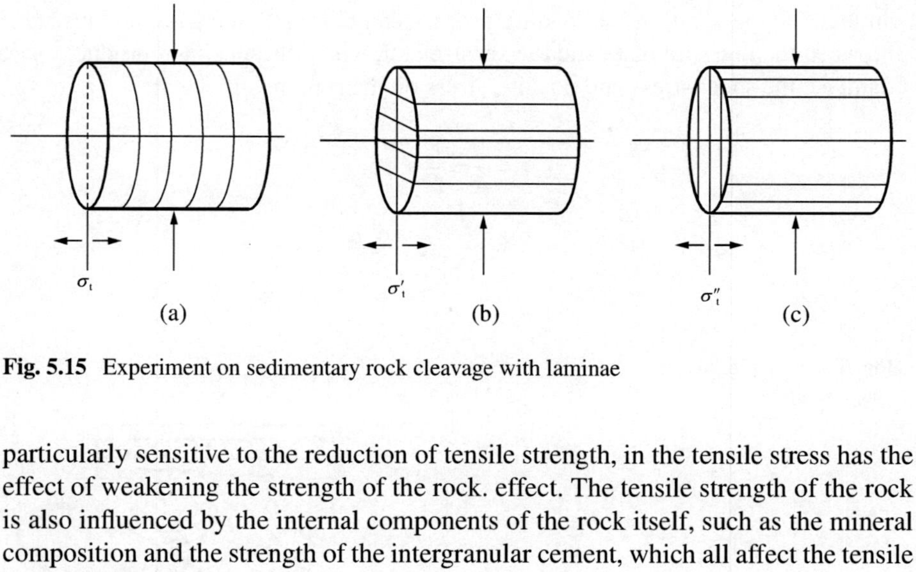


**Fig. 5.15** Experiment on sedimentary rock cleavage with laminae

particularly sensitive to the reduction of tensile strength, in the tensile stress has the effect of weakening the strength of the rock. effect. The tensile strength of the rock is also influenced by the internal components of the rock itself, such as the mineral composition and the strength of the intergranular cement, which all affect the tensile strength of the rock.

In addition, the tensile strength of rocks generally increases with increasing loading rate. The tensile strength of rocks decreases with increasing temperature, moisture and porosity. This conclusion is the same as for compressive strength, but the rate of increase or decrease is not the same.

## 5.4 Shear Strength of Rocks and Its Influencing Factors

Shear strength is generally defined in two ways: one is the maximum shear stress on the shear breaking surface of the rock when the specimen is subjected to a normal load; the other is defined as the maximum shear stress on the shear breaking surface in pure shear (i.e., without a normal load). The former takes into account the cohesion and internal friction contained in the rock during shear damage; the latter depends only on the cohesion. Therefore, some people also call the former as shear strength and the latter as shear strength. However, the former is mostly used at present, and the shear strength referred to in this section refers to the maximum shear stress on the shear breaking surface of the rock in the presence of a normal load.

### *5.4.1 Direct Shear Experiment*

The most widely used is the wedge-shaped simple shear, the main device as shown in Fig. 5.16, the rectangular column (10 cm × 10 cm × 15 cm) specimen placed in the shear, the pressure applied in the press for shear damage test. When the load P reaches a certain value, the specimen is sheared along the ab section, generally in the shear device is equipped with rollers between the upper and lower and pressure plate, and add slippery oil, in the loading process can eliminate the frictional resistance between the pressure plate and the shear meter, when the specimen produces shear damage, the shear stress and positive stress on the rupture surface are.

$$
\begin{aligned}
\tau &= \frac{\mathrm{T}}{\mathrm{A}} = \frac{\mathrm{P}}{\mathrm{A}} \sin\alpha \\
\sigma_{\mathrm{N}} &= \frac{\mathrm{N}}{\mathrm{A}} = \frac{\mathrm{P}}{\mathrm{A}} \cos\alpha
\end{aligned}
\tag{5.9}
$$

**Fig. 5.16** Simple shear gauge setup

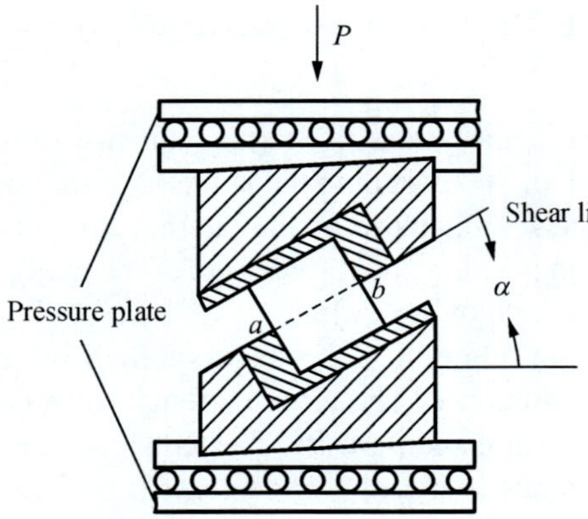

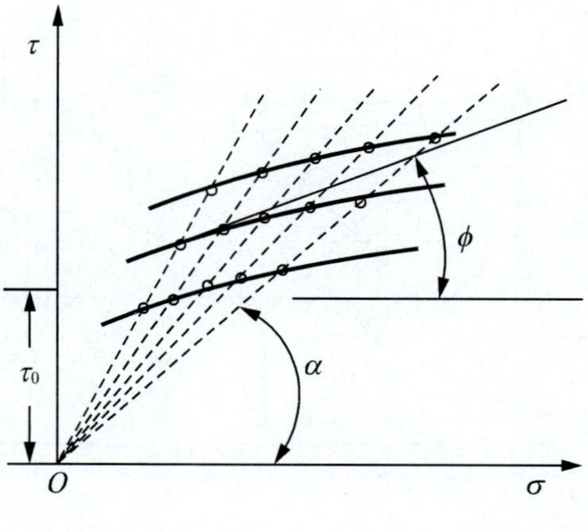


**Fig. 5.17** Rock shear strength curve

In the above equation, P is the load when the specimen produces shear damage, T is the shear force acting on the shear damage surface, N is the pressure acting on the shear damage surface, A is the area of the shear damage surface, and $\alpha$ is the angle between the horizontal plane and the shear damage surface.

For shear damage experiments, multiple rock specimens are used for the same rock, each at a different angle $\alpha$. When shear damage occurs, a pair of $\tau$ and $\sigma_n$ values can be obtained corresponding to each $\alpha$ value, and a series of points with different $\alpha$ values are plotted in the $\sigma \sim \tau$ coordinate system (i.e., a row of circles in Fig. 5.17), and a smooth curve is used to connect these points, and this curve is the strength curve of a certain rock when shear damage occurs (the three curves in Fig. 5.17 represent the strength curves of three rocks (the strength curves of three rocks). It reflects two parameters of the shear strength of the rock, namely the cohesion (the intersection of the curve with the $\tau$ axis) and the angle of internal friction (the angle between the tangent line of the curve and the intersection of the $\tau$ axis and the $\sigma$ axis).

To make the alpha angle change, this is generally accomplished by adding wedge shims with the same inclination angle and opposite direction to the upper and lower ends of the shear gauge, as shown in Fig. 5.18. The $\alpha$ value is usually made to vary in the range of 40° to 60°. If the $\alpha$ angle is too small, the positive stress on the shear surface is too large to destroy the shear surface as intended, and if the $\alpha$ angle is too large it will cause tensile stresses and there may be a danger of tipping the shear device.

### 5.4.2 Rock Triaxial Experiments

Rock triaxial experiments are performed using the triaxial tester shown in Fig. 5.19, where the rock specimen is placed in a pressure chamber and a certain circumferential pressure ($\sigma$) is applied$_3$, followed by a vertical pressure ($\sigma_1$), until the rock breaks. This gives the values of $\sigma_1$ and $\sigma_3$ at the time of rock destruction. A damage stress circle can then be drawn in the $\sigma - \tau$ coordinate system. A series of different values

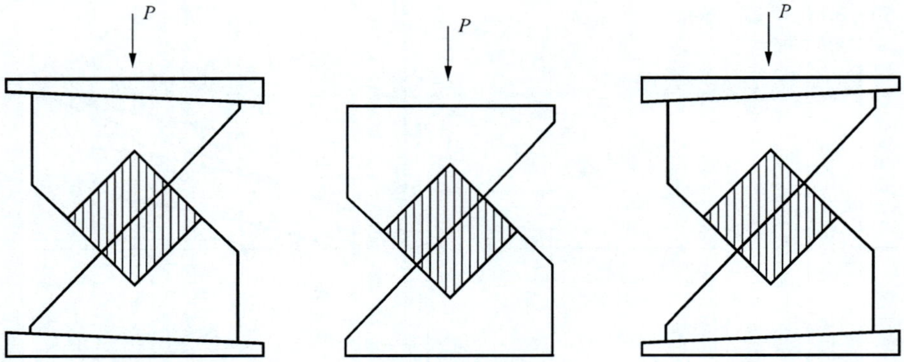


**Fig. 5.18** Device for changing the shear angle

of $\sigma_1$ and $\sigma_3$ can be obtained by performing damage experiments with specimens of the same rock at different circumferential pressures $\sigma_3$ and vertical pressures $\sigma_1$, and a set of damage stress circles can be drawn. The envelope of this set of damage stress circles is the rock shear strength curve. The vertical coordinate of any point on the envelope in Fig. 5.20 represents the shear strength $\tau_\mathrm{f}$ along the shear rupture surface under a certain surrounding pressure and vertical compressive stress; the angle between the tangent line and the horizontal coordinate of any point represents the angle of internal friction on the corresponding shear surface; the intercept between the tangent line and the vertical coordinate is the cohesive force $\tau_0$ at the shear rupture surface.

Experiments show that the rock envelope is generally quadratic at higher envelope pressures as shown in Fig. 5.20. It can be seen from the figure that both the angle of internal friction $\varphi$ and cohesion $\tau_0$ of the rock vary with the possible shear damage surface at different envelope pressures. In other words the values of $\varphi$ and $\tau_0$ vary with the magnitude of the confining pressure. For shear damage at higher confining pressures, the angle of internal friction $\varphi$ becomes smaller while the rock cohesion $\tau_0$ increases, and conversely at lower confining pressures the angle of $\varphi$ becomes larger while $\tau_0$ decreases.

Sometimes, in the case of low perimeter pressure, to simplify the experiment, the uniaxial compressive strength and uniaxial tensile strength damage stress circles are approximated, and the common tangent of the two circles is drawn, and this common tangent is used as an approximate substitute for the envelope of the rock (i.e., the strength curve), as shown in Fig. 5.21. Its shear strength can be used as follows.

$$\tau_f = \tau_0 + \sigma \tan\phi \tag{5.8}$$

A bilinear envelope can also be approximated instead for the high and low confining pressure zones, using different $\tau_0$ and $\varphi$ in these two regions. Table 5.6 shows the applicable ranges of $\tau_0$ and $\varphi$ for several rocks.

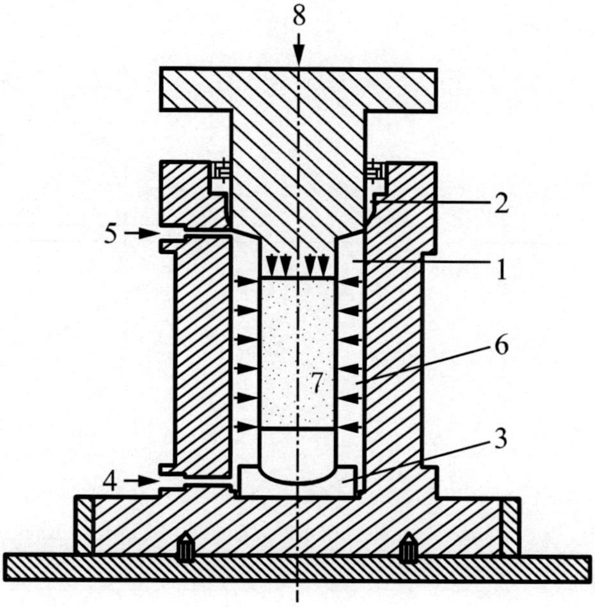


**Fig. 5.19** Servo-controlled conventional 3-axial experimental system. 1—pressure chamber; 2—sealing ring; 3—spherical base; 4—pressure input port; 5—exhaust port; 6—confining pressure; 7—test sample; 8—applying vertical pressure

**Fig. 5.20** Shear strength curves for triaxial experiments

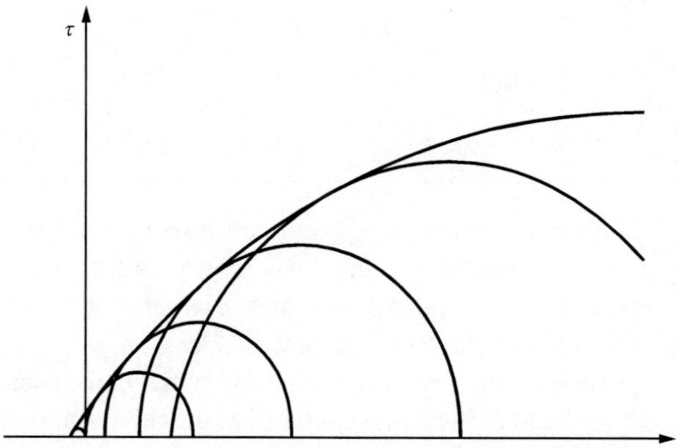


It should be noted here that all of the above methods assume that the coefficient of internal friction f is a constant regardless of low or high confining pressure, but for soft rocks (e.g., shale, mudstone, etc.), even at very low confining pressures, a nonlinear envelope is shown.

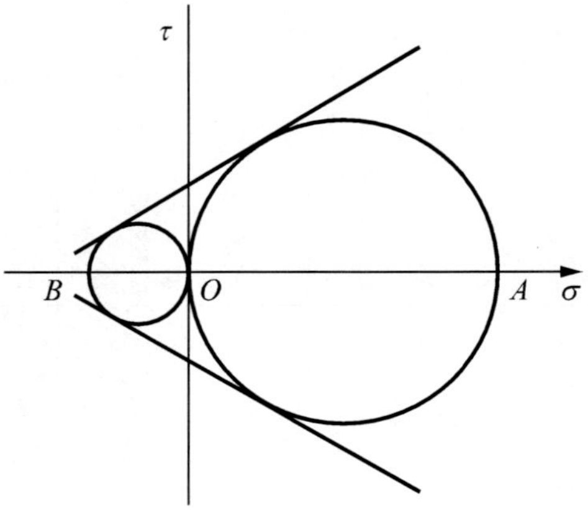


**Fig. 5.21** Simplified linear envelope

**Table 5.6** $\tau_0$ and $\varphi$ values for several rocks

| Rock type | Range of application/MPa | Cohesion $\tau_0$/MPa | Angle of internal friction $\varphi$ |
|---|---|---|---|
| Gray-green bauxite | ≤70 | 26 | 51°30′ |
| | >70 | 80 | 23°20′ |
| Purple-red bauxite | ≤50 | 23 | 59°40′ |
| | >50 | 135 | 27°20′ |
| Red sandstone | ≤100 | 26 | 44°30′ |
| | >100 | 80 | 28° |
| Marble | ≤19.5 | 8.5 | 66° |
| | >19.5 | 40 | 29°20′ |
| Dolomitic chert | ≤40 | 25 | 62° |
| | >40 | 70 | 37° |

The size of the rock specimen also has an effect on the shear strength, which generally decreases as the cross-section of the rock increases. Therefore, it is recommended that the ratio of height to diameter of the rock specimen should be 2.0–2.5, within which the value of $\tau_0$ varies less.

Because the strength is not only a mechanical property of the medium itself, but also related to the stress state inside the medium, such as direct shear test, although it is forced to produce shear rupture on the predetermined shear surface, but sometimes the predetermined shear surface may not be the easiest direction to shear, which is related to the stress distribution on the shear surface, triaxial test results show that the shear strength curve is almost the same as the rupture stress circle envelope. Therefore, theoretically it seems to reflect its shear strength, in fact, most rocks reach the strength limit, the rock shows anisotropic characteristics, the rock has a certain stress gradient inside, so to really understand the shear strength of the rock, there are still some difficulties.

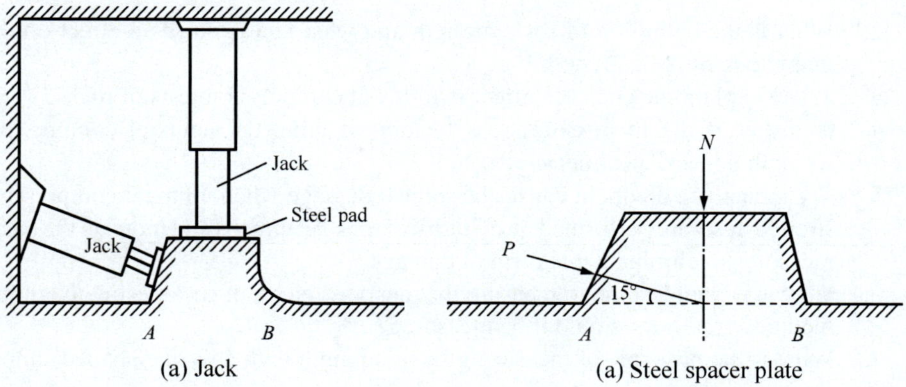


**Fig. 5.22** Schematic diagram of the field rock shear experiment

## 5.4.3 Rock Shear Experiments at the Mine Site

In the mining field, the shear strength of the rock at the mine site is currently determined mainly by the direct shear method, the apparatus of which is shown in Fig. 5.22a. Generally, a convex rock block is cut in a flat lane with a bottom area of 1 square meter, a jack is installed on the rock to apply a certain pressure, and an inclined jack is also arranged on its side with an angle of $\alpha = 15°$ gradually increasing the load until the rock block is sheared along the *AB* surface, and the shear stress and positive stress on the surface are.

$$\tau = \frac{P \cos 15°}{A}$$
$$\sigma = \frac{N + P \sin 15°}{A} \tag{5.9}$$

The above experiment is repeated for blocks with the same rock properties using different vertical pressures *N*. This results in a set of $\tau$, $\sigma$ values. A shear strength curve can be obtained in the $\tau$-$\sigma$ coordinate system.

Without considering the influence of other factors, the order of magnitude of the various strengths is

Triaxial compressive strength > uniaxial compressive strength > shear strength > uniaxial tensile strength.

The triaxial compressive strength of the rock is the highest, while the uniaxial tensile strength is the lowest. The uniaxial compressive strength is 10–50 times the uniaxial tensile strength, while it is about 3–10 times the shear strength.

### Exercises

1. Into which categories can rock damage be classified based on the percentage of strain before damage, briefly explain.

2. What is the definition of rock strength and what factors have an effect on the magnitude of rock strength?
3. Try to explain the end-face effect in uniaxial compression tests of rocks.
4. Analysis of the main extraneous factors affecting the uniaxial compressive strength of rock specimens.
5. Try to analyze the main causes of conical damage when uniaxial compressive strength tests are performed on cylindrical rock specimens and indicate the basic measures to eliminate this form of damage.
6. Why is it troublesome to measure the tensile strength of rocks using the direct method? And how should it be measured?
7. What is the principle of measuring shear strength with a wedge-shaped simple shear gauge?
8. What is the principle of triaxial testing to determine shear strength?
9. Experiments on the shear strength of marble, when the positive stress is $\sigma_{n1} =$ 6 MPa and $\sigma_{n2} =$ 10 MPa when the corresponding shear stresses are $\tau_{n1} =$ 19.2 MPa and $\tau_{n2} =$ 22 MPa. When this rock is tested in triaxial compressive test, when the enclosing pressure $\sigma_3 =$ 0 the strength of the rock is $\sigma_c =$ 100 MPa The strength of the rock is When the circumferential pressure is $\sigma_3 =$ 6 MPa time, try to find its peak compressive strength?
10. The following table shows the results of triaxial compression experiments on cylindrical specimens with a perimeter pressure of 10 MPa and a pore pressure of 0.

| Total axial force/kN | Specimen height/mm | Specimen diameter/mm | Total axial force/kN | Specimen height/mm | Specimen diameter/mm |
|---|---|---|---|---|---|
| C.00 | 100.84 | 50.20 | 190.62 | 100.56 | 50.25 |
| 19.89 | 100.80 | 50.20 | 191.90 | 100.54 | 50.25 |
| 39.60 | 100.77 | 50.20 | 180.22 | 100.52 | 50.26 |
| 63.40 | 100.74 | 50.20 | 137.56 | 100.49 | 50.26 |
| 88.67 | 100.71 | 50.21 | 115.79 | 100.48 | 50.27 |
| 116.18 | 100.68 | 50.21 | 101.93 | 100.43 | 50.28 |
| 144.68 | 100.65 | 50.22 | 97.97 | 100.40 | 50.28 |
| 162.38 | 100.63 | 50.22 | 96.98 | 100.37 | 50.28 |
| 185.23 | 100.58 | 50.24 | | | |

Try to estimate the yield strength, peak strength, residual strength, tangential modulus of elasticity at 50% peak strength and tangential Poisson's ratio at 50% peak strength.

## References

Burshtein LS. Effect of moisture on the strength and deformability of sandstone. J Min Sci. 1969;5(5):573–6.

Cai M. Rock mechanics and engineering. Beijing: Science Press; 2002.

Chen Ziguang. Mechanical properties of rocks and tectonic stress fields. Beijing: Geological Press, 1986.

China National Standardization Administration Committee. GB/T 23561 methods for determination of physical and mechanical properties of coal and rocks. Beijing: China Quality Inspection Press; 2010.

Gao L. Mechanics of mine rock mass. Beijing: Metallurgical Industry Press; 1979.

Griggs D, Handin J. Rock deformation. Geological Society of America; 1960.

Heard HC. Transition from brittle fracture to ductile flow in Solenhofen limestone as a function of temperature, confining pressure, and Geological Society of America Memoirs 1960;79(1):193–226.

Hudson JA, Harrison JP. Engineering rock mechanics. Beijing: Science Press; 2008.

ISRM. The complete ISRM suggested methods for rock characterization, testing and monitoring: 1974–2006. International Society for Rock Mechanics, Commission on Testing Methods; 2007.

Jaeger JC, Cook NGW. Fundamentals of rock mechanics. London: CHAPMAN and HALL; 1979.

Kramadibrata S, Jones IO. Size effect on strength and deformability of brittle intact rock. In: The 2nd international workshop on scale effects in rock masses, Lisbon, Portugal; 1993, p. 277–84.

Lei G. Mining rock mechanics. Beijing: Metallurgical Industry Press; 1979.

Li X. Mechanical properties of rock masses. Beijing: Coal Industry Press; 1983.

Mogi K. Effect of the triaxial stress system on the failure of dolomite and limestone. Tectonophysics. 1971;11(2):111–27.

Schwartz AE. Failure of rock in the triaxial shear test. In: The 6th US symposium on rock mechanics. American Rock Mechanics Association; 1964.

Vutukuri VS. Mechanical properties of rocks, vol. 1. Clausthal, W. Germany: Trans. Tech. Publications; 1974.

Vutukuri VS. The effect of liquids on the tensile strength of limestone. Int J Rock Mech Min Sci Geomech Abstracts. Pergamon 1974;11(1):27–9.

You M. Rock mechanical properties. Beijing: Coal Industry Press; 2007a.

You M. Mechanical properties of rocks. Beijing: Geological Press; 2007b.

# Chapter 6
# Rock Strength Failure Criterion



## 6.1 Coulomb Failure Criterion

Failure to rocks can usually be divided into brittle failure, which is failure that occurs with very little deformation, and ductile failure, which is failure that deformation reaches a significant degree. The brittle or ductile failure of rock is not only affected by stress and strain state, but also controlled by temperature, confining pressure, strain rate and other factors. But most rock failure criterions are considered to be related only to stress or strain states. In the state of unidirectional stress, rock failure can be obtained directly from rock mechanics experiments. For example, the maximum stress during failure is used as the basis for rock failure. However, in general, rocks are in a three-dimensional stress state, and rock failure is often related to the magnitude of the three principal stresses and the distance between them. So a failure criterion must be sought that can be applied to all stress states.

The Coulomb failure criterion is by far the most commonly used and simplest criterion in rock mechanics (Cai 2002).

According to this criterion, rock shear failure occurs along one plane. It is not only related to the magnitude of the shear stress on that plane, but also to the normal stress on that plane. The rock does not failure along the plane where the maximum shear stress acts, but along one of the planes where its shear stress and normal stress reach their most unfavorable combination. That is.

$$|\tau_f| = \tau_0 + f\sigma_n \tag{6.1}$$

$|\tau_f|$ is the shear strength along the rock shear plane

$\tau_0$ is the intrinsic shear strength or cohesion of the rock, $f\sigma_n$ is the frictional resistance at the shear plane, and $\sigma_n$ is the normal stress at the shear plane.

$f$ is the coefficient of internal friction, $f = \tan\varphi$.

Taking $\sigma$,$\tau$ as the horizontal and vertical axes of the coordinate system, the above equation is a linear equation as shown below (Fig. 6.1).



Y. Cheng et al., *Foundations of Rock Mechanics in Oil and Gas Engineering*,
https://doi.org/10.1007/978-981-99-1417-3_6

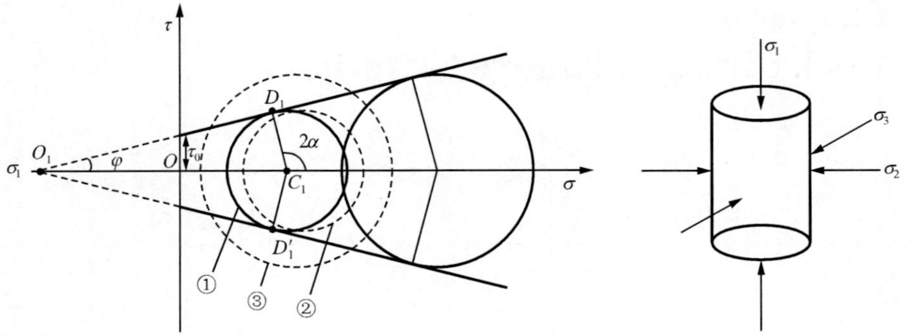


**Fig. 6.1** Schematic of the Coulomb criterion

When the stress circle drawn by the stress state at a point in the rock is tangent to that straight line, it indicates that shear failure is at a critical state. The direction of the shear plane can be determined by $D_1$ and $D_1'$ points where the stress circle is tangent to the shear strength line. If the stress circle drawn by the stress state at a point in the rock is between those two straight lines and not tangent to them, the rock is in an undamaged state. If the stress circle intersects the straight line of the shear strength, the rock has damaged and sliding occurs along the shear plane.

It can be seen that the two points $D_1$ and $D_1'$ where the ultimate stress circle is tangent to the shear strength straight line indicate the critical state where a set of conjugate shear failure planes will occur in the rock. From the Fig. it is seen that the shear stress on this set of shear planes is not the maximum shear stress. The angle between the normal direction outside the shear plane and the maximum principal stress $\sigma_1$ can be seen from the stress diagram as,

$$2\alpha = 90^\circ + \varphi \tag{6.2}$$

$$\alpha = 45^\circ + \frac{\varphi}{2} \tag{6.3}$$

Angle to minimum stress $\sigma_3$:

$$\alpha' = 45^\circ - \frac{\varphi}{2} \tag{6.4}$$

It follows that the maximum compressive stress $\sigma_1$ bisects the acute angle caught by the conjugate shear plane; the minimum principal stress $\sigma_3$ bisects the obtuse angle caught by the conjugate shear plane.

The Coulomb failure criterion can also be expressed in terms of principal stresses $\sigma_1$,$\sigma_3$, with stresses on the shear plane as

$$\sigma_n = \frac{1}{2}(\sigma_1 + \sigma_3) + \frac{1}{2}(\sigma_1 - \sigma_3)\cos 2\alpha$$

$$|\tau_f| = \frac{1}{2}(\sigma_1 - \sigma_3)\sin 2\alpha \tag{6.5a}$$

Thus:

$$\tau_0 = |\tau_f| - f\sigma_n = -\frac{f}{2}(\sigma_1 + \sigma_3) - \frac{1}{2}(\sigma_1 - \sigma_3)(f\cos 2\alpha - \sin 2\alpha) \tag{6.5b}$$

In the above equation, $\alpha$ is the angle between the normal direction of the shear failure plane and $\sigma_1$, also called the fracture angle, seen as Figs. 6.2 and 6.3.

Taking the derivative of (6.5b) on $\alpha$ and making its derivative zero, the extreme point is

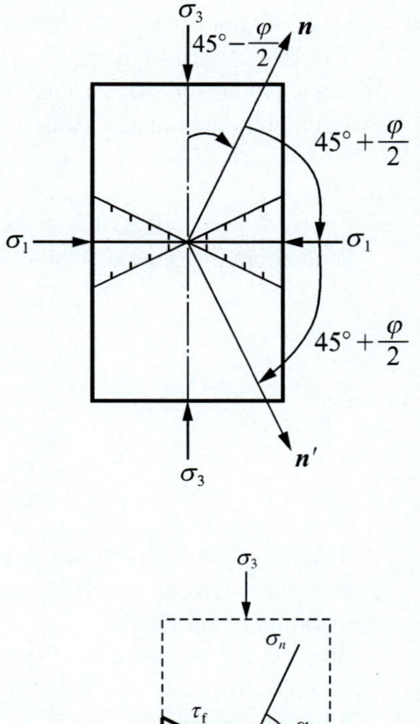

**Fig. 6.2** Fracture angle of rock

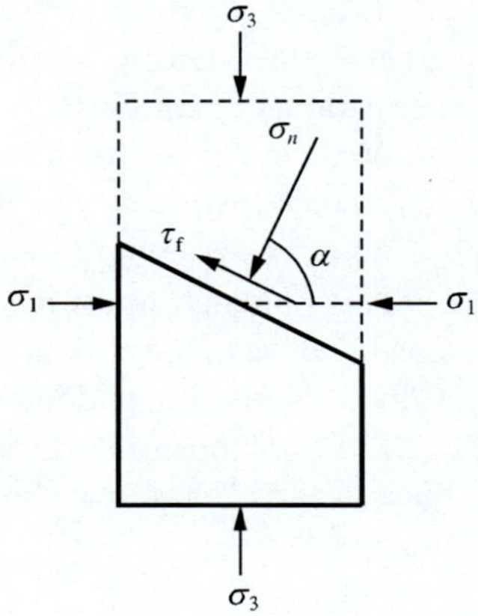

**Fig. 6.3** Relationship between stress on shear plane and principal stress

$$\tan 2\alpha = -\frac{1}{f} \tag{6.6}$$

Since $2\alpha$ is located between $\frac{\pi}{2}$ and $\pi$, then

$$\sin 2\alpha = -\tan 2\alpha \Big/ \sqrt{1 + \tan^2 2\alpha} = (f^2 + 1)^{-\frac{1}{2}}$$

$$\cos 2\alpha = -f \Big/ \sqrt{1 + \tan^2 2\alpha} = -f(f^2 + 1)^{-\frac{1}{2}}$$

Thus:

$$\tau_0 = \frac{\sigma_1}{2}[(f^2 + 1)^{1/2} - f] - \frac{\sigma_3}{2}[(f^2 + 1)^{1/2} + f] \tag{6.7}$$

The above equation is the Coulomb failure criterion expressed by $\sigma_1$ and $\sigma_3$. If the principal stresses $\sigma_1$ and $\sigma_3$ meet the above equation, then shear failure will occur.

When uniaxial tensile failure occurs, $\sigma_1 = 0$, $\sigma_3 = -\sigma_t$, then

$$2\tau_0 = \sigma_t[(f^2 + 1)^{1/2} + f] \tag{6.8}$$

When uniaxial compression failure occurs,$\sigma_1 = \sigma_c$, $\sigma_3 = 0$, then

$$2\tau_0 = \sigma_c[(f^2 + 1)^{1/2} - f] \tag{6.9}$$

So,

$$\frac{\sigma_c}{\sigma_t} = \frac{(f^2 + 1)^{1/2} + f}{(f^2 + 1)^{1/2} - f} \tag{6.10}$$

It follows that the criterion is applicable to materials with compressive strength greater than tensile strength, such as rocks at lower confining pressure and temperature conditions.

When $\sigma_c / \sigma_t = 10$, $f = 1.5$, Equivalent to hard magmatic rocks.

When $\sigma_c / \sigma_t = 6$, $f = 1.0$, Equivalent to strong sedimentary rocks.

When $\sigma_c / \sigma_t = 4$, $f = 0.7$, Equivalent to unstable sedimentary rocks.

This ratio, as shown by laboratory experiments, is relatively close to the experimental value of the rock in a brittle state. Table 6.1 shows the experimental parameter values of cohesion and internal friction angle for typical rocks.

The Coulomb criterion can also be expressed in the following various expressions, such as the Mohr circle shown above (Fig. 6.1), in $\Delta O_1 D_1 C_1$

$$\sin\varphi = \frac{\overline{C_1 D_1}}{\overline{O_1 C_1}} \tag{6.11}$$

where

**Table 6.1** $\tau_0$, $\varphi$ values of typical rock

| Rock | $\tau_0$/MPa | $\varphi/^\circ$ | $f$ |
|---|---|---|---|
| Shale | 3–30 | 15–30 | 0.25–0.6 |
| Sandstone | 8–40 | 35–50 | 0.7–162 |
| Limestone | 10–50 | 35–50 | 0.7–1.2 |
| Marble | 15–30 | 3550 | 0.7–1.2 |

$$\overline{C_1D_1} = \frac{1}{2}(\sigma_1 - \sigma_3) \tag{6.12}$$

$$\overline{O_1C_1} = \frac{1}{2}(\sigma_1 + \sigma_3) + \tau_0 ctg\varphi \tag{6.13}$$

So,

$$\sin\varphi = \frac{\sigma_1 - \sigma_3}{\sigma_1 + \sigma_3 + 2\tau_0 \text{ctan}\varphi} \tag{6.14}$$

Collated from,

$$\sigma_1 = \sigma_3.\frac{1 + \sin\varphi}{1 - \sin\varphi} + 2\tau_0\frac{\cos\varphi}{1 - \sin\varphi} \tag{6.15}$$

Equation (6.15) is an alternative form of the Coulomb criterion expressed by $\sigma_1$ and $\sigma_3$ which can also be written as.

$$\sigma_1 = \sigma_3.\frac{1 + \sin\varphi}{1 - \sin\varphi} + 2\tau_0\sqrt{\frac{1 + \sin\varphi}{1 - \sin\varphi}} \tag{6.16}$$

According to the trigonometric constant,

$$\frac{1 + \sin\varphi}{1 - \sin\varphi} = \tan^2(45^\circ + \frac{\varphi}{2}) \tag{6.17}$$

$$\tan(45^\circ + \frac{\varphi}{2}) = \frac{1}{\tan(45^\circ - \frac{\varphi}{2})} \tag{6.18}$$

So Eq. (6.15) can again be written as

$$\sigma_1 = \sigma_3 \tan^2(45^\circ + \frac{\varphi}{2}) + 2\tau_0 \tan(45^\circ + \frac{\varphi}{2}) \tag{6.19}$$

Or

$$\sigma_3 = \sigma_1 \tan^2(45^0 - \frac{\varphi}{2}) - 2\tau_0 \tan(45^\circ - \frac{\varphi}{2}) \tag{6.20}$$

Both of the above equations are alternative forms of the Coulomb criterion, and whenever the principal stresses in the rock satisfy either form of the expression above, the rock is about to shear fracture.

If the pore pressure P is present in the rock, it only reduces the positive stress in any section and has no effect on the shear stress in that section according to the definition of effective stress. Thus, the Coulomb criterion can be written as

$$|\tau_f| = \tau_0 + f(\sigma_n - p) \tag{6.21}$$

The above equation indicates that in the $\sigma - \tau$ coordinate system, the stress circle is simply moved to the left by a distance whose magnitude is equal to the magnitude of the pore pressure, but the pore pressure does not change the radius of the stress circle. Thus, under the influence of pore pressure, the stress circle, which was originally in a stable state, is shifted to the left by a distance $P$. Then it may be tangent to the shear strength, making it in the limit state, and the rock will be destroyed. This fact indicates that the pore pressure P reduces the strength of the rock as well as makes the rock susceptible to brittle failure because the pore pressure counteracts the effect of the confining pressure.

## 6.2 Mohr Failure Criterion

From the results of triaxial experiments, the rock shear failure line is approximately an oblique straight line when the confining pressure is low, but it converts to a curve when the confining pressure is high. Both $\varphi$ and $\tau_0$ change with the magnitude of the confining pressure. When the confining pressure is higher, the internal friction angle $\varphi$ decreases but the cohesion $\tau_0$ increases, and the converse is true when the confining pressure is lower. Mohr failure criterion states that when shear failure occurs in a certain section, the shear stress $\tau_f$ in that section must increase to a critical value, and that value $\tau_f$ depends on the positive stress in that section, i.e.,

$$\tau_f = f(\sigma_n) \tag{6.22}$$

This functional relationship can be determined experimentally. That is, for a group of rock specimens with the same properties, the triaxial failure tests of rocks are carried out under different principal stresses $\sigma_1$ and $\sigma_3$ respectively. In this way, a set of ultimate stress circles can be derived, and the failure criterion of the rock can be established from the envelope of this set of ultimate stress circles, as shown in Fig. 6.4.

It should be stated that this criterion does not consider the effect of the intermediate principal stress $\sigma_2$.

The shape of the envelope of rocks can be divided into two categories, as shown in Fig. 6.5. One type of envelope in the high confining pressure region, the curve gradually bends toward the horizontal coordinate axis $\sigma$, tending to the horizontal

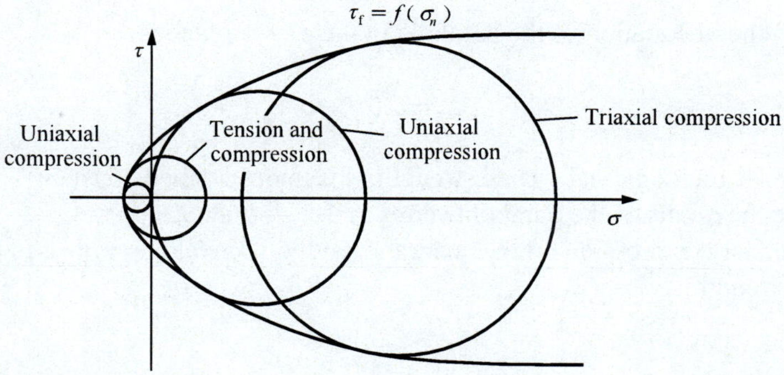


**Fig. 6.4** Mohr envelope

asymptote, as shown in Fig. 6.5a. The rock with many pores, relatively loose, large compressibility and good ductility belongs to this type, such as: coal, clayey shale and other ductile rocks. The other type of envelope is skewed to the sides in the high confining pressure region without bending to the horizontal axis $\sigma$, which is a non-contracting type of envelope, as shown in Fig. 6.5b. Rocks with denser structures, such as sandstone, limestone, granite and other brittle rocks, belong to this type (Fig. 6.6).

To describe the failure curve of a rock, the shear strength curve of the rock is often approximated by envelopes of parabolic and hyperbolic.

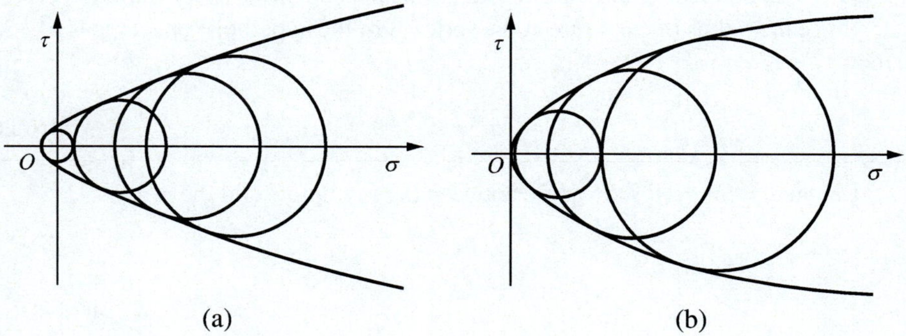


**Fig. 6.5** Two types of envelopes for rocks

**Fig. 6.6** Parabolic type envelope

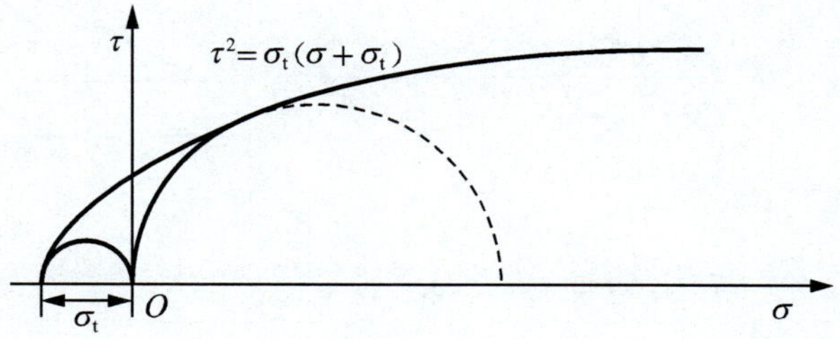

The general equation of the parabola in the $\tau - \sigma$ plane is

$$\tau^2 = \lambda(\sigma + \sigma_t) \tag{6.23}$$

where, $\sigma_t$ is rock uniaxial tensile strength, $\lambda$ is undetermined constant

Since the radius of the parabola vertex is $\frac{\lambda}{2} = \frac{\sigma_t}{2}$,then $\lambda = \sigma_t$.

The general expression for the quadratic parabolic envelope expressed by uniaxial tensile strength is

$$\tau^2 = \sigma_t(\sigma - \sigma_t) \tag{6.24}$$

Hyperbolic envelope is able to take into account the fracture factors in the rock mass. Due to the presence of fractures in the rock mass, the strength of the rock mass is much reduced, especially the tensile strength, but still has some compressive and shear strength. Therefore, according to the degree of fracture development in the rock mass, the envelope can be appropriately shifted towards the compression region. If the rock mass completely loses its tensile strength, the envelope can be tangent to the vertical coordinate at the origin *O*, as shown in Fig. 6.7.

The hyperbolic envelope parameter equation is

$$\begin{aligned} \sigma + \sigma_t &= a(\cosh 2t - 1) \\ \tau &= b \sinh 2t \end{aligned} \tag{6.25}$$

where *t* is coefficient, *a* and *b* are constants that depend on the shape of the hyperbola.

Since the radius of curvature at the vertex B of the hyperbolic envelope is $b^2 / 2a$, then

$$\sigma_t = 2b^2 / a \tag{6.26}$$

Let $\tan\eta = b/a$, $\eta$ is the angle between the asymptote and the $\sigma$ axis.

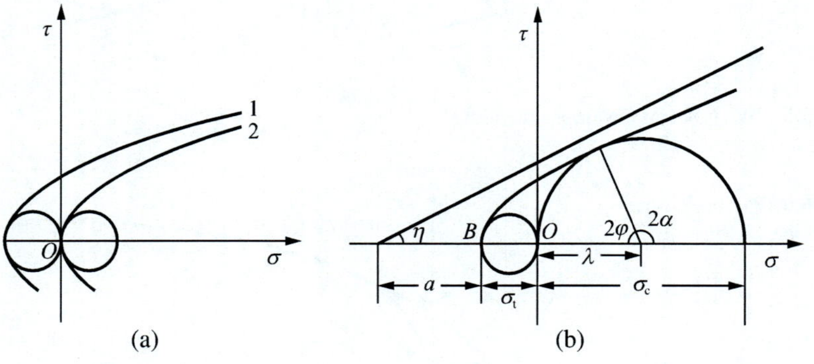


**Fig. 6.7** Hyperbolic envelope

So $\sigma_t / \tan \eta = 2b, b = \frac{\sigma_t}{2 \tan \eta}$

$$\sigma_t / \tan^2 \eta = 2a$$

$$a = \frac{\sigma_t}{2 \tan^2 \eta}$$

Taking the above equation into Eq. (6.24),

$$\begin{aligned} \sigma + \sigma_t &= \frac{\sigma_t}{2 \tan^2 \eta}.(cosh2t - 1) \\ \tau &= \frac{\sigma_t}{2 \tan \eta} \sinh 2t \end{aligned} \tag{6.27}$$

The condition that the uniaxial compressive strength stress circle is tangent to the hyperbolic envelope was used to determine $\tan \eta$. And by further derivation it was obtained that

$$\begin{aligned} \tau &= \left[\tan^2 \eta(\sigma + \sigma_t)^2 + \sigma_t(\sigma + \sigma_t)\right]\gamma^2 \\ \tan \eta &= \frac{1}{2}\sqrt{\frac{\sigma_c}{\sigma_t} - 3} \end{aligned} \tag{6.28}$$

## 6.3 Failure Criterion for Rock Formations with Weak Planes

There are a large number of structural planes (cleavages, joints or faults) in the rock mass. Due to geological action, there are often weak intercalated layers on these structural planes. Their strength is much lower than the strength of the rock mass itself. This makes it possible for the rock mass to generate shear slip along the weak planes (Jaeger and Cook 1979; Hudson and Harrison 2008). The criterion for determining whether a rock mass has produced slip along a weak plane is similar to the Coulomb failure criterion, namely.

$$\left|\tau_f\right| = \tau_u + \sigma_n \tan \varphi_u \tag{6.29}$$

where, $\tau_f$ is the shear strength of the weak plane, $\tau_u$ is the intrinsic shear strength of the weak plane, $\varphi_u$ is the internal friction angle of the weak plane, $\sigma_n$ is the normal stress acting on the weak plane.

For the two-dimensional problem, the relationship between the normal and shear stresses at any point on the weak plane and the principal stresses $\sigma_1$ and $\sigma_3$ is given by

$$\begin{cases} \sigma_n = \frac{1}{2}(\sigma_1 + \sigma_3) + \frac{1}{2}(\sigma_1 - \sigma_3)\cos 2\alpha \\ \left|\tau_f\right| = \frac{1}{2}(\sigma_1 - \sigma_3)\sin 2\alpha \end{cases} \tag{6.30}$$

where, $\alpha$ is the angle between the outer normal to the weak plane and the maximum principal stress $\sigma_1$.

Since the average normal stress and the maximum shear stress are

$$\begin{aligned} \sigma_m &= \frac{1}{2}(\sigma_1 + \sigma_3) \\ \tau_{\max} &= \frac{1}{2}(\sigma_1 - \sigma_3) \end{aligned} \tag{6.31}$$

Then Eq. (6.29) can be written as.

$$\begin{cases} \sigma_n = \sigma_m + \tau_{\max}\cos 2\alpha \\ \left|\tau_f\right| = \tau_{\max}\sin 2\alpha \end{cases} \tag{6.32}$$

Substituted into the failure criterion Eq. (6.29)

$$\tau_{\max} \cdot [\sin 2\alpha - \tan\varphi_u \cos 2\alpha] = \tau_u + \sigma_m tg\varphi_u \tag{6.33}$$

Let $\tan\delta = \sin\varphi_u \csc(2\alpha - \varphi_u)$,then the above equation can be written as

$$\tau_{\max} = (\sigma_m + \tau_u \cot\varphi_u)\tan\delta \tag{6.34}$$

The above equation is the shear slip criterion generated by the weak plane.

Substituting Eq. (6.33) into the above equation

$$\sigma_1 = [(2\tau_u + 2f_u\sigma_3)/(1 - f_u\cot\alpha)\sin 2\alpha] + \sigma_3 \tag{6.35}$$

where, $f_u$ is the internal friction coefficient of the weak plane.

Let $k = \sigma_3 / \sigma_1$, then

$$\sigma_1 = 2\tau_u \cot\varphi_u / [(1 - k)\sin(2\alpha - \varphi_u)\csc\varphi_u - (1 + k)] \tag{6.36}$$

Equations (6.35) and (6.36) are both alternative forms of the slip criterion along a weak plane.

From Eq. (6.34), it can be seen that $\sigma_1 - \sigma_3$ converges to $\infty$, if $\alpha$ converges to $\pi/2$ or $\varphi_u$. Therefore when $\varphi_u < \alpha < 90°$, it is possible to slip along the weak plane. If we take the derivative of $\sigma_1$-$\sigma_3$ with respect to $\alpha$, and make it zero, we get that $\sigma_1 - \sigma_3$ on the weak plane at $\tan 2\alpha = -1/f_u$ reaches the extreme value, i.e.,

$$(\sigma_1 - \sigma_3)_{\max} = 2(\tau_u + f_u\sigma_3)\left[\left(f_u^2 + 1\right)^{\frac{1}{2}} + f_u\right] \tag{6.37}$$

The angle $\alpha$ between the normal to the weak plane and the maximum principal stress $\sigma_1$ is

$$\tan 2\alpha = -\frac{1}{\tan \varphi_u} = \tan(90^\circ + \varphi_u) \quad (6.38)$$

That is, when $\alpha = 45^\circ + \varphi_u/2$, it is most conducive to the sliding of the weak plane.

As can be seen from Fig. 6.8, The *PQR* straight line represents the failure criterion for the weak plane. If the stress on the weak plane in the rock mass is at point *D* on the stress circle, it is obvious that it is not possible to produce slip on this weak plane. However, if the stress is located at any point on the *QR* arc of the stress circle, it corresponds to the orientation within which the weak plane has produced slip, and the points *Q* and *R* on the stress circle indicate that the weak plane corresponding to that orientation is in a critical failure state.

Metamorphic rocks or rocks with bedding are anisotropic (transverse isotropy). There is a big difference in strength between parallel and perpendicular planes, and the shear strength is lower in the direction parallel to the bedding.

Figure 6.9 shows the experimental information for the four rocks and it can be seen that the strength is controlled by the angle between the principal stress and the weak plane. When $\alpha = 90^\circ$ and $\alpha = 0^\circ$, both have higher strength. When $\alpha = 30^\circ$–$45^\circ$, it has the lowest strength.

As the confining pressure increases, the strength of the rock all increases relatively. Therefore, the rock formation with anisotropic characteristics can be considered as a set of parallel weak planes, so that the failure criterion of weak plane can be applied to the anisotropic rock formation.

$$|\tau_f| = \tau_w + \tan \varphi_w \sigma_n \quad (6.39)$$

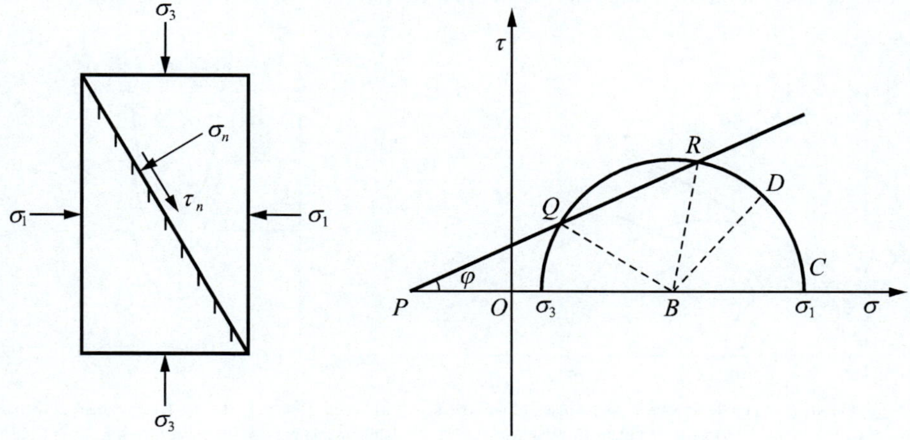


**Fig. 6.8** Slip along a weak plane

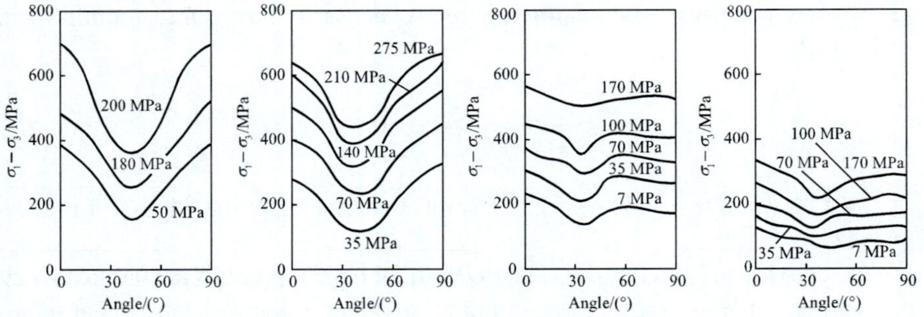


**Fig. 6.9** Shear strength curves for different angles of maximum principal stress and bedding plane

where, $\tau_w$ is the cohesion at the weak plane, $\varphi_w$ is the internal friction angle at the weak plane.

The angle between the orientation normal to the layer most favorable to shear slip and the maximum principal stress $\sigma_1$ is still $45° + \varphi_w/2$.

Here there is also a possibility that if the stresses at the rock layer do not satisfy the criterion for layer slip, but may produce shear failure along the plane intersecting the layer. In this case the Coulomb criterion expressed by the intrinsic shear strength $\tau_0$ of the rock and the internal friction angle $\varphi$ can still be applied (Fig. 6.10).

The straight line *CD* in Fig. 6.10a corresponds to the shear strength of the rock, i.e., the Coulomb criterion $|\tau_f| = \tau_0 + f\sigma_n$. The straight line AB corresponds to the shear strength along the bedding in the formation, i.e.,$|\tau_f| = \tau_w + f_w\sigma_n$, but $\tau_0 > \tau_w$ and $f > f_w$.

In what state will the stress in the formation cause slip failure along the bedding plane or failure along the section intersected with the bedding plane? To illustrate this, assume that $\sigma_3$ remains constant while gradually increasing $\sigma_1$. When $\sigma_1$ reaches a certain value $\sigma_{min}$ (point *E*), the stress circle is tangent to *AB*. And the rock layer in this stress state first slips on the plane where the angle between the outer normal

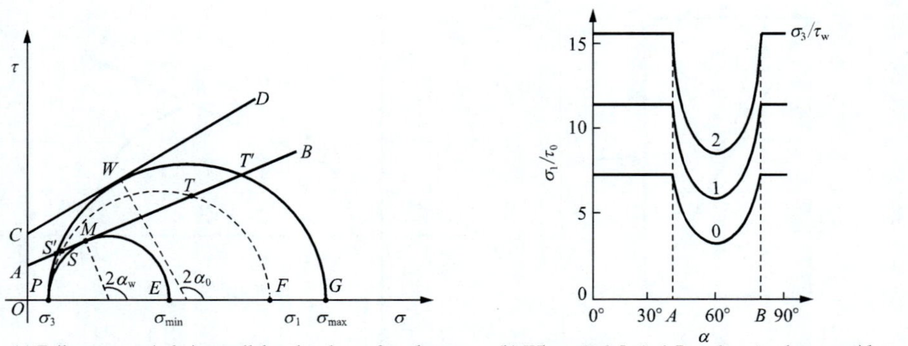


(a) Failure stress circle in parallel to the plane of weakness and in the plane intersecting the plane of weakness

(b) When $f_w$=0.5, $f_w$=0.7, $\tau_0$=2, $\sigma_1/\tau_0$, changes with $\alpha$, and the numbers on the curve are $\sigma_3/\tau_w$

**Fig. 6.10** Influence of bedding plane angle on rock failure direction and the maximum stress $\sigma_1$ is $\alpha_w$. If the maximum principal stress $\sigma_1$ is greater than the stress at point F, the stress circle of its stress state is as the dashed circle in Fig. 6.10a. Then the group of bedding planes corresponding to the stress of the arc of $ST$ on the dashed circle produces slip, while the group of weak planes corresponding to the stress of the arc of $PS$ and $TF$ on the dashed circle cannot produce slip along the bedding planes. If $\sigma_1$ increases to $\sigma_{\max}$ (point $G$), the stress state corresponding to the stress circle tangent to the line $CD$ and cut to the line $AB$, two slip states are possible.

If the stresses at bedding plane are within the orientation range indicated by the arcs PS and T'G, shear slip can only be produced in the plane obliquely intersecting the bedding plane in the direction whose normal direction is at an angle $\alpha_0$ with the maximum principal stress $\sigma_1$.

If the stress at the bedding plane is within the range of the arc $ST'$, but the stress is not necessarily at the point $W$, then slip should be generated along the bedding plane. So the range of slip along the bedding plane is such that if $\sigma_3$ remains constant, a finite range of $\sigma_{\min} \le \sigma_1 \le \sigma_{\max}$ produces slip along the bedding plane.

Figure 6.10b represents the three curves of the maximum principal stress $\sigma_1$ with the variation of the angle $\alpha$, when $\sigma_3/\tau_w$ is constant 0, 1, and 2 respectively under the conditions of $f_w = 0.5$ of the layer, $f = 0.7$ and $\tau_0 = 2\tau_w$ of the rock mass. Where the horizontal lines indicate the failure criterion of the rock mass and the curves indicate the failure criterion of the weak plane. If $\sigma_{1/}\tau_w$ lies above these lines then the specimen may failure along the weak plane or failure along the section that intersects the weak plane. If below these lines, no failure occurs. If the weak plane is oriented at an angle outside the $AB$ range, no failure will occur along the weak plane. Figure 6.11 represents the relationship between the principal stress difference $\sigma_1$-$\sigma_3$ and the shear fracture angle $\alpha_0$.

To determine whether jointed rock formation is in a stable state at a given stress state still approximates the Coulomb criterion, $|\tau_f| = \tau_0 + \sigma_n \tan\varphi$. If $\tau_0 = 0$ means that cohesion has been lost between the joint planes or anisotropic rock layers, then:

$$|\tau_f| = \sigma_n \tan\varphi \tag{6.40}$$

Therefore, the angle between the joint plane and the maximum principal stress $\sigma_1$ is $\beta = \pi/2 + \alpha$, $\alpha$ is the angle between the direction normal to the outer joint

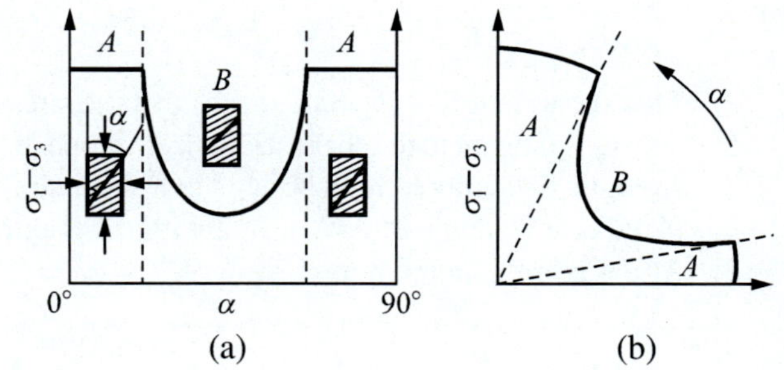


**Fig. 6.11** Effect of the angle between the principal stress and the weak plane on the strength

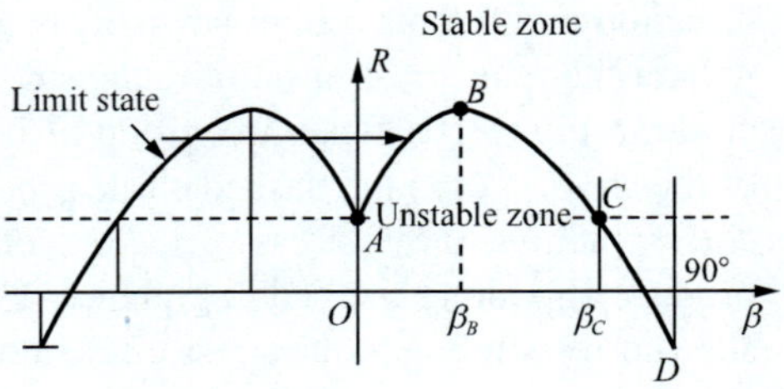


**Fig. 6.12** Stability conditions for rock masses with bedding planes

plane and $\sigma_1$, so that

$$\begin{aligned} \sigma_n &= \frac{\sigma_1+\sigma_3}{2} - \frac{\sigma_1-\sigma_3}{2}\cos 2\beta \\ |\tau_n| &= \frac{\sigma_1-\sigma_3}{2}\sin 2\beta \end{aligned} \tag{6.41}$$

To wit

$$(\sigma_1-\sigma_3)\sin 2\beta = (\sigma_1-\sigma_3)\tan\varphi_0 - \frac{\sigma_1-\sigma_3}{2}\cos 2\beta\tan\varphi_0$$

Or

$$R = \frac{\sigma_1+\sigma_3}{\sigma_1-\sigma_3} = \frac{\sin(2\beta+\varphi_0)}{\sin\varphi_0} \tag{6.42}$$

If the above conditions are met, then a jointed rock mass or anisotropic formation is in a limiting stable state, and the above equation $R = R(\beta)$ can be represented by Fig. 6.12.

(1) At the point $A$, $\beta_A = 0$. From Eq. (6.42), if $\sigma_3 = 0$, then $R = 1$. It indicates that the joint plane is in a critical state of brittle failure under the action of the unidirectional compressive stress $\sigma_1$ at the parallel joint plane.

At the point $B$ in the figure, $\beta_{\mathrm{B}} = 45°- \varphi_0/2$, then using Eq. (6.42), we can see that

$$R = \frac{\sigma_1+\sigma_3}{\sigma_1-\sigma_3} = \frac{\sin 90°}{\sin\varphi_0} = \frac{1}{\sin\varphi_0} \tag{6.43}$$

(2) This shows that $R > 1$, which means that the stress circle under the action of $\sigma_1$, and $\sigma_3$ is tangent to the shear strength envelope of the rock without consolidation strength. The jointed rock body is in a shear slip critical state when the jointed plane is at an angle of $45° - \varphi_0/2$ with the maximum principal stress.$\sigma_1$
(3) At the Point $C$ in the figure, $\beta_C = 90° - \varphi_0 = 2\beta_B$, by Eq. (6.42)

$$R = \frac{\sigma_1+\sigma_3}{\sigma_1-\sigma_3} = \frac{\sin(180° - \varphi_0)}{\sin\varphi_0} = 1 \tag{6.44}$$

It indicates that when $\sigma_3$ is zero, the principal stress $\sigma_1$ is at an angle of 90°− $\varphi_0$ to the joint plane and the jointed rock is in a critical state of unconfined lateral stress shear slip.

As the $\beta$ value varies from 90° − $\varphi_0$ to 90°, R tends to a negative value, indicating a steady state.

It follows that a jointed rock mass, whether in a unidirectional stress or two-way stress state, will produce an unstable state if it is within the ABC curve; outside the curve it is a stable state.

## 6.4 Griffith Criterion

Griffith (1921) established a brittle failure criterion in terms of the fundamental physical properties of the medium. He found from experiments on the strength of glass that the actual strength obtained was often 1/100–1/1000 of the theoretical value. Griffith explained as: the difference is due to the existence of tiny cracks randomly distributed within the medium. When the load reaches a certain value, the phenomenon of stress concentration occurs near the end of the crack most favorable to failure in it. The fracture begins to expand when the tensile stress at the end of the fracture is greater than or equal to the tensile strength at that point. This criterion assumes that the fracture is an elliptical hole (called a Griffith fracture) and that adjacent fractures do not affect each other. From this basis, the failure criterion for planar problems is derived based on the theory of elastic mechanics (Li 1983; Chen 1986).

### *6.4.1 Griffith Criterion Derivation*

Assume that an elliptical fracture exists in the rock with an arbitrary long-axis orientation, and the angle with $\sigma_1$ is $\beta$, subject to two-way compressive stresses $\sigma_1$ and $\sigma_3$. The fracture stress model is shown in Fig. 6.13.

According to the stress transformation equation, $\sigma_1$ and $\sigma_3$ convert to $\sigma_x$, $\sigma_y$ and$\tau_{xy}$, seen as Eq. (6.44).

$$\begin{aligned} \sigma_x &= \sigma_1 \cos^2 \beta + \sigma_3 \sin^2 \beta \\ \sigma_y &= \sigma_1 \sin^2 \beta + \sigma_3 \cos^2 \beta \\ \tau_{xy} &= \frac{\sigma_1 - \sigma_3}{2} \sin 2\beta \end{aligned} \tag{6.45}$$

If the equation of the ellipse is $x = a\cos\alpha$, $y = b\sin\alpha$, and the relationship between the eccentric angle α and the coordinate is shown in Fig. 6.13b. $\theta$ is the magnitude angle at the study point, $\tan\theta = y/x = m\tan\alpha$. Where $m = b/a$, the shear stress around the ellipse is given by

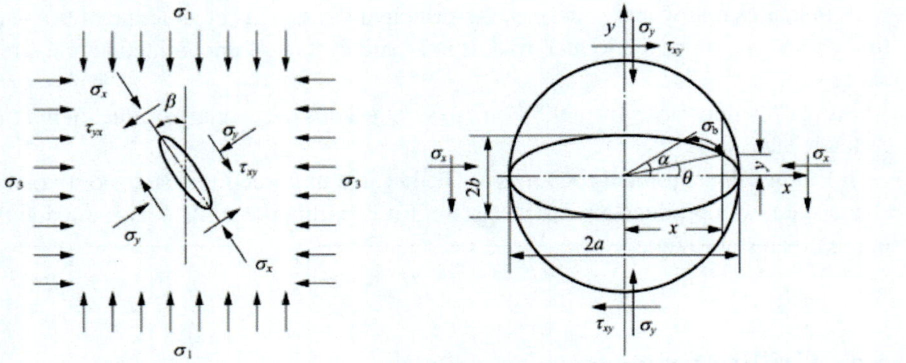


(a) The fracture force system (b) Stress system for calculating tangential stress around ellipse

**Fig. 6.13** Griffith fracture stress model

$$\sigma_b = \frac{\sigma_y\left[m(m+2)\cos^2\alpha - \sin^2\alpha\right]}{m^2\cos^2\alpha + \sin^2\alpha} + \frac{\sigma_x\left[(1+2m)\sin^2\alpha - m^2\cos^2\alpha\right]}{m^2\cos^2\alpha + \sin^2\alpha} + \frac{\tau_{xy}\left[2(1+m)^2\sin\alpha\cos\alpha\right]}{m^2\cos^2\alpha + \sin^2\alpha} \quad (6.46)$$

Since the fracture is flat and $m = b/a$ is small, when $\alpha \to 0$, $\sin\alpha \to \alpha$, $\cos\alpha \to 1$, omitting the second order infinitesimal $\sin^2\alpha$, the above equation simplifies to

$$\sigma_b = \frac{2\left(m\sigma_y - \alpha\tau_{xy}\right)}{m^2 + \alpha^2} \quad (6.47)$$

$\sigma_b$ is a function of $\alpha$. If $\sigma_b$ is differentiated with respect to $\alpha$ and made zero, the condition for maximum shear stress to occur is obtained as

$$\frac{d\sigma_b}{d\alpha} = -\frac{4m\alpha\sigma_y}{\left(m^2+\alpha^2\right)^2} - \frac{2\tau_{xy}}{m^2+\alpha^2} + \frac{4\alpha^2\tau_{xy}}{\left(m^2+\alpha^2\right)^2} = 0$$

That is,

$$\sigma_y = \frac{\alpha^2 - m^2}{2m\alpha}\tau_{xy} \quad (6.48)$$

Substituting Eq. (6.47) into Eq. (6.48) yields the extreme value $\sigma_\beta$ of $\sigma_b$

$$m\sigma_\beta = \frac{1}{2}[(\sigma_1+\sigma_3)(\sigma_1-\sigma_3)\cos 2\beta] \pm \frac{1}{2}[2(\sigma_1^2+\sigma_3^2) - 2(\sigma_1^2-\sigma_3^2)\cos 2\beta]^{1/2} \quad (6.49)$$

Then the shear stress extremes can be represented by the stress components $\sigma_y$, and $\tau_{xy}$.

Organize Eq. (6.48) into the expression for $\frac{1}{\alpha}$,

$$\frac{1}{\alpha^2} + \frac{2\sigma_y}{m\tau_{xy}} \cdot \frac{1}{\alpha} - \frac{1}{m^2} = 0$$

So, $\frac{1}{\alpha} = -\frac{1}{m\tau_{xy}}\left[\sigma_y \pm \left(\sigma_y^2 + \tau_{xy}^2\right)^{\frac{1}{2}}\right]$.

Substituting the above equation into Eq. (6.49), then

$$m \cdot \sigma_\beta = \sigma_y \pm \left(\sigma_y^2 + \tau_{xy}^2\right)^{\frac{1}{2}} \tag{6.50}$$

If expressed in terms of principal stresses, then substituting Eq. (6.45) into Eq. (6.50) gives

$$m\sigma_\beta = \frac{1}{2}[(\sigma_1 + \sigma_3) - (\sigma_1 - \sigma_3)\cos 2\beta] \pm \frac{1}{2}\left[2\left(\sigma_1^2 + \sigma_3^2\right) - 2\left(\sigma_1^2 - \sigma_3^2\right)\cos 2\beta\right]^{1/2} \tag{6.51}$$

From the above equation, if $\sigma_1$, and $\sigma_3$ is known and $\sigma_\beta$ is a function of $\beta$. If the extreme value of $\sigma_\beta$ is found, the most dangerous angle $\beta$ can be found and further the fracture criterion can be found.

Let $\frac{\mathrm{d}\sigma_\beta}{\mathrm{d}\beta} = 0$, i.e.,

$$m\frac{\mathrm{d}\sigma_\beta}{\mathrm{d}\beta} = (\sigma_1 - \sigma_3)\sin 2\beta \left\{1 \pm \frac{\sigma_1 + \sigma_3}{2\left[\frac{1}{2}\left(\sigma_1^2 + \sigma_3^2\right) - \frac{1}{2}\left(\sigma_1^2 - \sigma_3^2\right)\cos 2\beta\right]^{1/2}}\right\} = 0 \tag{6.52}$$

Then

$$\sin 2\beta = 0 \tag{6.53}$$

or $1 \pm \frac{\sigma_1 + \sigma_3}{2\left[\frac{1}{2}\left(\sigma_1^2 - \sigma_3^2\right) - \frac{1}{2}\left(\sigma_1^2 - \sigma_3^2\right)\cos 2\beta\right]^{1/2}} = 0$

That is,

$$\cos 2\beta = \frac{\sigma_1 - \sigma_3}{2(\sigma_1 + \sigma_3)} \tag{6.54}$$

When the direction of the crack meets Eqs. (6.52) and (6.53), $\sigma_\beta$ of the fracture is the extreme value. If this extreme value is $\sigma_0$, substitute (6.52) and (6.53) into Eq. (6.51) to find the extreme value $\sigma_0$.

When $\sin 2\beta = 0$, $\beta = 0°$ or 90°, i.e.,$\cos 2\beta = \pm 1$, substituting into Eq. (6.51), we have

$$m\sigma_0 = 2\sigma_3,\ 0,\ 2\sigma_1,\ 0 \tag{6.55}$$

When $\cos 2\beta = (\sigma_1 - \sigma_3)/2(\sigma_1 + \sigma_3)$

$$m\sigma_0 = \frac{(3\sigma_1 + \sigma_3)(\sigma_1 + 3\sigma_3)}{4(\sigma_1 + \sigma_3)} \tag{6.56}$$

Or

$$m\sigma_0 = -\frac{(\sigma_1 - \sigma_3)^2}{4(\sigma_1 + \sigma_3)} \tag{6.57}$$

From the above it can be seen that $m\sigma_0$ has six extreme values, where the maximum tensile stress reaches the tensile strength $\sigma_t$ and the rock is destroyed.

In the unidirectional tension state, $\sigma_3 < 0$ and $\sigma_1 = 0$. Substitute them into Eqs. (6.56)–(6.58) gets extreme values $2\sigma_3$, 0, 0, 0, $\frac{3}{4}\sigma_3$, $-\frac{1}{4}\sigma_3$($\sigma_3$ is negative), and obviously $2\sigma_3$ is the maximum tensile stress. In this way, when the tensile stress reaches the uniaxial tensile strength, the following formula always holds,

$$\begin{aligned} m\sigma_0 &= -2\sigma_t \\ m\sigma_0 &= 2\sigma_3 = 2(-\sigma_t) \\ \sigma_3 &= -\sigma_t \end{aligned} \tag{6.58}$$

At this point $\cos 2\beta = \pm 1$.

$\beta = 0°$ for dangerous directions.

In the unidirectional compression state, $\sigma_3 = 0$, $\sigma_1 > 0$. Substitute them into Eqs. (6.56)–(6.58) get

$$0, 0, 2\sigma_1, 0, \frac{3}{4}\sigma_1, -\frac{1}{4}\sigma_1$$

Since this theory assumes that rock failure is caused by tension, $-\frac{1}{4}\sigma_1$ is the maximum tensile stress, and this value is equal to the extreme value at which the rock breaks in unidirectional tensile rupture.

That is, $m\sigma_0 = -\frac{1}{4}\sigma_1 = 2(-\sigma_t)$.

So, $\sigma_1 = 8\sigma_t$.

That is, Griffith's criterion applies only to rocks with compressive strength 8 times the tensile strength. Substituting $\sigma_1 = \sigma_1$, $\sigma_3 = 0$ into Eq. (6.54) gives $\beta = 30°$. Therefore, under uniaxial compression, when the fracture strike is at an angle of 30° to the compressive stress, it is most conducive to fracture.

When biaxial stress is applied, because $|\cos 2\beta| < 1$.

So $\left|\frac{\sigma_1 - \sigma_3}{2(\sigma_1 + \sigma_3)}\right| < 1$.

i.e., $\sigma_1 + 3\sigma_3 > 0$ is the condition for forming biaxial stress.

Similarly when $m\sigma_0 = -2\sigma_t$, the rock breaks up and substituting into Eq. (6.57) gives

$$-\frac{(\sigma_1-\sigma_3)^2}{4(\sigma_1+\sigma_3)}=-2\sigma_t$$

$$\frac{(\sigma_1-\sigma_3)^2}{\sigma_1+\sigma_3}=8\sigma_t \tag{6.59}$$

If $\sigma_1+3\sigma_3<0$, a tensile stress condition, at which point $2\sigma_3$ is the maximum tensile stress. The rock breaks when $m\sigma_0=2\sigma_3=-2\sigma_t$, giving a failure criterion:

$$\sigma_3=-\sigma_t \tag{6.60}$$

Thus the Griffith failure criterion is

$$\begin{cases}(\sigma_1-\sigma_3)^2-8\sigma_t(\sigma_1+\sigma_3)=0 & (\sigma_1+3\sigma_3>0)\\ \sigma_3=-\sigma_t & (\sigma_1+3\sigma_3<0)\end{cases} \tag{6.61}$$

### 6.4.2 Modified Griffith Criterion

Since Griffith ignored the fact that fractures may produce closure under certain pressure in his assumptions, it caused some difference between the theoretical and measured values, and for this reason, corrections had to be made. That is, it is assumed that a certain frictional force will be generated between the crack faces when the crack is closed under pressure and it will affect the crack rupture. It is due to the presence of a positive stress $\sigma_n$ interacting between the crack faces when the crack is closed and a frictional force opposite to the direction of the crack sliding $\tau_f=f\sigma_n$ ($f$ is the coefficient of friction between the cracks). So this problem is similar to the case of frictionless cracks with $\sigma_y'=\sigma_y-\sigma_n$ and $\tau_{xy}'=\tau_{xy}-\tau_f$ acting at them. In a similar way to the derivation above, a modified Griffith criterion can be derived as follows.

$$\sigma_1[(1+f^2)^{1/2}-f]-\sigma_3[(1+f^2)^{1/2}+f]=4\sigma_t \tag{6.62}$$

When $\sigma_n=\frac{1}{2}(\sigma_1+\sigma_3)-\frac{1}{2}(\sigma_1-\sigma_3)\cos 2\beta>0$, Eq. (6.62) is established is established.

Because $\sigma_n>0$ indicates that the positive stress on the contact surface of the fracture is compressive, it means that the fracture is closed by compression.

If $\sigma_n<0$, indicates that the positive stress on the contact surface of the fracture is tensile, the fracture is not closed and the Griffith criterion should still be applied in this case.

Griffith's theory describes when cracks begin to rupture, but does not describe how cracks extend and propagate, and because the problem of crack extension and propagation is complex it will not be discussed here.

## Exercises

1. What is the strength failure criterion of rocks? Why is the rock strength failure criterion proposed?
2. The strength failure curve of a homogeneous rock is $\left|\tau_f\right| = \tau_0 + \sigma_n \tan(\varphi)$, where $\tau_0 = 40\text{MPa}$ and $\varphi = 30°$. Try to find the peak strength of the rock at confining pressure of 30 MPa and draw the orientation of the failure plane.
3. A rock specimen is subjected to uniaxial test. It is about to be failure when its compressive stress reaches 27.6 MPa. The angle between the failure plane and the plane of maximum principal stress is 60°. Assuming that the shear strength varies linearly with the normal stress, try to calculate

    (1) The shear strength on the plane where the normal stress is equal to zero.
    (2) The shear strength in that plane at an angle of 30° to the maximum principal stress plane in the above test.
    (3) Internal friction angle.
    (4) Normal stress and shear stress on the failure plane.

4. The rock specimen was subjected to a series of uniaxial tests to obtain an average uniaxial compressive strength of 20 MPa. And the same rock was subjected to a series of triaxial tests at a confining pressure of 60 MPa to obtain an average principal stress of 225 MPa. Please draw stress circles on the Mohr circle representing the two test results, and determine its internal friction angle and cohesion.
5. Assuming that the uniaxial tensile and uniaxial compressive strengths of a brittle limestone are known and that the triaxial strength meets the Coulomb-Navier linear failure criterion, try to analyze the internal friction angle of this limestone.
6. The rock mechanics experiments have also failed. The following information was obtained during the experiments: the first uniaxial compressive test did not yield uniaxial compressive strength, but the presence of a failure plane was recorded, with an angle of 20° to the specimen axis. The second specimen yielded uniaxial compressive strength of 85 MPa. Try to infer the cohesion and internal friction angle of the rock from this information.
7. The results of a triaxial experiment on a quartz sandstone are known as follows.

| | | | | | | | |
|---|---|---|---|---|---|---|---|
| $\frac{\sigma_1+\sigma_3}{2}$/MPa | – 6.65 | 100 | 135 | 160 | 200 | 298 | 435 |
| $\frac{\sigma_1-\sigma_3}{2}$/MPa | 6.65 | 100 | 130 | 150 | 180 | 248 | 335 |

Try to determine the relevant parameters of the Coulomb-Navier linear failure criterion and the Griffith failure criterion.

# References

Cai M. Rock mechanics and engineering. Beijing: Science Press; 2002.
Chen Z. Mechanical properties of rocks and tectonic stress fields. Beijing: Geological Press; 1986.
Hudson JA, Harrison JP. Engineering rock mechanics. Beijing: Science Press; 2008.
Jaeger JC, Cook NGW. Fundamentals of rock mechanics, London: CHAPMAN and HALL, 1979.
Li X. Mechanical properties of rock masses. Beijing: Coal Industry Press; 1983.

# Chapter 7
# In-Situ Stress States

## 7.1 Description of In-Situ Stress State

Rocks buried deep in the ground are subjected to stresses prior to engineering disturbances, and this stress is generally referred to as in situ stress. Because the rocks have undergone a long geological period and have undergone many complex tectonic movements, the in situ stress state of the rocks has become very complex (Jaeger and Cook 1979; Brady and Brown 1985; Cai 2002; Zoback 2007). To meet engineering needs, the in situ stress state is generally considered to consist of the overburden pressure (gravity) and two horizontal principal stresses. A reference coordinate system is established as shown in Fig. 7.1, with $xy$ forming the horizontal plane, $x$ direction being the direction of maximum horizontal principal stress, $y$ direction being the direction of minimum horizontal stress, and $z$ direction being the vertical direction.

The causes of the in-situ stress are complex and not yet well understood. A large amount of empirical data and theoretical studies from all over the world in recent decades have shown that the formation of the in-situ stress is mainly related to various dynamical processes in the Earth, including: plate boundary compression, mantle thermal convection, internal Earth stress, gravity, Earth rotation, magma intrusion and non-uniformity of the Earth's crust. Of these, the tectonic and gravitational stress fields are the main components of the present-day geostress field.

Heim (1912) was the first to assume that the crust is subjected to compressive stresses in the vertical direction ($z$-axis) by the gravity of the overburden of rocks at a certain depth in the crust.

$$\sigma_v = \int \rho(z) g dz \tag{7.1}$$

where $\rho(z)$ is the density at a burial depth of $z$ and g is the acceleration of gravity.

Due to the surrounding rock constraints, the rock at this point is not allowed to expand freely in all directions ($\varepsilon_{\mathrm{H}} = \varepsilon_{\mathrm{h}} = 0$), which inevitably causes equal compressive stresses in the horizontal direction (x, y axis), i.e.



Y. Cheng et al., *Foundations of Rock Mechanics in Oil and Gas Engineering*,
https://doi.org/10.1007/978-981-99-1417-3_7

**Fig. 7.1** In-situ stress state

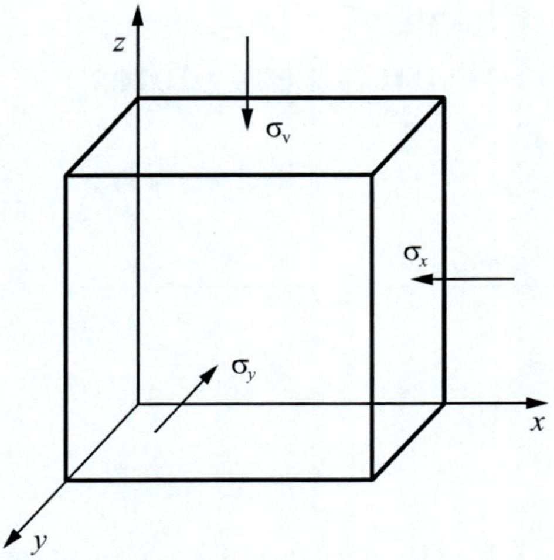


$$\sigma'_{\mathrm{H}} = \sigma'_h = \frac{\mu}{1-\mu}\sigma_v \tag{7.2}$$

For tectonic motions, they are generally very complex. It is difficult to express it in a specific expression. For simplicity, Huang (1986) assumes that the tectonic motion does not affect the vertical in-situ stresses and only generates the horizontal in-situ stress components ($\sigma''_{\mathrm{H}}$, $\sigma''_{\mathrm{h}}$) and is expressed as

$$\begin{aligned}\sigma'_{\mathrm{H}}&=\xi_1\sigma_{\mathrm{v}}\\ \sigma'_{\mathrm{h}}&=\xi_2\sigma_{\mathrm{v}}\end{aligned} \tag{7.3}$$

Thus, the total in situ stress state is expressed as:
The vertical Principal Stress (Overburden Pressure) $\sigma_v = \int \rho(z)\mathrm{d}z$.
Maximum horizontal principal stress $\sigma_{\mathrm{H}}=\left(\frac{\mu}{1-\mu}+\xi_1\right)\sigma_{\mathrm{v}}$.
Minimum horizontal principal stress $\sigma_{\mathrm{H}}=\left(\frac{\mu}{1-\mu}+\xi_2\right)\sigma_{\mathrm{v}}$.
Considering the effect of pore pressure $p_{\mathrm{p}}$, the in situ stress state is expressed as

$$\begin{aligned}\sigma_{\mathrm{v}}&=\int\rho(\mathrm{z})\mathrm{gdz}\\ \sigma_{\mathrm{H}}&=\left(\frac{\mu}{1-\mu}+\xi_1\right)(\sigma_{\mathrm{v}}-\alpha\mathrm{P_p})+\alpha\mathrm{P_p}\\ \sigma_{\mathrm{H}}&=\left(\frac{\mu}{1-\mu}+\xi_2\right)(\sigma_{\mathrm{v}}-\alpha\mathrm{P_p})+\alpha\mathrm{P_p}\end{aligned} \tag{7.4}$$

Equation (7.4) is generally referred to as the Huang Rongzun model.

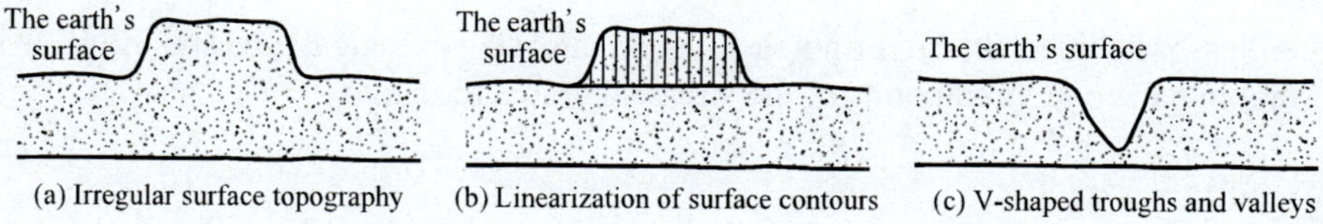


**Fig. 7.2** Effect of topography on ground stress

## 7.2 Factors Affecting the State of In Situ Stress

There are many factors affecting the in-situ stress state, and we describe below some of the factors that are easily understood by common.

### 7.2.1 *Surface Shape*

The previous discussion showed that for a flat surface, the average vertical stress component should be close to the depth stress. For irregularly shaped surfaces, as shown in Fig. 7.2, the stress state at any point can be considered to be a composite of the stress components due to the depth stress and the irregular distribution of the surface overload. The effect of the latter stress can be estimated by linearizing the surface profile as shown in Fig. 7.2b. Using elasticity theory to solve the half-space problem, the stress state can be estimated for a number of locations. An example is the area near the bottom of a V-shaped trough. This surface shape may produce a higher horizontal stress component than the vertical stress component at such locations. However, the effect of an irregular surface shape on the in situ stress state decreases rapidly with increasing depth below the surface at that point. This factor, therefore, is only relevant for near-surface geotechnical projects.

### 7.2.2 *Residual Stresses*

Residual stresses can be generated within the rock mass due to non-uniform cooling of the rock mass, or if the rock mass is uniformly cooled but has a different coefficient of thermal expansion than its adjacent rock units.

Various local mineral changes in the rock medium can also cause residual stresses to develop. Local recrystallization processes in the rock may produce volumetric strain. Changes in the content of mineral aggregates can also produce strain and residual stresses.

It is practically impossible at present to obtain a comprehensive thermodynamic history and detailed geological evolution of the various components of a rock formation, so the problem of residual stresses still constitutes a constraint that prevents either basic mechanical principles or exhaustive geological investigations from making accurate predictions of the stress state in a rock mass.

### 7.2.3 Envelope

Inclusions in rock masses are rock units that are created between layers of the original rock mass. Common inclusions are of an ejecta nature, such as walls, beds, and veins of minerals such as quartz and lapis lazuli. The presence of vertical inclusions in a rock body that are nearly flat in shape may affect the in situ stress state in two ways. First, if the inclusions are formed under the action of horizontal resistance resisting the surrounding pressure. Then a high stress component will act in the direction perpendicular to the plane of the envelope. The second possible effect of the envelope is related to the difference in the modulus of elasticity of the envelope and the surrounding rock. Any load in the system, for example, such a load originating from a change in the effective stress in the original rock mass, or from displacement in the medium caused by tectonic activity, produces a stress in the inclusion body that is different, either higher or lower, than the stress value in the original rock mass. A harder inclusion will be subjected to a higher state of stress, and vice versa. If the modulus of elasticity of the original rock and the inclusions are different, there will be a high stress gradient in the original rock near the inclusions. In contrast, the stress state in the inclusions themselves is more uniform.

### 7.2.4 Tectonic Stress

The earth's crust has undergone many complex crustal movements over a long geological period, thus producing extremely complex tectonic traces of different forms, orientations, properties, levels and sequences (Su et al. 2002). The tectonic system has gone through a long period of geological history from occurrence to formation, so the tectonic stress field is essentially a non-stable tectonic stress field that evolves with time. The tectonic stress field can be divided into paleotectonic stress field and present tectonic stress field by time period. The former refers to the tectonic forms left over from the geological period, while the latter refers to the stress fields that are still active in the present-day crust since the Cenozoic. In terms of its scope of study, the tectonic stress field can be divided into local tectonic stress field (stress distribution pattern of a tectonic unit), regional tectonic stress field (stress distribution pattern of a region) and global tectonic stress field. Due to the tectonic stresses, the in situ stress state is greatly changed. For the three common types of faults, the stress combinations at the time of fault formation are different, as shown in Fig. 7.3. For normal faults, the overburden pressure is the maximum principal stress, the maximum horizontal ground stress is along the fault strike, the minimum horizontal ground stress is perpendicular to the fault strike, and the dip angle of normal faults is steeper

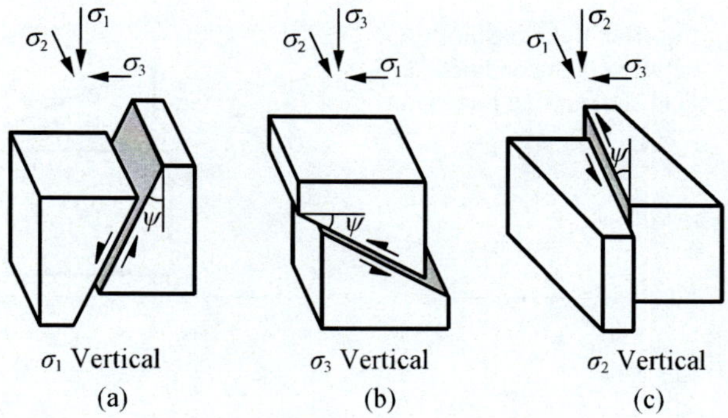


**Fig. 7.3** Effect of fault type on ground stress

according to Coulomb's criterion; for reverse faults, the overburden pressure is the minimum principal stress, the maximum horizontal in-situ stress is perpendicular to the fault strike, and the minimum horizontal in-situ stress is along the fault strike, and the dip angle of reverse faults is slower by the same reasoning; for strike-slip faults, the overburden pressure is the middle principal stress, the maximum horizontal in-situ stress is the maximum principal stress, the minimum horizontal in-situ stress is the minimum principal stress, the angle between the direction of the maximum horizontal ground stress and the fault strike is less than 45°, the angle between the minimum horizontal ground stress and the fault strike is more than 45°, and the fault surface is approximately vertical.

The formation of a fault produces a stress release that reduces the level of in-situ stress acting on the fault, but increases it again as tectonic movement continues. If a fault is formed a long time ago, new tectonic movements may occur after the formation of the fault, resulting in changes or shifts in in-situ stresses, so caution should be exercised in determining the order of magnitude of in-situ stress combinations based on fault morphology.

### *7.2.5 Fissure Groups and Discontinuity Surfaces*

The presence of fractures in the rock mass, either as a finite continuous group of joints or as a major tectonic surface through the rock formation, limits the equilibrium state of stresses in the medium. It makes the stress distribution in the rock mass more complex. For example, for a set of conjugate faults, the direction of maximum principal stress in the stress field before fault formation coincides with the acute two-sided angular bisector formed by the two fault faces, while the axis of minimum principal stress, coincides with the obtuse bisector, and the axis of intermediate principal stress coincides with the intersection of the two fault faces (Fig. 7.4). This interpretation is based on the damage pattern obtained from true triaxial compression experiments of the rock. However, this interpretation does not apply to the stress state after the fracture period. In fact, rock fracture is essentially a process of energy

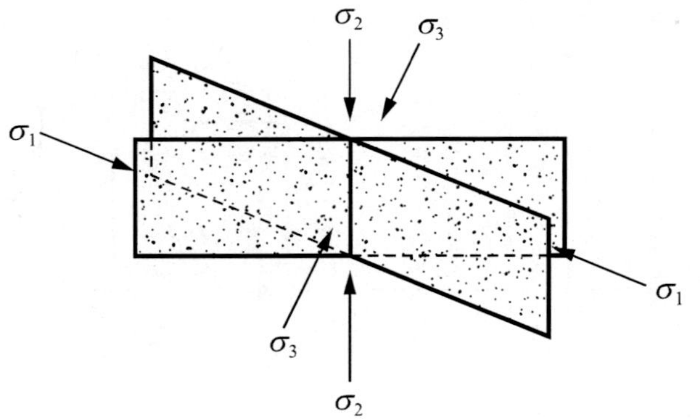


**Fig. 7.4** Relationship between fault conditions and field stresses that form faults

dissipation and stress redistribution. The stress state after fracture can be determined by the requirement to maintain equilibrium conditions at the fracture surface; it has little to do with the stress state before fracture. This brief discussion shows that it is extremely difficult to estimate the state of stress around a rock mass from the initial state. A method for determining the local stress tensor is necessary to successfully determine the in situ stress state.

## 7.3 Method for Determining In-Situ Stress

### *7.3.1 General Approach*

Since determining the original stress state of rock is a prerequisite for evaluating the stability of underground projects, much effort has been spent on developing stress measurement equipment and exploring methods of stress measurement. The measurement methods can be divided into two categories: field measurements and laboratory measurements. For field measurements, the various methods to date can be divided into two distinct categories in terms of the principle of the measurement method, although most of them of course use boreholes to approach the measurement site. One of the most common methods is to determine the strain in the borehole wall or other deformation of the borehole that results from the part of the borehole where the measuring instrument is placed in the set. If a sufficient number of stresses and strains are measured during this stress relief, then the six stress components of the field stress tensor can be derived directly from the measured data using the solution method of elasticity theory. The second method, represented by pressure pillow measurements and hydraulic fracturing, measures the annular positive stress components at specific locations on the borehole wall. At each location, the normal stress component is obtained by applying a pressure to the measurement slot that is balanced by the local normal stress component acting in the perpendicular direction of the slot. The circumferential stress at each measurement location may be directly related to the stress state at the test site prior to drilling, and the local value of the stress tensor can be determined directly if sufficient boundary stresses are determined at the perimeter of the wellbore. The main methods of indoor measurements are anelastic strain recovery (ASR), which uses the elastic after-effects of the rock to measure the strain release in different orientations of the core immediately after it is removed from the site, which is then used to calculate the in situ stress state. Differential strain analysis (DS), where the strain values of the core in three orthogonal directions are measured under pressurized conditions and then used to analyze the in-situ stress state. Acoustic Emission (AE), which determines the in-situ stress state of a core by measuring the Kessel effect in the core.

### 7.3.2 Three Directional Stress Gauges

Devices for direct or indirect determination of field stresses are photoelastic gauges, USBM borehole deformation gauges, bi-directional strain gauges, and tri-directional strain gauges. Leeman and Hayes (1966), and Worotnicki and Walton (1976) have described soft envelope strain, which is the simplest of the above instruments in principle, since only one stress relief is required to determine the full component of the stress tensor. As shown in Fig. 7.5a, the strain gauge consists of at least three strain flowers mounted on a deformable substrate or shell. The operation is shown in Fig. 7.5b–d. Select suitable epoxy or polyester resin to bond the strain gauge to the borehole wall. Stress relief in the vicinity of the strain gauge causes the strain flower to produce a strain equal in magnitude to the original strain in the borehole wall, but opposite in sign. Therefore, it is simple to determine the strain state of the borehole wall before the stress relief from the measured strains. Using these observations of borehole strain, and using the elastic properties of the rock and an expression for the stress concentration around the round hole, the local stress state of the rock prior to drilling can be deduced (Zhong et al. 2005; Yang et al. 2010).

The method for determining the in situ stress tensor components from borehole strain observations is derived from the solution to the problem of stress distribution around a circular hole in an object subjected to three direction stress (Leeman 1969; Leeman and Hayes 1966; Fairhurst 1986). Figure 7.6a represents the orientation of the stress measurement hole, which is determined by the dip angle $\alpha$ and the direction angle$\beta$of the tendency line in the overall coordinate system *xyz*. Relative to these axes, the surrounding field stress components (before the borehole) are $P_{xx}$, $P_{yy}$, $P_{zz}$, $P_{xy}$, $P_{yz}$, $P_{zx}$. Figure 7.6a also plots a local coordinate system *l, m, n* for the borehole. *n* is oriented parallel to the axis of the borehole, with the m axis in the horizontal plane (*x*, *y*). The field stress components $P_{ll}$, $P_{ln}$, etc. in the local coordinates of the borehole are easily transformed into components $P_{xx}$, $P_{xz}$, etc. in overall coordinates by means of the stress transformation equation and the following rotation matrix.

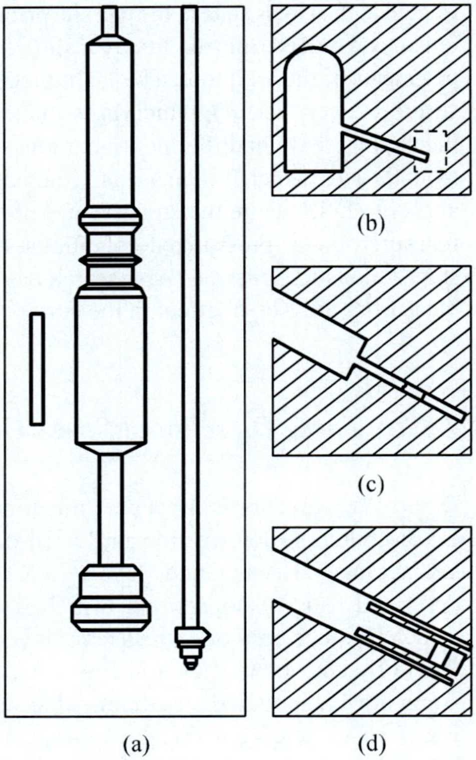


**Fig. 7.5** **a** Three-way strain gauge, **b–d** in stress relief

$$[R]=\begin{vmatrix} \lambda_{xl} & \lambda_{xm} & \lambda_{xn} \\ \lambda_{yl} & \lambda_{ym} & \lambda_{yn} \\ \lambda_{zl} & \lambda_{zm} & \lambda_{zn} \end{vmatrix} = \begin{bmatrix} -\sin\alpha\cos\beta & -\sin\beta & \cos\alpha\cos\beta \\ -\sin\alpha\sin\beta & -\cos\beta & \cos\alpha\cos\beta \\ cos\alpha & 0 & sin\alpha \end{bmatrix}$$

The location of the point on the borehole wall in Fig. 7.6b is represented by the angle $\theta$ in the lm plane, with $\theta$ counterclockwise positive, and for an isotropic elastic medium, the following relationship exists between the boundary stress at this point

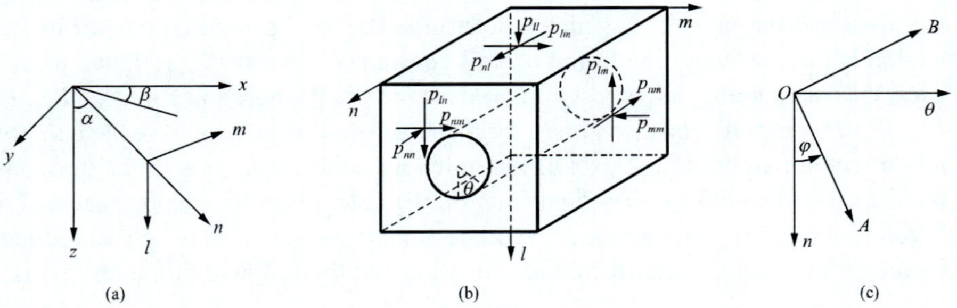


**Fig. 7.6** **a** Definition of local axes of the borehole, **b** field stress components in local coordinates of the borehole, position coordinate angle, **c** axes of the borehole wall

and the local field stress.

$$\begin{aligned}
\sigma_{rr} &= \sigma_r = 0 \\
\sigma_{\theta\theta} &= p_{ll}(1-\cos 2\theta) + p_{mm}(1+\cos 2\theta) - 4p_{lm}\sin 2\theta \\
\sigma_{nn} &= p_{nn} - 2\mu(-p_{ll}\cos 2\theta + p_{mm}\cos 2\theta - 2p_{lm}\sin 2\theta) \\
\tau_{\theta n} &= 2p_{mn}\cos\theta - 2p_{nl}\sin\theta
\end{aligned} \tag{7.5}$$

Equation (7.5) defines the nonzero boundary stress component $\sigma_{\theta\theta}, \sigma_{\theta n}, \sigma_{nn}$ with respect to the n, $\theta$ coordinate axes, with the n axis aligned with the direction of the borehole axis and the $\theta$ axis aligned with the orthogonal direction, which is the direction of the tangent to the borehole boundary in the l, m plane. Another right-handed right-angle coordinate system *OA, OB* can be introduced at the borehole boundary, as shown in Fig. 7.6c. Here the angle $\varphi$ represents the angle of rotation from the n, $\theta$ axis to the *OA, OB* axis. The normal component of the boundary stress along the *OA, OB* direction is given by the following equation.

$$\begin{aligned}
\sigma_A &= \frac{1}{2}(\sigma_{nn} + \sigma_{\theta\theta}) + \frac{1}{2}(\sigma_{nn} - \sigma_{\theta\theta})\cos 2\phi + \tau_{n\theta}\sin 2\phi \\
\sigma_B &= \frac{1}{2}(\sigma_{nn} + \sigma_{\theta\theta}) - \frac{1}{2}(\sigma_{nn} - \sigma_{\theta\theta})\cos 2\phi - \tau_{\theta n}\sin 2\phi
\end{aligned} \tag{7.6}$$

It is assumed that the OA direction in Fig. 7.6c coincides with the direction and location of the strain gage used to measure the strain state at the hole wall. Since the stress relief process is a planar stress state on the hole boundary, the following relationship between the measured positive strain components and the local boundary stress components is obtained.

$$\varepsilon_A = \frac{1}{E}(\sigma_A - \mu\sigma_B) \tag{7.7}$$

Substituting the expressions for $\sigma_A$, $\sigma_B$ Eq. (7.6) into Eq. (7.7) and then substituting Eq. (7.5) in the resulting expression, the relationship between the local strain state of the hole wall and the field stress is obtained as follows.

$$\begin{aligned}
E\varepsilon_A = p_{ll}&\left\{\frac{1}{2}[(1-\mu) - (1+\mu)\cos 2\phi] - \left(1-\mu^2\right)(1-\cos 2\phi)\cos 2\theta\right\} \\
&+ p_{mm}\left\{\frac{1}{2}[(1-\mu) - (1+\mu)\cos 2\phi] + \left(1-\mu^2\right)(1-\cos 2\phi)\cos 2\theta\right\} \\
&+ p_{nn}\frac{1}{2}[(1-\mu) + (1+\mu)\cos 2\phi] - p_{lm}2\left(1-\mu^2\right)(1-\cos 2\phi)\sin 2\theta \\
&+ p_{mn}2(1+\mu)\sin 2\phi\cos\theta - p_{nl}2(1+\mu)\sin 2\phi\sin\theta
\end{aligned} \tag{7.8a}$$

Or

$$a_1 p_{ll} + a_2 p_{mm} + a_3 p_{nn} + a_4 p_{lm} + a_5 p_{mn} + a_6 p_{nl} = b \tag{7.8b}$$

Equations (7.8a) and (7.8b) indicate that the strain state at a location on the hole wall, in the direction specified by the angle θ and φ, can be determined linearly from the field stress components. The coefficients of Eq. (7.8b) $a_i$ ($i = 1$ to 6) can be calculated directly from the location of the measurement point and the two directional angles and the Poisson's ratio of the rock. Therefore, if six independent observations are obtained for the strain state at six locations/directions on the hole wall, a joint set of six equations can be built. Writing,

$$[A][P] = [b] \tag{7.9}$$

Here [P] is a column vector consisting of the stress components $P_{ll}$, $P_{mm}$, $P_{nn}$, Plm, $P_{mn}$, $P_{nl}$. Provided that the location/orientation of the strain observation points is suitably chosen to ensure that the non-sickness coefficient matrix [A] is obtained, then the field stresses $P_{ll}$, $P_{lm}$, etc. can be solved directly from Eq. (7.9).

Actual designs of three-way strain gauges capable of providing more than six independent strain observations are usually available. From the redundant observations many equally valid field stress tensors can be solved (Brady et al. 1976). Within the zone of influence of the stress determination, these solutions can be used to determine the local average solution of the surrounding stress state. As with the various parameters that define the field stress tensor, the reliability of the measured values of the stress state is limited.

### 7.3.3 Pressure Pillow Measurement

Stress measurements with strain gauge equipment are usually required in small diameter boreholes with a volume of rock to be measured of approximately $10^{-3}$ $m^3$. If a larger diameter cylinder wall can be taken as the measurement site, then a larger volume of rock can be tested. In the case of human-accessible chambers, the stress state of the wall is measured directly, without testing the strain state. This also eliminates the need to determine or estimate the modulus of deformation of the rock mass (Hudson and Cooling 1988). As a method of measuring the stress state at the boundary of the refuge wall, the pressure pillow method is particularly attractive because it is a zero method, i.e. this measurement method seeks to recover the initial local stress state at the test site. This method is inherently more accurate than those methods that require disturbance of the initial state to be measured.

In order to determine the field stresses successfully by the pressure pillow method, three conditions need to be satisfied, namely

(a) The surfaces of the chambers forming the test site must be less disturbed.
(b) The geometry of the chamber shall be such that the solution representing the relationship between the far-field stress and the boundary stress has a closed form.
(c) The rock mass properties must be elastic because the displacement must be recoverable when the stress increment causing the displacement changes in the opposite direction.

Requirements (a) and (c) effectively preclude the use of a chamber formed by conventional blasting as a test site. Cracking associated with blasting and other transient effects may cause extensive disturbance of the elastic stress distribution in the rock and may cause inelastic displacement of the rock during the measurement. Requirement (b) specifies that the chamber must have a simple geometry, with chambers of circular cross-section being the most convenient.

Figure 7.7 describes the practical use of a pressure pillow. The pressure pillow consists of two parallel metal plates, approximately 300 mm square, welded together at the edges of the two plates. A one-way conduit is connected to the hydraulic pump. Two needles, selected for use with a DEMEC deformometer or other similar deformometer, are mounted to the rock surface so that a measurement field is established perpendicular to the direction of the groove axis. The distance $d_0$ between the two needles is to be measured. A series of interlap rock boreholes are drilled to form the measurement slot. The distance between the two displacement measurement points is reduced. The pressure pillow is cemented into the slot with mortar and the two measurement points will move in opposite directions when the pressure pillow is pressurized with oil or water. It is assumed that when the distance between the two displacement monitoring points returns to the initial value $d_0$, the pressure acting on the pressure pillow is equal to the normal stress component that was already present in the rock before the slot was opened and is perpendicular to the direction of the axis of the slot. This assumption causes some error, mainly due to the boundary effect of the pressure pillow, but this error can be corrected by a reasonable calibration of the pressure pillow.

The biggest disadvantage of this system is that it requires at least six measurements from different directions and at six different locations, so it is inevitable that these measurement points have to be arranged on the cavern side walls. Inevitably, at each test point its actual stress is different. Therefore, in order to reasonably interpret the measurements, it is necessary to know as much as possible about the distribution of stresses around the test hole.

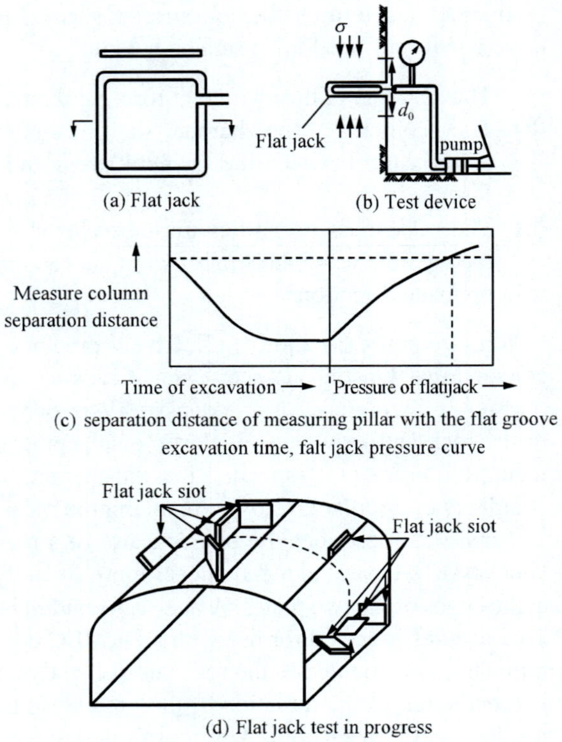


**Fig. 7.7** Pressure pillow test (Kim and Franklin 1987)

### *7.3.4 Hydraulic Fracturing*

Hydraulic fracturing is the only method capable of measuring stresses deep in the Earth's crust. This method is a routine process method in oil and gas drilling. It is performed by pumping fluid through a mud pump or frac truck at the casing shoe. The purpose is to check the quality of the cementing operation and determine the minimum fracture pressure of the formation below the casing shoe. It is also possible to seal a bare-hole section at any depth with a packer and then pump fluid through the frac truck. With the increase of fluid pumping volume, the borehole pressure increased. Two parameters, pump volume and pressure, were recorded.

#### **7.3.4.1 Test Method**

The test method is described below as an example of a bare-hole section hydraulic fracturing test, as shown in Fig. 7.8. The top and bottom of the well section to be tested are sealed with packers and filled with high-pressure fluid in between, and the fluid pumping volume and tubing pressure are recorded during the test until the

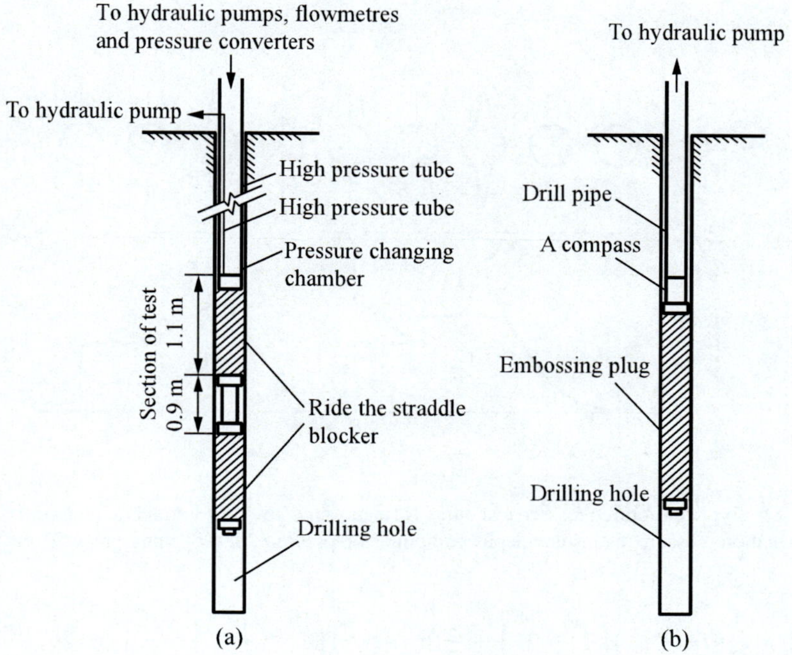


**Fig. 7.8** Schematic diagram of the hydraulic fracturing test tool Illustrated description of **a** hydraulic fracturing tool, **b** embossing plug

surrounding rock is fractured by tension. The orientation and geometric characteristics of the fracture can be observed after the test by means of an embossed plug or downhole television.

During pressurization, the effective stress in the tangential or vertical direction of the borehole may become tensile stress, and when this tensile stress reaches the tensile strength of the formation, the borehole ruptures. The pressure at this point is called the fracture pressure, and when a fracture is formed in the borehole wall, the pressure at which the surrounding rock is further split continuously is called the propagation pressure $P_{\text{pro}}$. When the fracture expands to three times the diameter of the borehole, the pump is stopped and the hydraulic system is shut down, forming an "instantaneous stopping" pressure $P_{\text{s}}$. If the surrounding rock has good permeability, the pressure in the fracture will gradually decay after the pump is stopped to The formation pore pressure $P_{\text{p}}$, Fig. 7.9 shows a typical fracture pressure test curve.

#### 7.3.4.2 Fundamentals

As shown in Fig. 7.10, for a vertical well, the stress distribution pattern in the borehole envelope can be represented by the following equation.

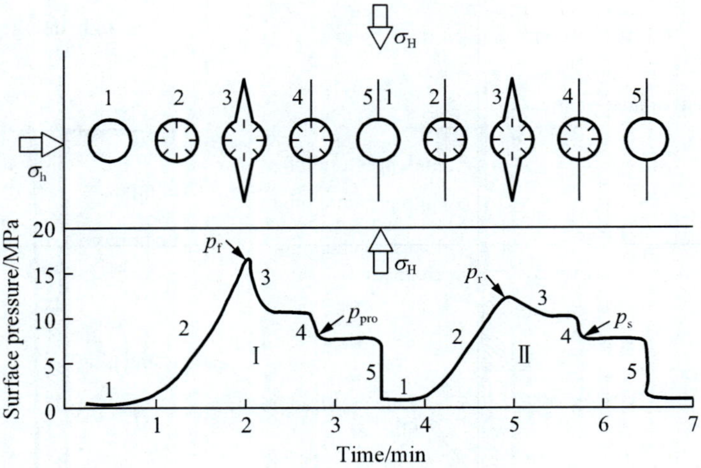


**Fig. 7.9** Typical rupture pressure test curve (Liu and Zhai 2005) Pf initial cracking pressure, Ppro propagation pressure, ps instantaneous pumping stop pressure, pr reopening pressure

$$\sigma_r = \frac{1}{2}(\sigma_H + \sigma_h)\left(1 - \frac{a^2}{r^2}\right) + \frac{1}{2}(\sigma_H - \sigma_h)\left(1 - \frac{4a^2}{r^2} + \frac{3a^4}{r^4}\right)\cos 2\theta + P_m\frac{a^2}{r^2} \tag{7.10}$$

$$\sigma_\theta = \frac{1}{2}(\sigma_H + \sigma_h)\left(1 + \frac{a^2}{r^2}\right) + \frac{1}{2}(\sigma_H - \sigma_h)\left(1 - \frac{3a^4}{r^4}\right)\cos 2\theta - p_m\frac{a^2}{r^2} \tag{7.11}$$

where $\sigma_{\mathrm{H}}$ is the maximum horizontal in situ stress component, $\sigma_{\mathrm{h}}$ is the minimum horizontal in situ stress component, $P_{\mathrm{m}}$ is the borehole fluid column pressure.

At the well wall ($r = a$), the upper two equations change to

$$\begin{aligned}\sigma_r &= p_m \\ \sigma_\theta &= (\sigma_H + \sigma_h) - 2(\sigma_H - \sigma_h)\cos 2\theta - p_m\end{aligned} \tag{7.12}$$

**Fig. 7.10** Borehole mechanics model

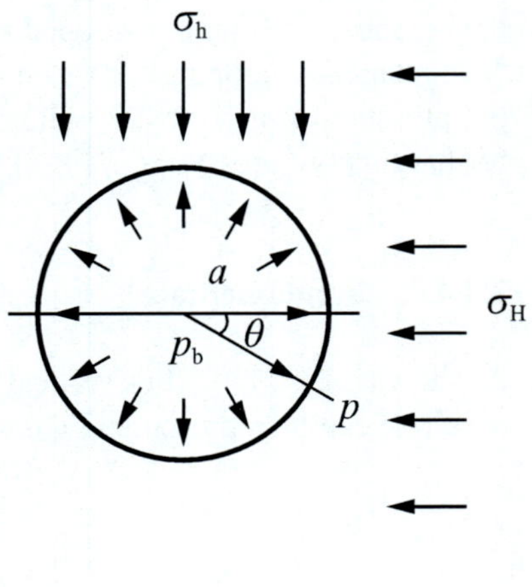

When $\theta = 0$, $\sigma_\theta$ obtains the minimum value, i.e.,

$$\sigma_\theta = 3\sigma_h - \sigma_H - p_m \tag{7.13}$$

When a well wall fractures,

$$\sigma_\theta = -\sigma_t \tag{7.14}$$

$\sigma_t$ is the tensile strength of the strata, so that we have

$$\sigma_H = 3\sigma_h - p_f + \sigma_t \tag{7.15}$$

When pore pressure $p_p$ is present within the formation, Eq. (7.15) is rewritten as

$$\sigma_H = 3\sigma_h - p_f + \sigma_t - \alpha p_p \tag{7.16}$$

The formation is ruptured and then repressurized, the pressure of reopening fracture is $p_r$.

$$\sigma_H = 3\sigma_h - p_r - \alpha p_p \tag{7.17}$$

From Eqs. (7.16) and (7.17), we have.

$$\sigma_t = p_f - p_r \tag{7.18}$$

The instantaneous stopping pump pressure Ps is the borehole fluid column pressure in equilibrium with the minimum horizontal ground stress, and is therefore generally considered to be.

$$\sigma_h = p_s \tag{7.19}$$

If the formation is horizontally fractured, this indicates that the vertical stress is the minimum principal stress and that the cracking conditions are different from the former as shown in Fig. 7.11. For the hydraulic case with $P_m$ inside the fracture, Jaeger and Cook (1979) gives the following expression.

When $\sigma_H + 3\sigma_v > 4\,pm$

$$(\sigma_H - \sigma_v)^2 - 8\sigma_t(\sigma_H + \sigma_v - 2p_m) = 0 \tag{7.20}$$

When $\sigma_H + 3\sigma_v < 4\,pm$

$$\sigma_v - p_m + \sigma_t = 0 \tag{7.21}$$

For the case with pore pressure, Eq. (7.21) is often used to establish the relationship for horizontal fracture initiation, when

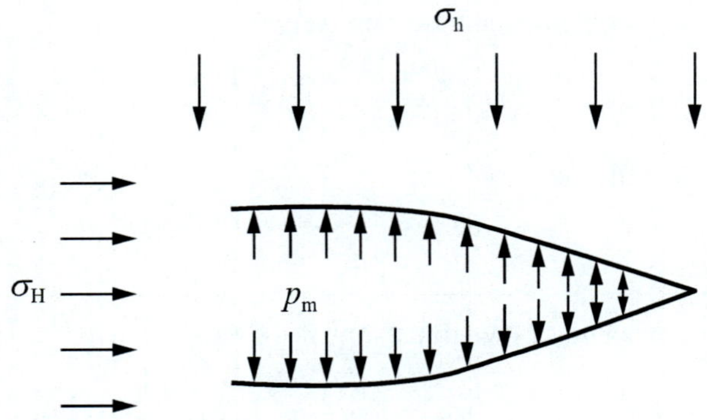


**Fig. 7.11** Horizontal fracture schematic

$$\sigma_v = p_m - \sigma_t + \alpha P_p \tag{7.22}$$

Wang et al. (2005) used hydraulic fracturing to measure the in situ stress in the Yalong River at the proposed dam site of the South-North Water Transfer West Line. The measurement depths ranged from 139.35 to 154.96 m. The treatment pressure curves for five measurement points are given in Fig. 7.12. The fracture orientation was observed by impressions after treatment (as shown in Fig. 7.13). The hydraulic fracture surface appears symmetrically on both sides of the borehole wall in an approximately vertical manner. The fracture surface is relatively regular, extends long, and runs through almost the entire fractured section, which accurately reflects the maximum horizontal in situ stress direction.

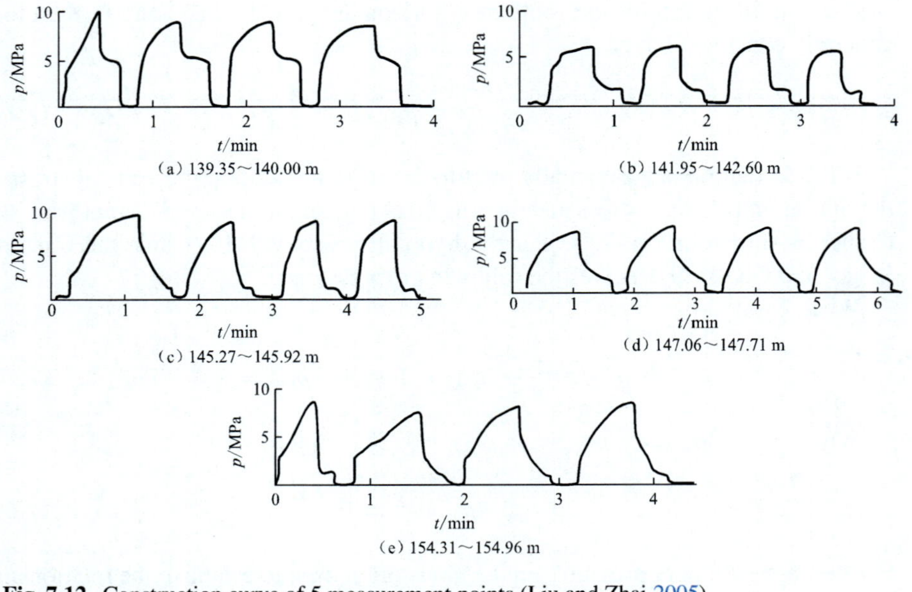


**Fig. 7.12** Construction curve of 5 measurement points (Liu and Zhai 2005)

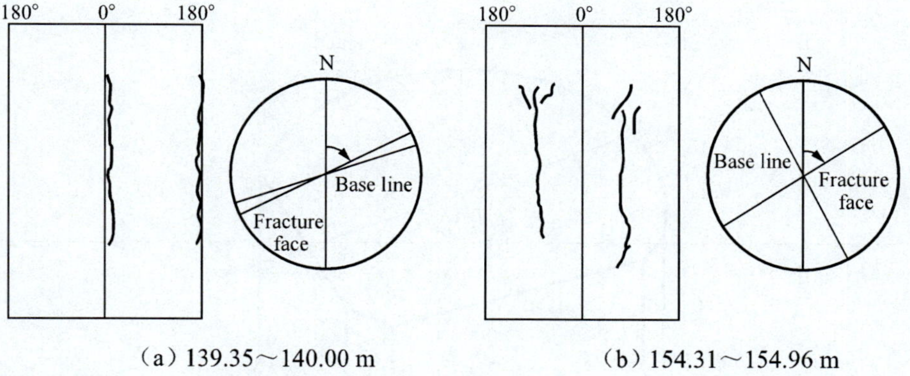


**Fig. 7.13** Crack orientation shown in the impression

### *7.3.5 Differential Strain Method*

The differential strain method is a method of determining three-dimensional stress directions and stress values in the field through indoor core experiments.

The rocks in the formation are under three-direction in-situ stress and are in compression. When coring is performed, the core is removed from the stress environment and an elastic stress release occurs, which is accompanied by maximum deformation in the direction of maximum stress and minimum deformation in the direction of minimum stress. The core, on which stress release has occurred, is machined into a specimen of the desired shape, with at least three planes orthogonal to each other, and strain gauges are attached as required, as shown in Fig. 7.14. The cores were sealed and placed in a high-pressure vessel and restoratively loaded under stress conditions of $\sigma_1 = \sigma_2 = \sigma_3$. During the loading process, the strain components in the three directions appeared to differ, with the largest strain component in the direction of the maximum original stress; the smallest strain component in the direction of the minimum original stress. By testing the difference between the strains in each direction of the above core, the state of stress to which this core is subjected in the formation can be known.

For each strain test at each selected confining pressure value nine strain values can be given, and a single subscript is used to represent these measured strain values. From these nine strains, six strain components describing the strain state at that moment can be calculated, and from these six strain values, the principal strain value at that moment can be calculated, as shown in Fig. 7.15.

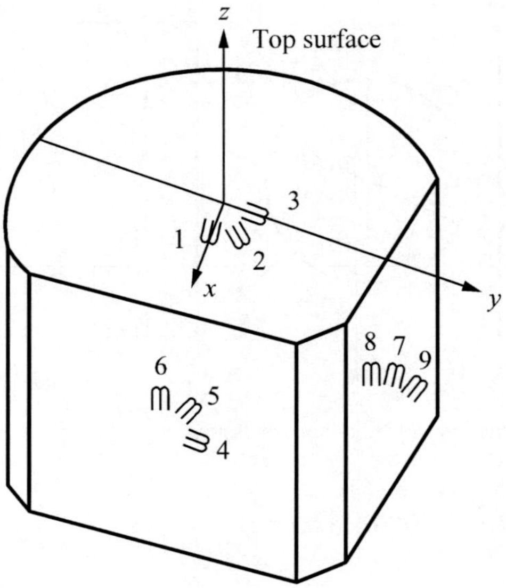


**Fig. 7.14** Schematic diagram of the principle of differential strain testing

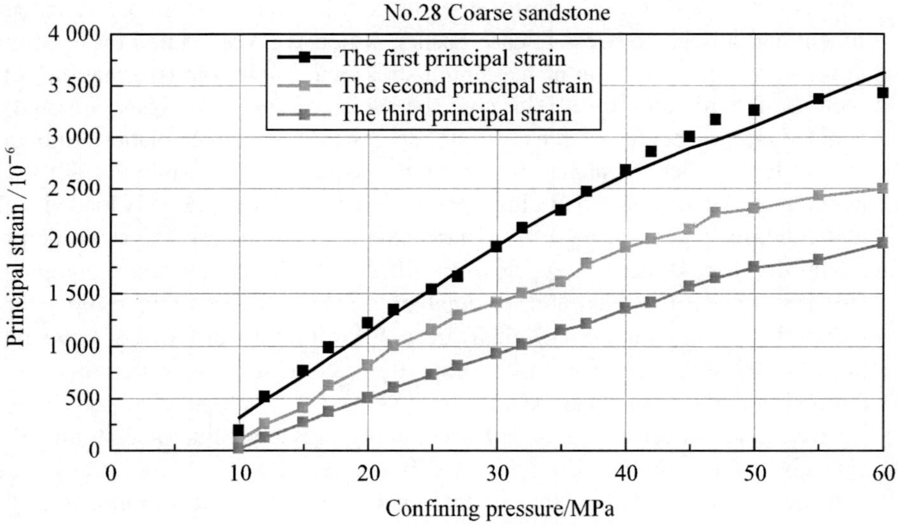


**Fig. 7.15** Plot of the three principal strains versus the enclosing pressure

$$\left.\begin{aligned}\varepsilon_{xx} &= \varepsilon_1 \\ \varepsilon_{yy} &= \varepsilon_3 \\ \varepsilon_{zz} &= \varepsilon_6 \\ \varepsilon_{xy} &= \varepsilon_2 - (\varepsilon_1 + \varepsilon_3)/2 \\ \varepsilon_{xz} &= \varepsilon_8 - (\varepsilon_7 + \varepsilon_9)/2 \\ \varepsilon_{yz} &= \varepsilon_5 - (\varepsilon_4 + \varepsilon_6)/2\end{aligned}\right\} \quad (7.23)$$

The three principal stress components (overburden pressure, maximum horizontal in-situ stress, and minimum horizontal in-situ stress) can be obtained by determining the principal stress from the principal strain according to the generalized Hooke's law. If the core is oriented, the orientation of the maximum horizontal in-situ stress can be determined from the marker line position.

## 7.4 Basic Laws of In-Situ Stress Distribution

Extensive research on ground stress has been carried out in recent decades due to the needs in the fields of seismic and geotechnical engineering. Some basic laws are outlined below.

### *7.4.1 In-Situ Stress Is a Relatively Stable Unsteady Stress Field*

Ground stress is a three-way unequal compressive stress field dominated by horizontal stresses in the vast majority of areas. The magnitude and direction of the three principal stresses vary with space and time, and thus it is a non-stationary stress field. The variation of ground stress in space is evident on a small scale, and the magnitude and direction of ground stress may be different from a point to another point tens of meters away from each other. However, for a certain region as a whole, the variation of ground stress is not significant. For example, in northern China, the dominant direction of the ground stress field is north-west to nearly east–west in the main compressive stress.

In some areas with active seismic activity, the magnitude and direction of the ground stress change significantly with time. Before the earthquake, it is in the stress accumulation stage, and the stress value keeps rising, while the earthquake enables the release of the concentrated stress, and the stress value suddenly drops significantly. The main stress direction will change significantly during the earthquake, and will return to the pre-earthquake state some time after the earthquake.

### 7.4.2 *Measured Vertical Stress Is Essentially Equal to the Overlying Rock Pressure*

Statistical analysis of measured vertical stresses σ v worldwide shows that in the depth range of 25–2700 m, σ v increases linearly and is roughly equivalent to the gravity calculated from an average capacity γ equal to 27 kN $m^{-3}$. However, the measurements in some areas deviate to some extent, and in addition to some of the above deviations that may be attributed to measurement errors, plate movement, magma convection and intrusion, dilation, and uneven expansion can also cause vertical stress anomalies, as shown in Fig. 7.16. The figure shows the pattern of variation of σ v with depth H for all countries of the world as summarized by Hoek and Brown (1980). Figure 7.17 shows the pattern of ground stress distribution in North China as measured by the National Seismological Bureau. The resulting expression is.

$$\begin{aligned} \sigma_H &= 0.7 + 0.023ll \\ \sigma_h &= -0.5 + 0.018H \\ \sigma_v &= 0.021H \end{aligned} \tag{7.24}$$

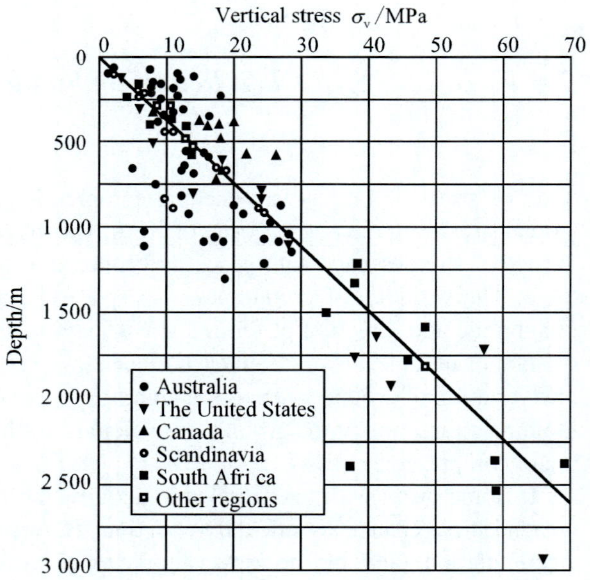


**Fig. 7.16** The variation pattern of vertical stress σ v with depth in various countries of the world (Hoek and Brown 1980)

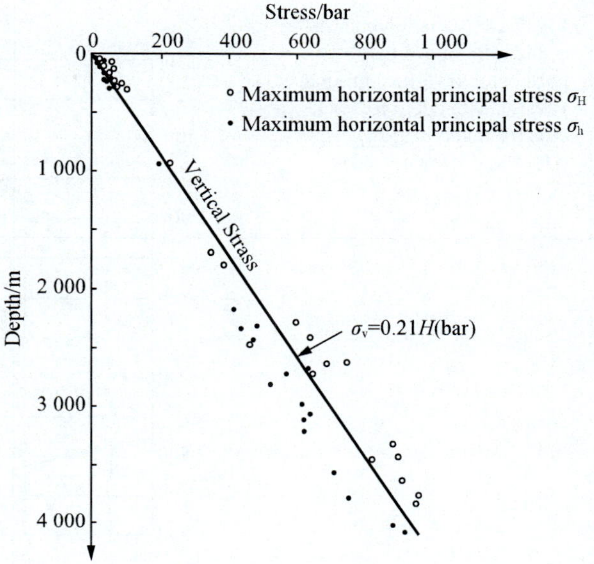


**Fig. 7.17** Variation of in-situ stress with depth in North China

## *7.4.3 The Horizontal Stress Distribution Is More Complex*

Measured data show that in the vast majority of areas there are two principal stresses located in the horizontal or near horizontal plane, and their angle with the horizontal plane is generally not greater than 30°. $\sigma_H/\sigma_v$ ranges from 0.5 to 5.5, and in many cases the ratio is greater than 2. If the average $\sigma_{h,av}$ of the maximum horizontal principal stress $\sigma_H$ and the minimum horizontal principal stress $\sigma_h$ is compared with $\sigma_v$, summarizing the current worldwide in-situ stress measurements, the value of $\sigma_{h,av}/\sigma_v$ is generally 0.5–5.0 and most are 0.8–1.5, as shown in Fig. 7.18. This indicates that the average horizontal in-situ stress is also generally greater than the vertical stress in the shallow crust. Vertical stresses are in most cases minimum principal stresses, in a few cases intermediate principal stresses, and in only a few cases maximum principal stresses. This again suggests that horizontal tectonic movements such as plate movements and collisions control the formation of shallow crustal in-situ stresses.

The ratio of average horizontal to vertical stress decreases with depth, hitting different areas and changing at very different rates. Hoek E. and Brown E. T. regressed the following equation based on the results shown in Fig. 7.18 to represent the range of variation with depth for $\sigma_{h,av}/\sigma_v$.

$$\frac{100}{H} + 0.3 \leq \frac{\sigma_{h,av}}{\sigma_v} \leq \frac{1500}{H} + 0.5$$

where $H$ is the depth in m.

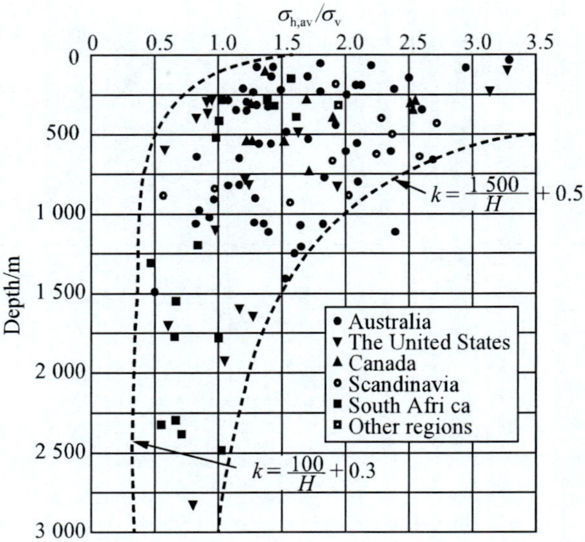


**Fig. 7.18** Variation pattern of the ratio of average horizontal to vertical ground stresses with depth for each country in the world (Hoek and Brown 1980)

Figure 7.18 shows that the values of $\sigma_{h,av}/\sigma_v$ are fairly scattered at modest depths, and that the range of variation in this value gradually narrows and concentrates near 1 as depth increases, suggesting the possibility of hydrostatic stress states deep in the crust.

The maximum horizontal in-situ stress and minimum horizontal in-situ stress also show a linear increase with depth. Unlike the overburden pressure, the constant term in the linear regression equation for the horizontal principal stress is larger than the value of the constant term in the linear regression equation for the overburden pressure, reflecting the fact that significant horizontal stresses still exist near the surface in some areas, and Stephenson et al. gave linear equations for the maximum and minimum horizontal ground stresses with depth for the Fennoscandian paleo-continent based on empirical results.

$$\sigma_H = 6.7 + 0.0444H$$
$$\sigma_h = 0.8 + 0.0329H$$

where $H$ is the depth in m.

## 7.4.4 *Performance Characteristics of High Stress Areas*

The level of in situ stress in the area can be determined using some of the phenomena revealed by the exploration works.

(1) Core pancaking is related to the difference in in-situ stress. If the drill core is burnt and destroyed piece by piece, this is the product of high in situ stress, core

pancaking is mainly related to the difference in in situ stress, the greater the difference in stress perpendicular to the drilling direction, the more serious the pancaking will be.

(2) It leads to instability of subsurface structures such as boreholes, and due to the presence of high stress zones, brittle destruction of rocks can occur during drilling and stripping of strain energy accumulated in the rocks due to sudden release. Shrinkage can occur for soft rock formations.

## Exercises

1. Briefly describe the concept of in situ stress and what are the main influencing factors affecting in situ stress?
2. Briefly understand the episodic and multi-period characteristics of regional tectonic movements and what are the main factors influencing present-day tectonic stresses?
3. A formation with a burial depth of 3000 m is known to have a modulus of elasticity of 20,000 MPa and a Poisson's ratio of 0.25. The average density of its overburden is 2.35 g/cm$^3$, the pore pressure equivalent density of the formation is 1.2 g/cm$^3$, and the coefficient of tectonic motion is $\xi_1 = 0.25$ and $\xi_2 = 0.35$. What is the magnitude of the in situ stress in this formation based on the Huang Rongtun model?
4. A coal seam is buried at a depth of 475 m, with a thickness of 1.5 m, a formation pressure gradient of 0.005 MPa/m, a modulus of elasticity of 5000 MPa and a Poisson's ratio of 0.35. The maximum surface injection pressure in the second cycle was 2.50 MPa, with a cumulative injection volume of 0.057 m$^3$ and a return flow rate of 0 m$^3$, and a shut-in time of 30 min. The maximum surface injection pressure in the fourth cycle was 3.30 MPa, with a cumulative injection volume of 0.223 m$^3$ and a return flow rate of 0 m and a shut-in time of 40 min. The in-situ stress test pressure vs. time curve is shown below. The G-function method was used to analyze the closure pressure and rupture pressure of the formation and further determine the maximum and minimum horizontal ground stress of the formation (Fig. E1).
5. Stresses in the granite rock were measured in a vertical borehole using hydraulic fracturing and were tested at two depths. One test depth was 500 m and the other test depth was 1000 m. The results of the tests are shown in the table below. Assuming that the tensile strength of the rock is 10 MPa, try to find the magnitude of the ground stress at the two test sites.

| Depth/m | Rupture pressure/MPa | Instantaneous pump stop pressure/MPa |
|---|---|---|
| 500 | 14.0 | 8.0 |
| 1000 | 24.5 | 16.0 |

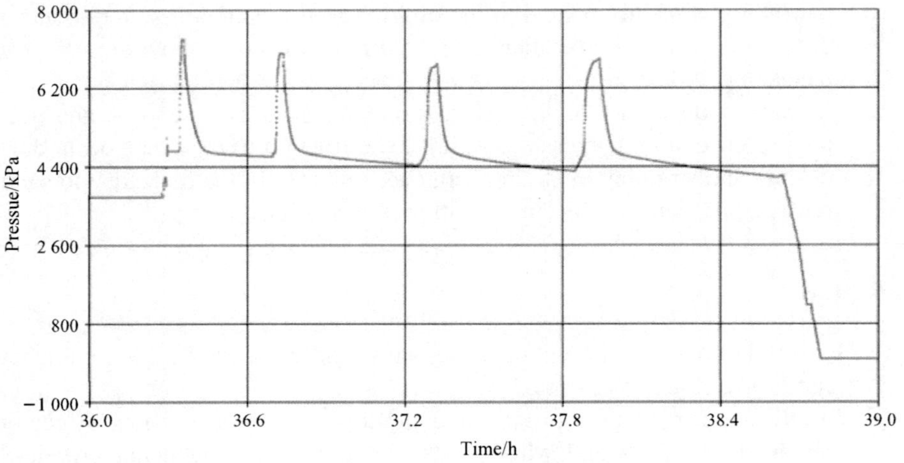


**Fig. E1** Measured pressure versus time curves for in situ stress testing of coal seams

# References

Brady BHG, Brown ET. Rock mechanics for underground mining. London: George Allen & Unwin; 1985.

Brady BHG, Friday RG, Alexander LG. Stress measurement in a bored raise at the Mount Isa Mine. "Advances in Stress Measurements" Proceedings of the ISRM symposium. Sydney. 1976;12–16.

Cai M. Rock mechanics and engineering. Science Press; 2002.

Fairhurst C. In situ stress determination - an appraisal of its significance in rock mechanics. In: Stephansson (ed) Proceedings of the international symposium on rock stress and rock stress measurements. Stockholm, Sept. Lulea: Centek Publishers; 1986, p. 3–17.

Hoek E, Brown ET. Underground excavation in rock. London: Institute of Mining and Metallurgy; 1980. p. 382–95.

Huang R. A new prediction method of formation fracture pressure. Oil Drilling & Production Technology; 1986.

Hudson JA, Cooling CM. In situ rock stresses and their measurement in the UK-Part 1, the current state of knowledge. Int J Rock Mech Min Sci Geomech Abstr. 1988;25(6):363–70.

Jaeger JC, Cook NGW. Fundaments of rock mechanics. London: Chapman and Hall; 1979.

Kim K, Franklin JA. Suggested methods for rock stress determination. Int J Rock Mech Min Sci Geomech Abstr. 1987;24(1):53–73.

Leeman ER. The CSIR "doorstopper" and triaxial rock stress measurement instruments. The CSIR "doorstopper" and triaxial rock stress measuring instruments. In: Proceedings of the international symposium on determination of stresses in rock masses. Lisbon: Laboratorio National de Engenharia Civil; 1969, p. 578–616.

Leeman ER, Hayes DJ. A technique for determining the complete state of stress in rock using a single borehole. In: Proceedings of the first congress of the international society of rock mechanics. 1966;2:17–24.

Liu C, Zhai C. Mining disturbance and simulation of stratigraphic space stress field. Yellow River Water Conservancy Press; 2005.

Su SR, Huang RQ, Wang ST. Effects of fracture tectonics on the ground stress field and its engineering applications. Science Press; 2002.

Wang X, Chen S, et al. Study on the geological conditions of South-North Water Diversion West Line Project. Yellow River Water Conservancy Press; 2005.

Worotnicki G, Walton RJ. Triaxial 'Hollow Inclusion' gauges for the determination of rock stresses in situ. In: Proceeding ISRM Symposium Investigation of stress in rock and advances in shear measurement. Sydney: supplement, 1976;1–8.

Yang Y, Zhou G, Li Y, et al. In-situ stress measurement and support technology of caving roadway along gob. Beijing: China Coal Industry Press; 2010.

Zhong Z, Chen Y, Liu Y, Zhou M. Three-dimensional ground stress measurement techniques for deep rock masses. In: Mine pressure and topping management, 2005;3:80–5.

Zoback MD. Reservoir geomechanics. London: Cambridge University Press; 2007.

# Chapter 8
# Mechanics of Wellbore Stability

## 8.1 Causes and Hazards of Wellbore Instability

### *8.1.1 Causes of Wellbore Instability and Research Methods*

In oil and gas drilling, borehole stability problems are a common problem worldwide. The direct economic losses caused by this amount to hundreds of millions of dollars every year. Therefore, many research institutes in the world are devoted to the field of wellbore stability.

Before drilling, the deeply buried rock formation is in equilibrium by the overburden pressure, maximum horizontal in-situ stress, minimum horizontal in-situ stress and pore pressure. After opening the borehole, the rock inside the well is taken away and the rock of the well wall loses its original support and is replaced by the drilling fluid hydrostatic pressure. Under this new condition, the stress in the rock surrounding the borehole will be redistributed, resulting in a high stress concentration near the wellbore, and if the rock strength is not large enough, instability of the wellbore will occur. By adjusting the drilling fluid density, the stress state of the wellbore confining rock can be changed to achieve the purpose of stabilizing the wellbore.

If the drilling fluid density is too low, the stress at the wellbore will exceed the shear strength of the rock and shear damage will occur (manifested as borehole collapse and expansion or yield shrinkage), at which time the critical borehole hydrostatic pressure is defined as the collapse pressure $P_c$; if the drilling fluid density is too high, tensile stress will be generated in the wellbore, and when the tensile stress reaches the tensile strength of the rock, tensile damage will occur (formation of fractures leading to well leakage), at which time the critical borehole fluid hydrostatic pressure is defined as the formation fracture pressure $P_f$. Shear failure is further divided into two types: one is brittle damage, resulting in well diameter expansion, which can cause difficulties in subsequent construction such as cementing and logging. This type of damage usually occurs in hard and brittle formations, but borehole enlargement may also occur in weakly cemented sandstone formations due to erosion. Another



Y. Cheng et al., *Foundations of Rock Mechanics in Oil and Gas Engineering*,
https://doi.org/10.1007/978-981-99-1417-3_8

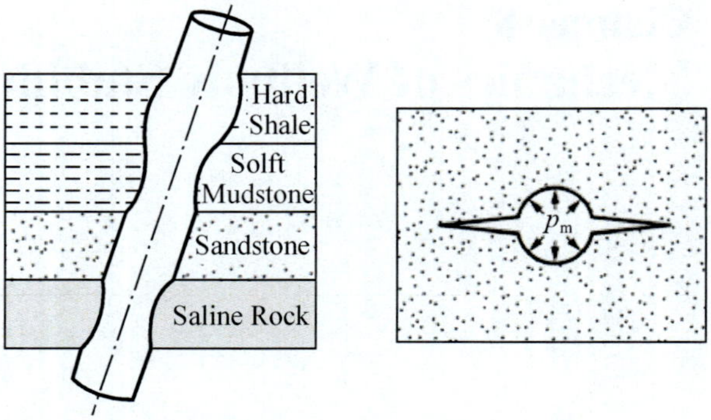

(a) Collapse failure (b) Tensile failure

**Fig. 8.1** Schematic diagram of the geometry of borehole destabilization damage in various lithological formations

type of shrinkage occurs in soft mudstone, sandstone, and rock salt formations where the drilling fluid density is too low and the wall stresses can exceed the yield limit of the rock, producing borehole shrinkage. Some limestone formations may also reproduce this phenomenon at depth. It is important to keep slicing the wellbore when this phenomenon is encountered in engineering, otherwise stuck drilling pipe will occur. Tensile damage produces hydraulic fracturing, which can lead to well leakage and, in severe cases, blowout. A schematic diagram of the geometry of borehole destabilization damage in different lithological formations is given in Fig. 8.1.

It can be seen that, in essence, the stability of the wellbore is ultimately expressed in the stress state of the rock surrounding the borehole as compared to the strength of the rock damage. If the stress in the borehole wall exceeds the rock strength envelope, the wall is to be destroyed; otherwise the wall is stable. However, there are many factors that affect the stress state and damage of the borehole envelope, making the problem very complex. In summary, the influencing factors can be divided into four major categories: (1) geomechanical factors, in situ stress state, formation pore pressure, in situ temperature, geological structure characteristics, etc. These factors are unchangeable, and they can only be accurately recognized. (2) Comprehensive properties of rocks, strength and deformation characteristics of rocks, porosity, water content, clay content, composition and compaction, etc. (3) Comprehensive properties of drilling fluids, chemical composition, nature of continuous phases, composition and type of internal phases, types of additives associated with continuous phases, index of the degree of strength reduction of the formation by drilling fluids, and maintenance of the drilling fluid system, especially for mud shales and mud cemented sandstones, where the influence of drilling fluids on their physical and mechanical properties is very significant. (4) Other engineering factors, including open borehole time, bare borehole length, borehole profile parameters (well depth, well deviation angle, azimuth angle), borehole fluid column pressure surge and suction, etc. These factors and parameters interact and influence each other, making the wellbore stability problem very complex.

There are still difficulties in accurately determining the pattern of influence of the factors, mainly due to the following reasons: (1) There are few methods of directly observing the morphology of the wellbore wall, and it is difficult to know exactly what is happening at depths of several kilometres downhole. (2) The mechanical properties of the rock encountered by drilling vary over a wide range. (3) The state of in situ stress is difficult to determine accurately. (4) The physicochemical interaction between the drilling fluid and the formation is very complex, leading to the problem of borehole instability, which is a world-class problem.

There are two main approaches to studying wellbore stability: one is drilling fluid chemistry and the other is rock mechanics. The study of wellbore stability from the aspect of drilling fluid chemistry has a long history, mainly studying the mechanism of mud shale hydration and swelling, searching for chemical additives and drilling fluid systems that inhibit mud shale hydration and swelling, minimizing the negative impact of drilling fluid on the formation, and keeping the wellbore stable (Chenevert 1970; Bol 1986). The rock mechanics study mainly includes the determination of in situ stress state, determination of rock mechanical properties, borehole confining rock stress analysis and stability analysis, and finally determining the reasonable drilling fluid density to keep the borehole stable (Bradley 1979; Aadnoy and Chenevert 1987; Cheng and Huang 1993; Deng et al. 2008).

Two tendencies should be avoided in the study of such complex problems, namely agnosticism, where the problem is considered too complex to be studied. The other is overly theoretical research, building more complex mechanical models in an attempt to describe the mechanical behaviour of the formation in detail, but the models are too complex to accurately determine the relevant parameters, making the application of the models difficult. It is now more feasible to couple drilling fluid chemistry and rock mechanics, to collect as much information as possible on borehole conditions (e.g., when and in what manner the borehole is complicated), to estimate the mechanical properties of the rock as accurately as possible, and to determine which parameters play a major role. Working with drilling engineers to test and analyze wellbore stability problems and establishing a combination of example-based analytical evaluations and theoretical simulations can be quite useful in solving formation stability problems during drilling.

### *8.1.2 Hazards of Unstable Well Walls*

In the long-term exploration and development process of major oil fields in China, the problem of wellbore instability has been more prominent. For example, the main manifestations in the Bohai Bay area are the hydration and swelling of the mud shale formation of Guatao and Minghuazhen formation strata, which causes shrinkage and stuck drilling accidents; the chipping off blocks of the mud shale formation of the Dongying, Shahejie and Kongdian Formation strata, which causes well diameter enlargement, collapse and stuck drill pipe, low quality of electrical logging, unqualified cementing and other engineering problems or accidents; some special formations such as: biological tuff, fractured basalt and soft sandstone also cause wellbore collapse and well leakage Complications, these accidents seriously delay the drilling cycle, significantly increase the drilling cost and bring adverse effects to the follow-up work, sometimes even make the well scrapped.

This chapter focuses on analyzing the mechanism of wellbore destabilization damage from the perspective of rock mechanics and gives specific steps to solve for wellbore stability.

## 8.2 Stress Distribution of Confining Rock of Vertical Well

To carry out mechanical analysis of wellbore stability, one of the first tasks is to determine the stress distribution law of the wellbore confining formation, the mechanical properties of the formation vary and the analysis method used varies. In this section, the law of stress distribution in a vertical well is given as an example of a porous linear elastic formation.

### *8.2.1 Stress Distribution Model of Vertical Well*

Since the borehole diameter is much smaller than the borehole axis, the vertical borehole model can be simplified to a plane strain problem. Figure 8.2 is a schematic diagram of a vertical borehole mechanical model with a maximum horizontal in-situ stress $\sigma_H$ acting at infinity in the $x$ direction, a minimum horizontal in-situ stress $\sigma_h$ acting at infinity in the $y$ direction, a hydrostatic pressure of drilling fluid $P_m$ acting inside the borehole, a formation pore pressure $p_p$ acting inside the formation, and a layer of ideal mud cake forming on the wellbore wall to isolate the drilling fluid from the formation fluid. Because the formation is linearly elastic, following the principle of superposition, the mechanical model of Fig. 8.2 can be decomposed into a superposition of two submodels shown in Fig. 8.3, a unidirectional pressurized large-plate small-hole stress concentration model (Fig. 8.3a) and a thick-walled barrel model (Fig. 8.3b). Both sub-models are classical elastic mechanics problems with analytical solutions.

**Fig. 8.2** Mechanical model of a vertical well

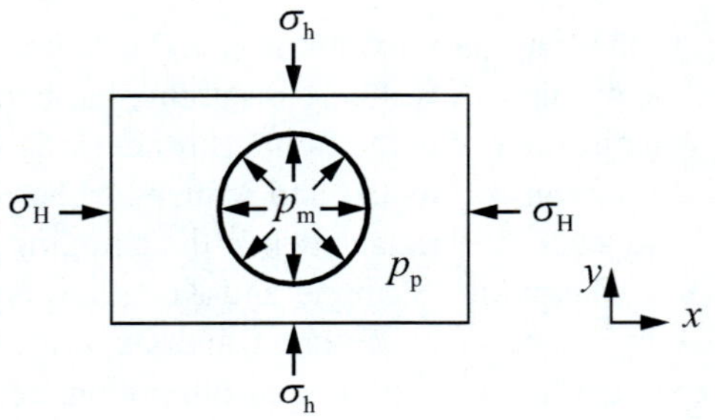

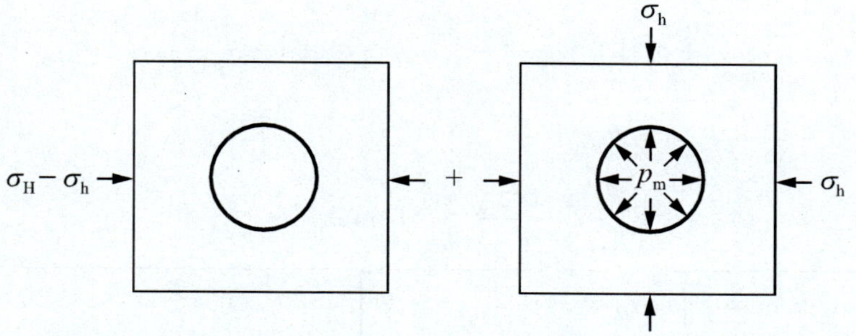


(a) Infinite plate small hole stress concentration model (b) Thick-walled cylinder under pressure model

**Fig. 8.3** Decomposition of a straight borehole model

According to the principle of elastic mechanics, the solution of the stress concentration model for small holes in infinite plates under unidirectional pressure is

$$\begin{aligned}
\sigma_r^1 &= \frac{1}{2}(\sigma_H - \sigma_h)\left[1 - \frac{a^2}{r^2} + \left(1 - \frac{4a^2}{r^2} + \frac{3a^4}{r^4}\right)\cos 2\theta\right] \\
\sigma_\theta^1 &= \frac{1}{2}(\sigma_H - \sigma_h)\left[1 + \frac{a^2}{r^2} - \left(1 + \frac{3a^4}{r^4}\right)\cos 2\theta\right] \\
\tau_{r\theta}^1 &= \frac{1}{2}(\sigma_H - \sigma_h)\left(1 + \frac{2a^2}{r^2} - \frac{3a^4}{r^4}\right)\sin 2\theta
\end{aligned} \tag{8.1}$$

The solution of the thick-walled cylinder model is given by

$$\begin{aligned}
\sigma_r^2 &= \left(1 - \frac{a^2}{r^2}\right)\sigma_h + \frac{a^2}{r^2}P_m \\
\sigma_\theta^2 &= \left(1 + \frac{a^2}{r^2}\right)\sigma_h + \frac{a^2}{r^2}P_m \\
\tau_{r\theta}^2 &= 0
\end{aligned} \tag{8.2}$$

Summing Eqs. (8.1) and (8.2) and considering the effect of formation pore pressure, the effective stress distribution in the confining rock of the vertical well is obtained as

$$\begin{aligned}
\sigma_r' = {} & \frac{1}{2}(\sigma_H + \sigma_h)\left(1 - \frac{a^2}{r^2}\right) + \frac{1}{2}(\sigma_H - \sigma_h)\left(1 + \frac{3a^4}{r^4} - \frac{4a^2}{r^2}\right)\cos 2\theta \\
& + \frac{a^2}{r^2}P_m - \alpha P_p
\end{aligned}$$

$$\sigma'_\theta = \frac{1}{2}(\sigma_H + \sigma_h)\left(1 + \frac{a^2}{r^2}\right) - \frac{1}{2}(\sigma_H - \sigma_h)\left(1 + \frac{3a^4}{r^4}\right)\cos 2\theta$$
$$- \frac{a^2}{r^2}P_m - \alpha P_p$$
$$\tau_{r\theta} = \frac{1}{2}(\sigma_H - \sigma_h)\left(1 + \frac{2a^2}{r^2} - \frac{3a^4}{r^4}\right)\sin 2\theta$$
$$\sigma'_z = \sigma_v - \mu\left[2(\sigma_H - \sigma_h)\left(\frac{a^2}{r^2}\right)\cos 2\theta\right] - \alpha P_p \tag{8.3}$$

where: $\sigma_{\mathrm{H}}$ and $\sigma_{\mathrm{h}}$ respectively are the maximum and minimum horizontal in-situ stresses, $\sigma_{\mathrm{v}}$ is vertical in-situ stress, i.e., the overburden pressure, $\sigma_r$ is the radial stress, $\sigma_\theta$ is the circumferential stress, $\sigma_z$ is the axial stress, $p_{\mathrm{p}}$ is the original pore pressure in the formation, $p_{\mathrm{m}}$ is the borehole fluid hydrostatic pressure, $\alpha$ is the effective stress coefficient (Biot coefficient), $a$ is the radius of the borehole, $r$ is the radius of the study point, $\theta$ is the circular angle of the study point (rotated counterclockwise from the direction of maximum horizontal in-situ stress), $\mu$ is stratigraphic Poisson's ratio.

From Eq. (8.3), it can be seen that the stress distribution in the borehole confining is mainly influenced by the in-situ stress, formation pore pressure, borehole fluid hydrostatic pressure, and the location of the analysis point.

At the borehole wall (r = a), the stress distribution is simplified as

$$\sigma'_r = P_m - \alpha P_p$$
$$\sigma'_\theta = (\sigma_H + \sigma_h) - 2(\sigma_H - \sigma_h)\cos 2\theta - P_m - \alpha P_p$$
$$\sigma'_z = \sigma_v - 2\mu(\sigma_H - \sigma_h)\cos 2\theta - \alpha P_p$$
$$\tau_{r\theta} = 0 \tag{8.4}$$

### 8.2.2 *The Stress Distribution in Vertical Wellbore Surrounding Rock*

From Eq. (8.3), it can be seen that the distribution of stresses in the confining formation of a vertical borehole depends first of all on the state of the original in-situ stresses. We know that if the type of controlling fault is different, the order of magnitude of the three in-situ stress components is different. It can also be seen that when the circumferential angle is equal to 0° (maximum horizontal in-situ stress direction) and 90° (minimum horizontal in-situ stress direction), the shear stress $\tau_{r\theta}$ is zero and the three normal stresses in the radial direction, circumferential direction and vertical direction of borehole confining are the principal stresses. The following is an example of a normal fault in-situ stress combination ($\sigma_{\mathrm{v}} > \sigma_H > \sigma_h$) first, to analyze the law of radius, circumferential angle, borehole pressure and other factors on the stress components of the borehole envelope. For a shale formation with a borehole depth of 3000 m, the overburden pressure is 73.5 MPa, the maximum horizontal in-situ stress is 58.5 MPa, the minimum horizontal in-situ stress is 52.5 MPa, the formation pore pressure is 34.5 MPa (equivalent density is 1.15 g/cm$^3$), the formation Biot coefficient is 0.55, the formation Poisson's ratio is 0.25, and the borehole radius is 0.1 m.

Figures 8.4 and 8.5 give the stress variation pattern of borehole confining formation with position coordinates for a lower borehole fluid hydrostatic pressure of 37.5 MPa (equivalent density of 1.25 g/cm$^3$). Figure 8.4 shows the variation of stress components of the borehole confining rock with position coordinates along the direction of maximum horizontal in-situ stress. The maximum shear stress is half of the difference between the axial and radial stresses at wellbore wall; away from the borehole wall, the radial stress gradually converges to the maximum horizontal in-situ stress, the circumferential stress gradually converges to the minimum horizontal in-situ stress, and the axial stress gradually converges to the overburden pressure, and the stress concentration basically disappears when $r/a = 10$. Figure 8.5 shows the variation of the stress component of the borehole confining rock with the position coordinates along the direction of the minimum horizontal in-situ stress. Again, the stress concentration is greatest at the borehole wall, but the maximum shear stress is half the difference between the circumferential and radial stresses, and is much greater than the maximum shear stress at the borehole wall at the maximum horizontal in-situ stress (from 33 to 48 MPa); away from the borehole wall and outward, the three stress components gradually converge to the original ground stress (effective stress).

Figures 8.6 and 8.7 give the stress variation pattern of borehole confining rock with position coordinates for higher borehole fluid hydrostatic pressure of 52.5 MPa (equivalent density of 1.75 g/cm$^3$). Figure 8.6 shows the variation of the stress component of the borehole confining rock with position coordinates along the direction of maximum horizontal in-situ stress; the circumferential stress increases with increasing radius in the near-well zone and slightly decreases with increasing radius

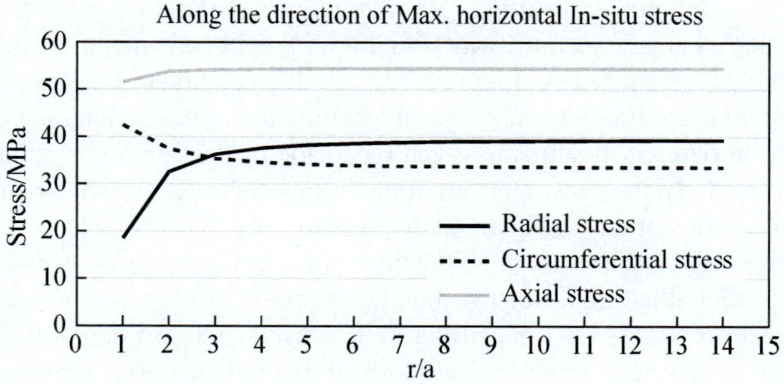


**Fig. 8.4** Variation pattern of stress components in the borehole perimeter rock with position coordinates along the direction of Max. horizontal in-situ stress at drilling fluid density of 1.25 g/cm$^3$

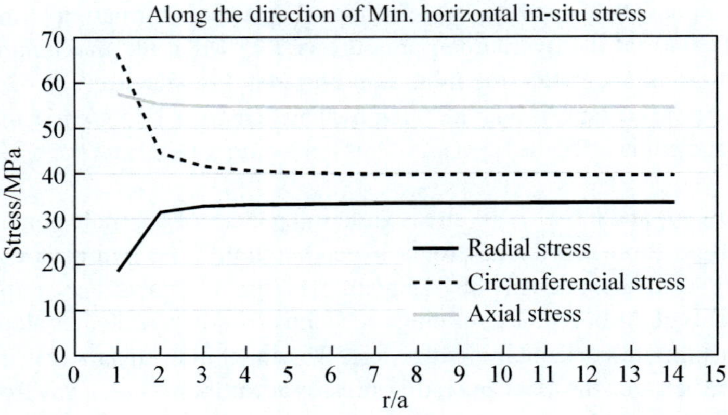


**Fig. 8.5** Variation pattern of stress components in the borehole perimeter rock with position coordinates along the direction of Min. horizontal in-situ stress at drilling fluid density of 1.25 g/cm$^3$

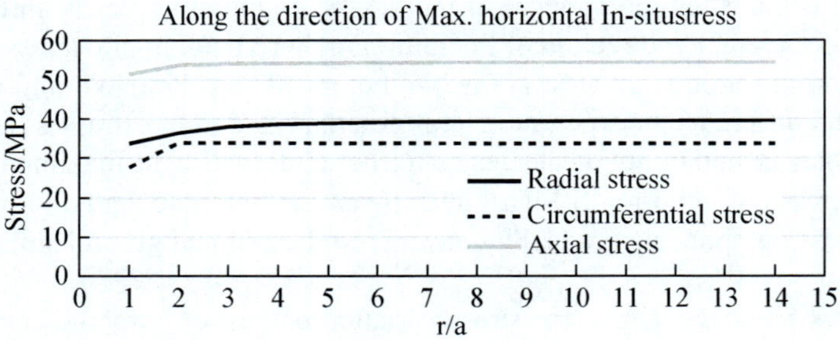


**Fig. 8.6** Variation pattern of stress components in the borehole confining rock with position coordinates along the direction of Max. horizontal in-situ stress at drilling fluid density of 1.75 g/cm$^3$

beyond 2r, which is significantly different from Fig. 8.4; and the stress concentration at the wellbore wall in Fig. 8.6 (maximum shear stress of 24 MPa) is much smaller than the case in Fig. 8.4 (maximum shear stress of 33 MPa); the stress concentration also disappears at a faster rate away from the well wall and outward than in the case of low fluid column pressure. Figure 8.7 shows the variation of the stress component of the borehole confining formation with position coordinates along the direction of minimum horizontal in-situ stress, and the shape of the curve is similar to that of Fig. 8.5. Similarly, the stress concentration disappears faster away from the well wall and outward than in the low hydrostatic pressure case.

Figure 8.8 gives the variation of the three stress components on the wellbore wall with circumferential angle for two borehole fluid hydrostatic pressures (a drilling fluid density of 1.25 g/cm$^3$, b drilling fluid density of 1.75 g/cm$^3$). It can be seen that the circumferential stress varies significantly with the circumferential angle at a certain borehole fluid column pressure, for Fig. 8.8a, the minimum value is 42 MPa at 0° and the maximum value is 66 MPa at 90°, for Fig. 8.8b, the minimum value

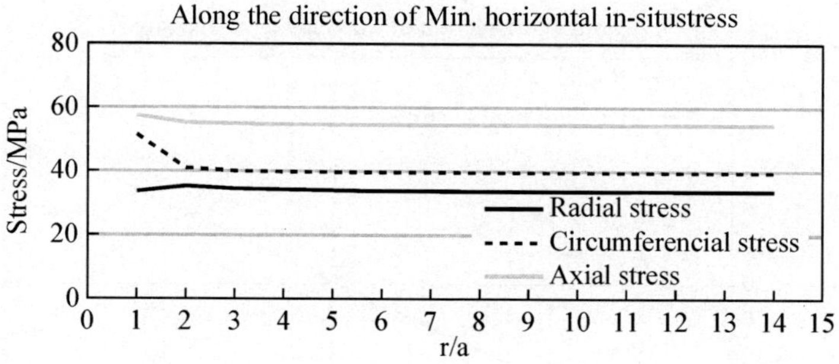


**Fig. 8.7** Variation pattern of stress components in the borehole confining rock with position coordinates along the direction of Min. horizontal in-situ stress at drilling fluid density of 1.75 g/cm$^3$

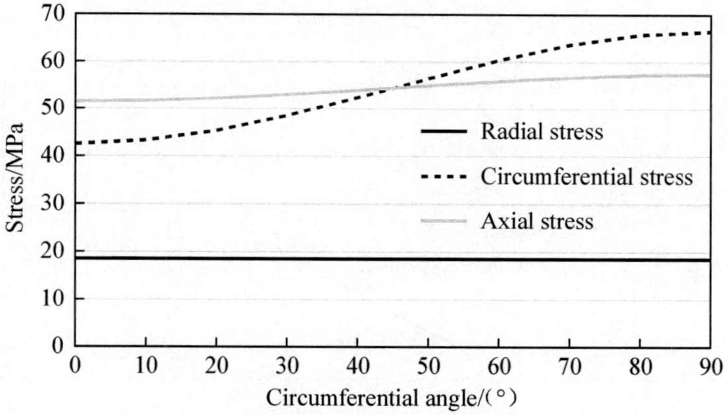


(a) Drilling fluid density of 1.25 g/cm$^3$

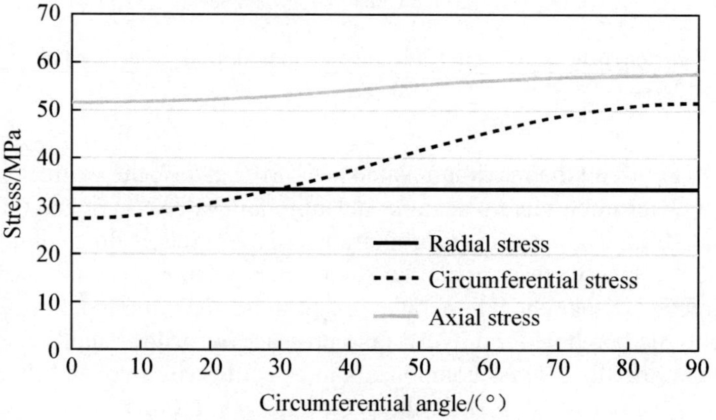


(b) Drilling fluid density of 1.75 g/cm$^3$

**Fig. 8.8** Relationship between the stress component of the well wall and the circumferential angle for different borehole fluid column pressures

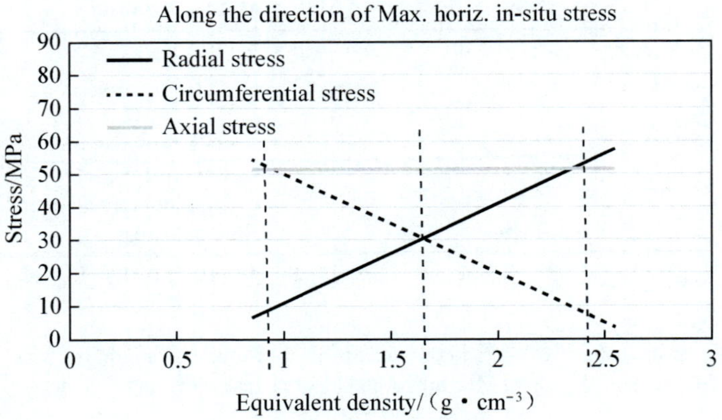


(a) Circumference angle 0°

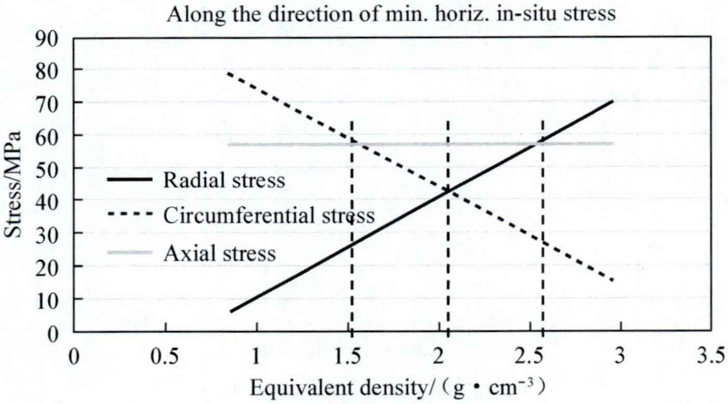


(b) Circumference angle 90°

**Fig. 8.9** Variation pattern of well wall stresses with drilling fluid density at 0° and 90° circumferential angles

is 27 MPa at 0° and the maximum value is 51 MPa at 90°; the vertical stress varies slightly with the circumferential angle, the minimum value is at 0° and the maximum value is at 90°; the radial stress is proportional to the borehole fluid column pressure.

Figure 8.9 gives the variation of the three stress components on the wellbore wall in the direction of maximum horizontal in-situ stress and minimum horizontal in-situ stress with the borehole fluid hydrostatic pressure (or drilling fluid density). It can be seen that the three stress components intersect two by two, and there are three intersection points $\rho_1$, $\rho_2$, $\rho_3$, at the position of $\rho_2$, the maximum shear stress is the smallest, and from the position of $\rho_2$ to the left or right, the maximum shear stress gradually increases, and the order of the three stress components changes. This results in four intervals: the first interval (left 1), the maximum principal stress is the circumferential stress, the middle principal stress is the axial stress, and the minimum principal stress is the radial stress; the second interval (left 2), the maximum principal stress is the axial stress, the middle principal stress is the circumferential stress, and the minimum principal stress is the radial stress; the third interval (left 3), the maximum principal stress is the axial stress, the middle principal stress is the radial stress, and the minimum principal stress is the circumferential stress; the fourth interval (left 4), the maximum principal stress is radial stress, the intermediate principal stress is axial stress, and the minimum principal stress is circumferential stress. Comparing the Fig. 8.9a, b, it can be seen that at low drilling fluid density, the maximum shear stress at the 90° position of the circumferential angle is much larger than the value at the 0° position of the circumferential angle; while the circumferential stress at the 0° position of the circumferential angle is the first to converge to zero as the drilling fluid density increases.

For the uniform horizontal ground stress case ($\sigma_{\mathrm{H}} = \sigma_{\mathrm{h}}$), the borehole confining formation stress distribution Eq. (8.3) simplifies to

$$
\begin{aligned}
\sigma_r &= \sigma_h - (\sigma_h - p_m)\frac{a^2}{r^2} - \alpha P_p \\
\sigma_\theta &= \sigma_h + (\sigma_h - p_m)\frac{a^2}{r^2} - \alpha P_p \\
\sigma_z &= \sigma_v - \alpha P_p
\end{aligned}
\tag{8.5}
$$

At the well wall ($r = a$), the stress component is

$$
\begin{aligned}
\sigma_r &= p_m - \alpha P_p \\
\sigma_\theta &= 2\sigma_h - p_m - \alpha P_p \\
\sigma_z &= \sigma_v - \alpha P_p
\end{aligned}
\tag{8.6}
$$

For the shale with a burial depth of 3000 m, under the action of uniform horizontal ground stress ($\sigma_{\mathrm{H}} = \sigma_{\mathrm{h}} = 52.5$ MPa) and lower drilling fluid density (1.25 g/cm$^3$), the variation of stress with radius in the surrounding rock of the borehole is shown in Fig. 8.10. It can be seen that the stress concentration phenomenon occurs only near the borehole, and when $r = 6a$, the stress concentration phenomenon basically disappears and the stress state returns to the original ground stress state.

For the uniform horizontal in-situ stress case, the stress components in the borehole confining rock are independent of the circumferential angle and are only related to the radius due to symmetry. The variation of the three stress components with drilling fluid density is given in Fig. 8.11, and similar to the non-uniform ground stress case, the order of magnitude of the three stress components is divided into the similar four intervals.

The stress distribution of the confining rock of the vertical borehole under the combination of normal fault-controlled in-situ stress is given above (Figs. 8.4 and 8.9), and the stress distribution of the confining rock of the vertical borehole under the combination of the 3000 m stratigraphic strike-slip fault-controlled in-situ stress ($\sigma_H > \sigma_{\mathrm{v}} > \sigma_h$), i.e., a maximum horizontal in-situ stress of 76.5 MPa, an overburden pressure of 73.5 MPa, and a minimum horizontal in-situ stress of 52.5 MPa, is

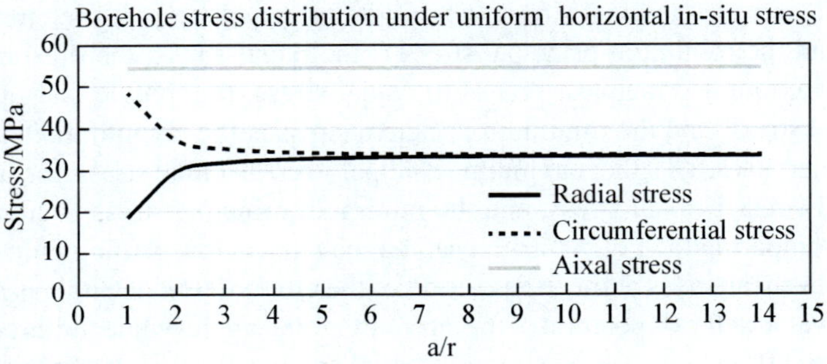


**Fig. 8.10** Variation of stress components in the borehole envelope with position

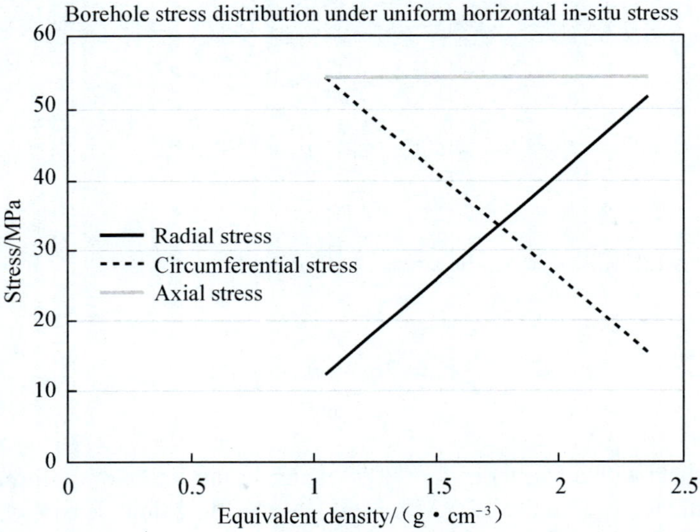


**Fig. 8.11** Relationship between stress components on the well wall and drilling fluid density

analyzed below. The pore pressure equivalent density of the formation is 1.15 g/cm$^3$, Biot coefficient is 0.55, Poisson's ratio of the formation is 0.25, and the radius of the borehole is 0.1 m. Figure 8.12 gives the variation of the three stress components on the wellbore wall with the drilling fluid density at the positions of 0° and 90° of the circumferential angle. It can be seen that when the circumferential angle is 0°, the circumferential stress gradually tends to 0 with the increase of drilling fluid density; when the circumferential angle is 90°, the three stresses intersect similar to Fig. 8.9b and are also divided into four intervals, the first interval (left 1), the maximum principal stress is the circumferential stress, the middle principal stress is the axial stress, and the minimum principal stress is the radial stress; the second interval (left 2), the maximum principal stress is the axial stress, the middle The third interval (left 3), the maximum principal stress is the axial stress, the intermediate principal

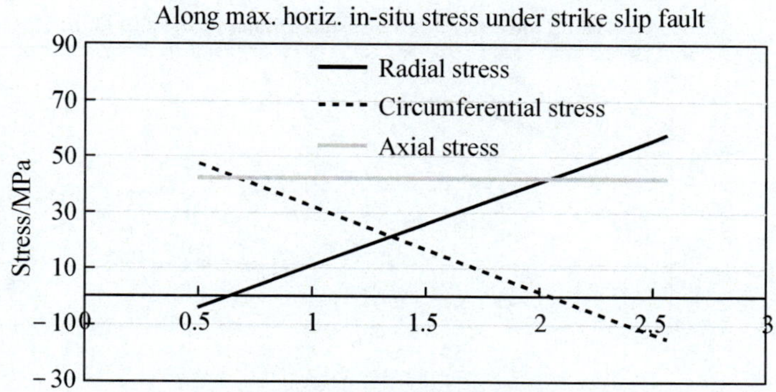


(a) Circumferential angle 0°

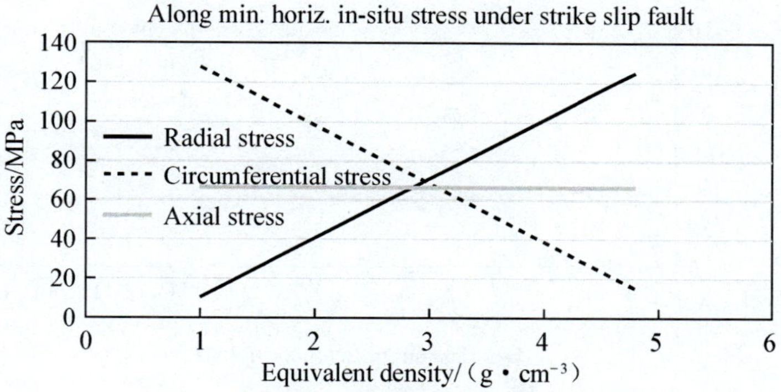


(b) Circumferential angle 90°

**Fig. 8.12** Relationship between wall stress and drilling fluid density in straight wells under walking slip fault control

stress is the radial stress, and the minimum principal stress is the circumferential stress; the fourth interval (left 4), the maximum principal stress is the radial stress, the intermediate principal stress is the axial stress, and the minimum principal stress is the circumferential stress. At this point $\rho_2$ is 2.9 g/cm$^3$, which is much higher than the case of normal faulting (see Fig. 8.9b, $\rho_2$ is 2.1 g/cm$^3$).

Figure 8.13 gives the variation pattern of the three stress components with radius for two positions with circumferential angles of 0° and 90° at a drilling fluid density of 1.25 g/cm$^3$. The circumferential stress in Fig. 8.13a appears to increase with increasing radius in the near-wellbore zone, and the radius $r/a$ exceeds 1.5 before it becomes decreasing with increasing radius; the radial and axial stresses increase monotonically with increasing radius. In Fig. 8.13b the radial stresses show a small decreasing phase in the near-wellbore zone, and the circumferential and axial stresses decrease monotonically with increasing radius.

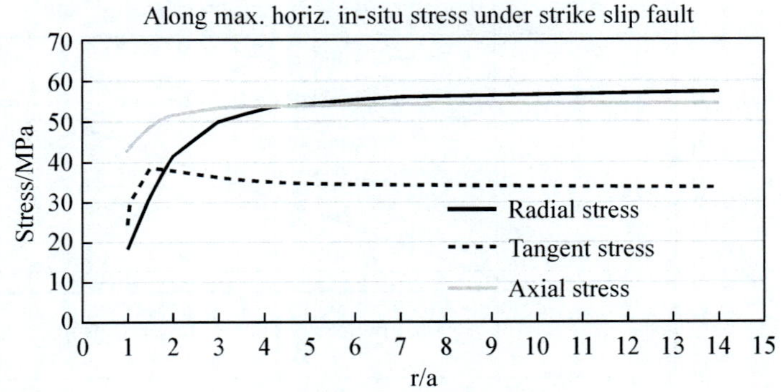


(a) The circumference angle is 0°

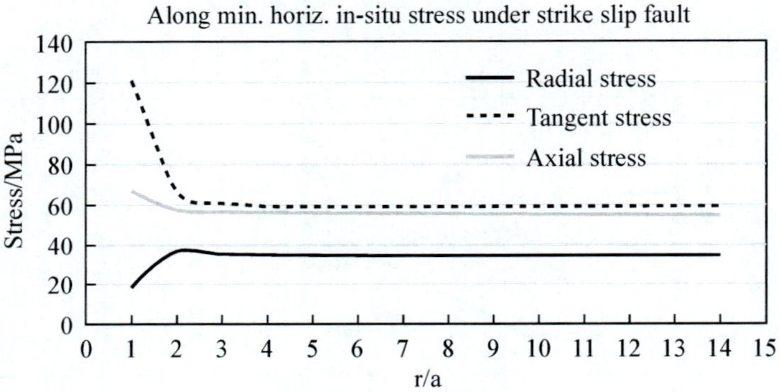


(b) The circumference angle is 90°

**Fig. 8.13** Variation pattern of surrounding rock stress components with radius at drilling fluid density of 1.25 g/cm $^{3}$

If the in-situ stress of the formation at a burial depth of 3000 m is controlled by the inverse fault ($\sigma_H > \sigma_h > \sigma_v$), its original in-situ stress state is: the maximum horizontal ground stress is 79.5 MPa, the minimum horizontal ground stress is 75.0 MPa, and the overburden pressure is 73.5 MPa. Still taking the above relevant parameters, the stress distribution of the surrounding rock of the vertical well is analyzed. Figure 8.14 gives the variation of the three principal stresses on the wellbore wall with the drilling fluid density at the positions of 0° and 90° of the circumferential angle. It can be seen that, similar to the case of the strike slip fault, the three stress components have three intersections, and the intersection of the radial and circumferential stresses in the direction of the minimum horizontal in-situ stress $\rho_2$ is about 2.75 g/cm$^3$.

The variation of stresses with radius in the borehole confining rock in the direction of maximum horizontal in-situ stress (circumferential angle 0°) and in the direction of minimum horizontal in-situ stress (circumferential angle 90°) is given in Fig. 8.15. It can be seen that all three stress components vary monotonically with position, with the greatest stress concentration at the borehole wall.

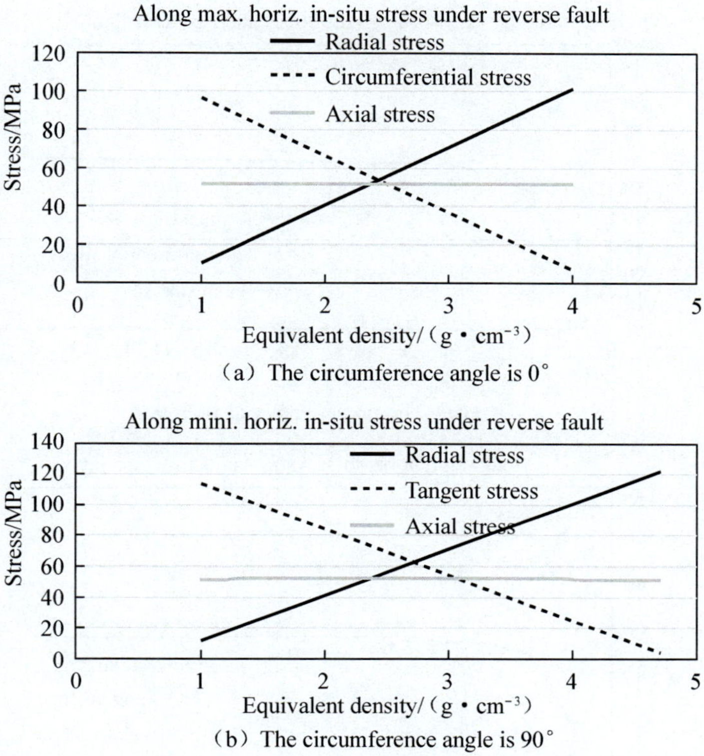


**Fig. 8.14** Relationship between well wall stress and drilling fluid density in straight wells under reverse fault control

Comparing Figs. 8.4 and 8.15 shows that under different original in-situ stress conditions and borehole fluid hydrostatic pressure, the radial stress or circumferential stress appears to vary non-monotonically with increasing radius in the direction of maximum horizontal in-situ stress and minimum horizontal in-situ stress, and the reasons for this are analyzed below. Equation (8.3) radial stress or circumferential stress to the radius for the first-order partial derivative in the direction of maximum horizontal in-situ stress is that

$$\begin{aligned}\frac{\partial \sigma_r}{\partial r} &= (\sigma_H + \sigma_h)\frac{a^3}{r^3} + (\sigma_H - \sigma_h)\left(-\frac{6a^5}{r^5} + \frac{4a^3}{r^3}\right) - p_m\frac{2a^3}{r^3} \\ \frac{\partial \sigma_\theta}{\partial r} &= -(\sigma_H + \sigma_h)\frac{a^3}{r^3} + (\sigma_H - \sigma_h)\frac{6a^5}{r^5} + p_m\frac{2a^3}{r^3}\end{aligned} \tag{8.7}$$

in the direction of minimum horizontal ground stress is that

$$\frac{\partial \sigma_r}{\partial r} = (\sigma_H + \sigma_h)\frac{a^3}{r^3} + (\sigma_H - \sigma_h)\left(\frac{6a^5}{r^5} - \frac{4a^3}{r^3}\right) - p_m\frac{2a^3}{r^3}$$

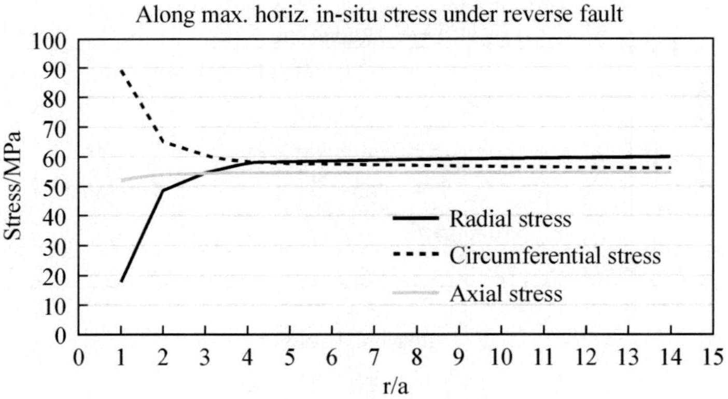


(a) The circumference angle is 0°

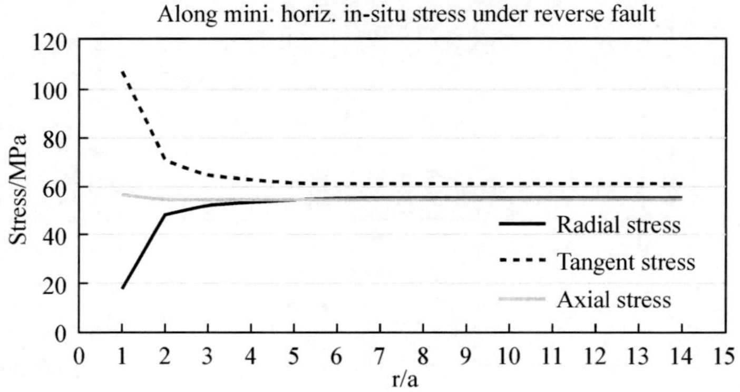


(b) The circumference angle is 90°

**Fig. 8.15** Variation of surrounding rock stress with radius in a straight borehole under reverse fault control

$$\frac{\partial \sigma_\theta}{\partial r} = -(\sigma_H + \sigma_h)\frac{a^3}{r^3} - (\sigma_H - \sigma_h)\frac{6a^5}{r^5} + p_m\frac{2a^3}{r^3} \tag{8.8}$$

Thus, the conditions under which the radial stress in the surrounding rock of a vertical borehole increases monotonically with increasing radius and converges to the horizontal in-situ stress in the direction of the maximum horizontal in-situ stress are

$$\frac{\partial \sigma_r}{\partial r} \geq 0 \tag{8.9}$$

The conditions under which the circumferential stress decreases monotonically with increasing radius and converges to horizontal ground stress are that

$$\frac{\partial \sigma_\theta}{\partial r} \leq 0 \tag{8.10}$$

At the wellbore wall, Eq. (8.9) becomes that

$$p_m \leq \frac{3\sigma_h - \sigma_H}{2} \tag{8.11}$$

At the wellbore wall, Eq. (8.10) becomes that

$$p_m \leq \frac{7\sigma_h - 5\sigma_H}{2} \tag{8.12}$$

The conditions for the monotonic variation of radial and circumferential stresses with position in the direction of minimum horizontal ground stress are the same as above.

At the wellbore wall, Eq. (8.9) becomes that

$$p_m \leq \frac{3\sigma_H - \sigma_h}{2} \tag{8.13}$$

At the wellbore wall, Eq. (8.10) becomes that

$$p_m \leq \frac{7\sigma_H - 5\sigma_h}{2} \tag{8.14}$$

Comparing the magnitudes of Eq. (8.11) with Eq. (8.13) and Eq. (8.12) with Eq. (8.14) shows that the radial and circumferential stresses in the surrounding rock of a vertical borehole are guaranteed to vary monotonically with radius as long as Eqs. (8.11) and (8.12) are satisfied.

## 8.3 Collapse and Rupture of Well Walls

### *8.3.1 Mechanisms of Well Wall Instability*

From Figs. 8.9, 8.12 and 8.14, it can be seen that the order of magnitude of the three principal stress components on the vertical wellbore wall will change due to different in situ stress states and different ranges of drilling fluid densities, making the well wall produce different damage patterns. It is also seen that the drilling fluid density corresponding to the intersection of radial and circumferential stresses $\rho_2$, the maximum shear stress is the smallest and the wellbore wall is the most stable; if the drilling fluid density used is less than $\rho_2$, the smaller the density, the greater the shear stress and the more unstable the well wall; if the drilling fluid density used is greater than $\rho_2$, the greater the density, the greater the shear stress and the more unstable the well wall. According to the Coulomb-Mohr strength damage criterion, the damage of the rock depends mainly on the maximum principal stress and the minimum principal stress, independent of the intermediate principal stress, and the angle between the normal outside the damage surface and the direction of the maximum principal stress is the rupture angle, and the direction of the damage surface is consistent with the intermediate principal stress; therefore, the damage of the wellbore wall will generally be manifested in the following four ways (Guenot 1990; Ewy and Cook 1990).

Mode A: Axial stress is intermediate stress
Mode B: circumferential stress is intermediate stress
Mode C: Radial stress is intermediate stress
Mode D: Tensile damage

These damage modes are depicted in Fig. 8.16; in modes A and B, the borehole wall collapses because the borehole pressure is too low; in modes C and D, shear or tensile damage occurs in the wall due to high borehole pressure.

**Mode A: Axial Stress Is Intermediate Stress**
No matter what kind of fault control the in-situ stress, if the drilling fluid density or borehole fluid hydrostatic pressure is too low (less than $\rho_1$), the following stress state will be generated on the wellbore wall in the direction of the minimum horizontal in-situ stress: circumferential stress is the maximum principal stress, radial stress is the minimum principal stress, Axial stress is the intermediate principal stress, and the maximum shear stress is very high, shear damage will occur, and the rupture surface is horizontal. At this time in the direction of the minimum horizontal in-situ stress borehole damage to form an elliptical borehole, resulting in the expansion phenomenon; and in the direction of the maximum horizontal in-situ stress no shear damage, the diameter of the wellbore is basically unchanged.

**Mode B: Circumferential Stress Is Intermediate Stress**
If the drilling fluid density or borehole fluid hydrostatic pressure increases ($\rho_1 < \rho_m < \rho_2$), different combinations of in-situ stresses will show different borehole damage characteristics. For positive fault-controlled ground stress conditions, if the formation strength is low, shear damage still occurs in the direction of minimum horizontal in-situ stress or maximum horizontal in-situ stress, when the rupture surface is a plumb plane along the radial direction, producing high steep fractures; for strike-slip fault-controlled in-situ stress conditions, the $P_1$ value is high, generally much higher than the formation pressure, and it is difficult to achieve this condition for equilibrium pressure drilling, and for reverse fault-controlled in-situ stress conditions, the This damage pattern does not occur in this borehole fluid hydrostatic pressure range.

**Mode C: Radial Stress Is Intermediate Stress**
If the drilling fluid density or borehole fluid hydrostatic pressure increases further ($\rho_2 < \rho_m < \rho_3$), the higher the drilling fluid density the higher the shear stress generated and the higher the risk of producing wellbore wall destabilization damage. At this point the rupture surface is a plumb plane perpendicular to the radial direction, and the wellbore wall damage is circumferential spalling off the block. Different combinations of in-situ stresses will exhibit different borehole damage characteristics. For

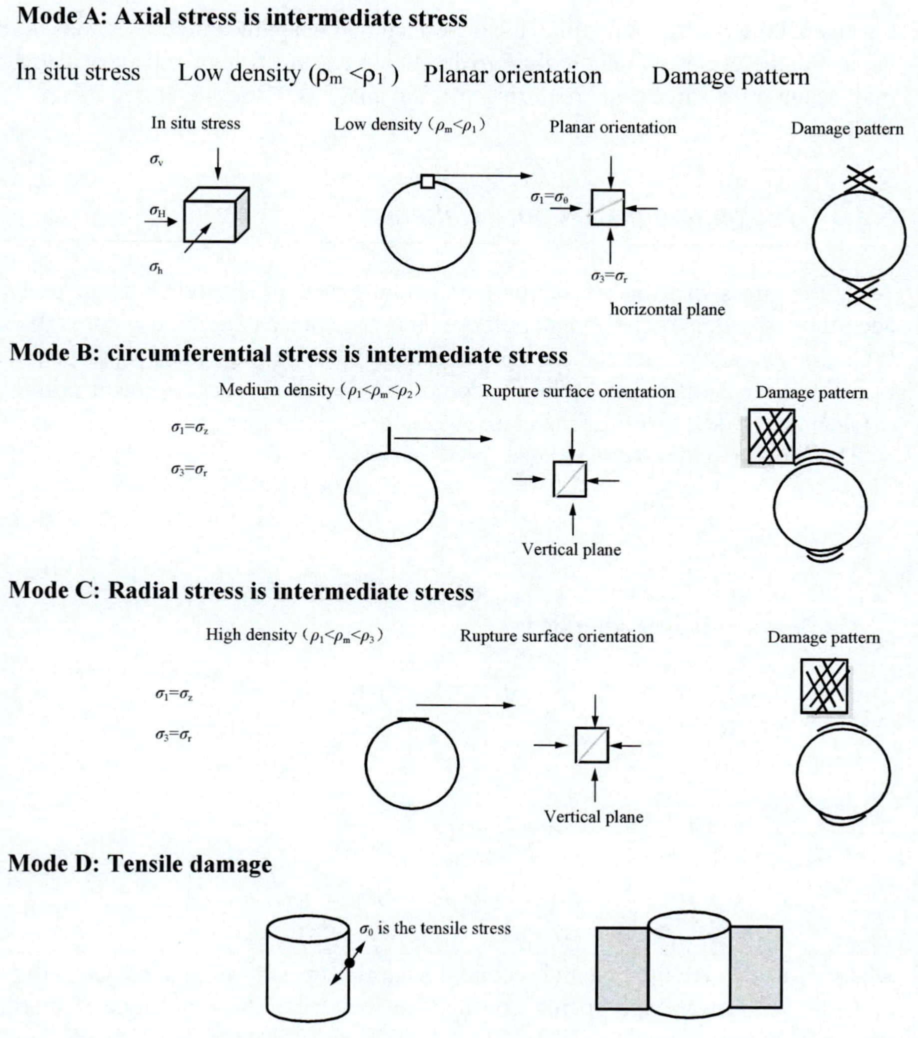


**Fig. 8.16** Damage pattern of well wall under different conditions

normal fault-controlled in-situ stress conditions, this phenomenon may occur if the formation pore pressure is very high. In contrast, for strike-slip fault and reverse-fault controlled in-situ stress combinations, this damage may occur in areas where the drilling fluid density is at $\rho_1 < \rho_m < \rho_2$.

**Mode D: Tensile Damage**

From Figs. 8.9, 8.12 and 8.14, it can be seen that regardless of the combination of in-situ stresses, after the drilling fluid density is high to a certain value, the circumferential stress in the direction of the maximum horizontal ground stress is the first to show tensile stress, which then produces tensile damage and forms a vertical fracture, which is the theoretical basis for borehole leakage or reservoir hydraulic fracturing during drilling. At the same time, it can be seen that when the circumferential stress is tensile, the general borehole fluid hydrostatic pressure is high and shear damage may occur at the same time, requiring specific analysis of the specific problem.

### 8.3.2 Judgement of Wellbore Collapse

After the stress distribution of the surrounding rock of the borehole is known, according to certain rock strength criterion, it is possible to determine whether destabilization damage occurs in the borehole wall, and the commonly used damage criterion for shear damage are Mohr–Coulomb criterion and Drucker–Prager criterion (Mclean and Addis 1990; Erling et al. 1992).

The Mohr–Coulomb criterion is

$$\tau_0 = \frac{\sigma_1}{2}[(f^2+1)^{1/2} - f] - \frac{\sigma_3}{2}[(f^2+1)^{1/2} + f] \tag{8.15}$$

$\tau_0$ is the cohesive force and f is the coefficient of internal friction.

The Drucker–Prager criterion is

$$\sqrt{J_2} = C_0 + C_1 J_1 \tag{8.16}$$

Of which

$$J_1 = \frac{1}{3}(\sigma_1 + \sigma_2 + \sigma_3);$$

$$J_1 = \frac{1}{6}\left[(\sigma_1 - \sigma_2)^2 - (\sigma_2 - \sigma_3)^2 - (\sigma_2 - \sigma_3)^2\right]$$

where $J_1$ and $J_2$ are the first and second deviatoric stress invariants, respectively.

$C_0$ and $C_1$ are the experimental constants related to cohesion and angle of internal friction (Mclean and Addis 1990).

Comparing the expression for the stress component of the wellbore wall Eq. (8.4) or Eq. (8.6) with the rock damage criterion Eq. (8.15) or Eq. (8.16), the critical borehole fluid hydrostatic pressure or drilling fluid density at which shear damage occurs in the wellbore wall can be found. This borehole pressure is called the collapse pressure of the formation and this drilling fluid density is called the collapse pressure equivalent density.

For the vertical well, uniform horizontal in-situ stress case, there are several possible damage scenarios depending on the relative magnitude of the principal stress components on the wellbore wall, considering first the case $\sigma_\theta > \sigma_z > \sigma_r$, which according to the Mohr–Coulomb criterion (8.15) has

$$P_c = P_m = \frac{2\sigma_h f_1 - (f_1 - f_2)\alpha P_p - 2\tau_0}{f_1 + f_2} \tag{8.17}$$

Of which $f_1 = \left(1+f^2\right)^{\frac{1}{2}} - f;\ \ f_2 = \left(1+f^2\right)^{\frac{1}{2}} + f$.

If the borehole fluid residence pressure falls below the value of Eq. (8.17), shear damage occurs in the borehole wall, producing uniform expanding.

As the borehole pressure increases (the drilling fluid density exceeds the uniform horizontal in-situ stress equivalent density), there will be $\sigma_r > \sigma_z > \sigma_\theta$, still according to the Mohr–Coulomb criterion, with

$$P_c = P_m = \frac{2\sigma_h f_2 + (f_1 - f_2)\alpha P_p + 2\tau_0}{f_1 + f_2} \tag{8.18}$$

The above equation shows that shear damage also occurs when the borehole fluid residence pressure is higher than Eq. (8.18), but this value tends to exceed the upper limit for tensile damage to occur in the borehole wall and will not actually exist. It can be seen that shear damage occurs in the wellbore wall, and theoretically there exists a low limit and a high limit to the drilling fluid density between which the borehole can remain stable.

For the vertical well, non-uniform horizontal in-situ stress case, the three principal stress components $\sigma_\theta$, $\sigma_z$, and $\sigma_r$ on the wellbore wall are related to the circumferential angle $\theta$. When $\theta = \pm 90°$, $\cos 2\theta = -1$, the shear stress is the largest. Due to different in situ stress states and different drilling fluid densities, the order of magnitudes of the three principal stress components $\sigma_\theta$, $\sigma_z$, and $\sigma_r$ on the wellbore wall exist in the following six cases, as shown in Table 8.1.

For Case 1, $\sigma_\theta \geq \sigma_z \geq \sigma_r$, the expression for the collapse pressure according to the Mohr–Coulomb criterion is

**Table 8.1** Well wall principal stress magnitude sequence and well wall damage pattern

| Statuses | Stress sequence $\sigma_1 \geq \sigma_2 \geq \sigma_3$ | Conditions of existence |
|---|---|---|
| 1 | $\sigma_\theta \geq \sigma_z \geq \sigma_r$ | In situ stresses controlled by positive faults, reverse faults, strike-slip faults and lower drilling fluid density |
| 2 | $\sigma_z \geq \sigma_\theta \geq \sigma_r$ | Positive fault-controlled in situ stress, low horizontal ground stress levels, lower drilling fluid density |
| 3 | $\sigma_z \geq \sigma_r \geq \sigma_\theta$ | Positive fault-controlled in situ stress, low horizontal ground stress levels, higher drilling fluid density |
| 4 | $\sigma_r \geq \sigma_z \geq \sigma_\theta$ | In situ stresses controlled by positive faults, reverse faults, strike-slip faults and very high drilling fluid densities |
| 5 | $\sigma_r \geq \sigma_\theta \geq \sigma_z$ | Inverse fault-controlled in situ stress, low strength soft strata, higher drilling fluid density; inverse fault-controlled in situ stress, very high horizontal ground stress levels, higher drilling fluid density (rare) |
| 6 | $\sigma_\theta \geq \sigma_r \geq \sigma_z$ | Inverse fault-controlled in situ stress, low strength soft strata, lower drilling fluid density; inverse fault-controlled in situ stress, very high horizontal ground stress levels, lower drilling fluid density (rare) |

$$P_c = \frac{(3\sigma_H - \sigma_h)f_1 - (f_1 - f_2)\alpha P_p - 2\tau_0}{f_1 + f_2} \tag{8.19}$$

For case 2, $\sigma_z \geq \sigma_\theta \geq \sigma_r$, the expression for the collapse pressure according to the Mohr–Coulomb criterion is

$$P_c = \frac{[\sigma_v + 2\mu(\sigma_H - \sigma_h)]f_1 - (f_1 - f_2)\alpha P_p - 2\tau_0}{f_2} \tag{8.20}$$

For Case 3, $\sigma_z \geq \sigma_r \geq \sigma_\theta$, the collapse pressure expression according to the Mohr–Coulomb criterion is:

$$P_c = \frac{2\tau_0 - [\sigma_v + 2\mu(\sigma_H - \sigma_h)]f_1 + (f_1 - f_2)\alpha p_p + (3\sigma_H - \sigma_h)f_2}{f_2} \tag{8.21}$$

For case 4, $\sigma_r \geq \sigma_z \geq \sigma_\theta$, the expression for the collapse pressure according to the Mohr–Coulomb criterion is

$$P_c = \frac{2\tau_0 + (3\sigma_H - \sigma_h)f_2 + (f_1 - f_2)\alpha P_p}{f_1 + f_2} \tag{8.22}$$

For case 5, $\sigma_r \geq \sigma_\theta \geq \sigma_z$, the expression for the collapse pressure according to the Mohr–Coulomb criterion is

$$P_c = \frac{2\tau_0 + [\sigma_v + 2\mu(\sigma_H - \sigma_h)]f_2 + (f_1 - f_2)\alpha P_p}{f_1} \tag{8.23}$$

For case 6, $\sigma_\theta \geq \sigma_r \geq \sigma_z$, the expression for the collapse pressure according to the Mohr–Coulomb criterion is

$$P_c = \frac{(3\sigma_H - \sigma_h)f_1 - [\sigma_v + 2\mu(\sigma_H - \sigma_h)]f_2 - (f_1 - f_2)\alpha P_p - 2\tau_0}{f_1} \tag{8.24}$$

If the Drucker-Prager criterion is used, a similar expression for the formation collapse pressure can be derived.

### 8.3.3 *Judgement of Wellbore Fracture*

In addition to the shear damage described above, another form of damage, tensile damage, exists for the borehole confining formation. It can be seen from Eq. (8.4) that as the borehole fluid hydrostatic pressure increases, the circumferential stress in the borehole wall gradually decreases and achieves a minimum value in the direction of the maximum horizontal ground stress (circular angle of 0°). Therefore, when the circumferential stress is equal to the tensile strength of the formation, tensile damage occurs in the wellbore wall, which is expressed as

$$\sigma_3 - \alpha P_p = -\sigma \quad (8.25)$$

where $\sigma_3$ is the minimum principal stress on the well wall, $\sigma_t$ is the tensile strength of the formation.

Comparing the expression for the stress component of the wellbore wall Eq. (8.4) or Eq. (8.6) with the rock damage criterion Eq. (8.25), the critical drilling fluid density at which tensile damage occurs in the wellbore wall can be found.

For vertical wells, the case of uniform horizontal in-situ stress with

$$P_f = 2\sigma_h + \sigma_t - \alpha P_p \quad (8.26)$$

For the case of vertical wells with non-uniform horizontal in-situ stress, $\sigma_\theta$ is minimum when $\theta = 0°$ or $180°$. There are

$$P_f = 3\sigma_h - \sigma_H + \sigma_t - \alpha P_p \quad (8.27)$$

That is, tensile rupture occurs first at this time in the direction of maximum horizontal in-situ stress in the wellbore wall, producing a vertical fracture. The critical borehole fluid hydrostatic pressure for wellbore wall rupture is called the fracture pressure of the formation and is determined by the above equation.

### *8.3.4 Factors Influencing Wellbore Stability*

In the actual drilling construction process, wellbore wall destabilization is related to many factors or parameters, such as formation lithology, in situ stress state, formation pore pressure, rock mechanical properties, drilling fluid properties, borehole opening time, temperature effect, operating characteristic parameters (agitation pressure or suction pressure), etc (Maury and Sauzay 1987; Santarelli 1987; Xu et al. 1999). These parameters are not independent of each other and can significantly reduce the formation strength if the drilling fluid performance is poor. The laws of in-situ stress, formation pore pressure and formation strength on wellbore stability have been given earlier. The effects of other factors on wellbore stability are discussed below.

**Formation Lithology**

Drilling practice shows that wellbore destabilization damage occurs mainly in mud shale formations, while for sandstone formations, even with very low strength, no downhole complications occur. Well diameter curves indicate that sandstone formations also appear to expanding, but due to the small size of the sandstone drop blocks and their rapid dispersion, they circulate to the surface (Santarelli et al. 1986; Guenot 1990). The main cause of wellbore destabilization in mud shale is the high clay mineral content of the mud shale and the interaction of the clay with the free water of the water-based drilling fluid, which generates swelling pressure and weakens the formation, hence the problem of time-dependent wellbore destabilization damage.

Mud shales are subdivided into two categories, soft mudstones and hard brittle shales, with different patterns of wellbore destabilization damage. Soft mudstones may show shrinkage and jamming, while hard brittle shales tend to show flaking off blocks and flaring and are periodic. In the case of fractured formations, large dropped chunks may also occur. Another formation prone to wellbore destabilization damage is the salt paste layer, especially under high temperature and pressure conditions, which has strong rheological properties, and after opening the borehole, if the drilling fluid density is not selected reasonably, the shrinkage jam may occur easily.

Although wellbore destabilization damage mainly occurs in mud shale formations, wellbore wall instability problems in other lithological formations have also occurred in different oil fields in China. For example, in the basalt formation of the Jidong oilfield, serious borehole destabilization damage problems have occurred, requiring higher drilling fluid densities to maintain borehole stability.

**Drilling Fluid Properties**

By opening the borehole, the original stress balance of the formation is broken and the drilling fluid hydrostatic pressure provides new wellbore support conditions, making drilling fluid density the most sensitive factor in wellbore stability analysis. However, in the conventional drilling process, the borehole fluid hydrostatic pressure is used to balance the formation pore pressure. It was only after the concept of formation collapse pressure was introduced in the 1980s that drilling technicians paid attention to the effect of drilling fluid density on wellbore stability. Even for areas where wellbore stability is not a serious problem, it is beneficial to increase the drilling fluid density appropriately. This is because borehole expansion can result in low quality logging or cementing. In most cases, drilling technicians are often reluctant to increase drilling fluid density because it may increase the risk of well leaks or differential pressure snags and reduce mechanical drilling rates. This requires the technician to take these factors into account and determine the optimum drilling fluid density based on regional experience.

In drilling engineering, traditional mud shale borehole collapse prevention is achieved through the chemical action of drilling fluids. Extensive drilling fluid chemical collapse prevention research has been conducted and a variety of drilling fluid admixtures and collapse prevention drilling fluid systems have been developed. For water-sensitive mudstones, drilling fluid chemistry is very important. The main mechanisms of hydration and swelling of mudstone are: surface hydration and osmotic hydration. Surface hydration is present in all clay types, where water molecules in the clay are bound by the clay lattice, but have a low capacity for volume expansion. Osmotic hydration is only present in montmorillonite-type clay minerals and causes strong volume expansion. The clay swelling causes dispersion of the formation into the drilling fluid, and this reactive particle rapidly reduces the size of the colloidal particles (less than 2 $\mu$m), making the performance of the drilling fluid difficult to maintain. At this point the drilling fluid must be diluted or flocculants added to allow the particles to agglomerate and then be screened out. This in turn requires the addition of further new chemicals, increasing the cost of the drilling fluid.

Oil-based drilling fluids are often used against shale formations with high montmorillonite content because of its stronger hydration inhibition. In particular, the addition of inorganic salts to the aqueous phase allows the water activity of the drilling fluid to be balanced with the water activity of the shale formation to prevent the free flow of water from the drilling fluid to the formation and improve the stability of the formation. This is because water transport changes the mechanical properties of the formation and makes the wellbore less stable. However, the high cost and serious environmental concerns of oil-based drilling fluids are not preferred.

Highly concentrated brine drilling fluids are also sometimes used to improve the inhibition of hydration of clay minerals. KCl solutions are most often used, and it is generally believed that the small size of $K^+$ makes it easy to enter the clay lamellae and acts as a barrier to further hydration. $Na^+$, on the other hand, is extremely hydrating and the use of sodium salts should be avoided.

Clay hydration is related to pressure, temperature, clay type, and solution components. During drilling, temperature and pressure can change, causing changes in the equilibrium water content of the shale. It is clear from the previous analysis that although the total stress in the formation near the wellbore is high, its effective stress magnitude is very limited, at which point it can lead to severe hydration swelling (Bol 1986). Hydration swelling also interacts with pore pressure transfer, and hydration swelling leads to an increase in pore diameter and higher permeability.

The pH in solution is another factor that affects the hydration and swelling of clays. The charge of clay minerals is pH dependent and as the pH increases, the negative charge of the clay minerals increases. Therefore the repulsion between the clay lamellae increases with increasing pH of the solution, which will result in weakening of the formation strength.

Polymers also have an inhibitory effect on shale hydration, but the mechanism has not been clearly explained. Bol (1986) concluded that polymers do not affect the equilibrium concentration of formation water, but can reduce the rate of hydration. In addition, high quality mudcake is particularly important for maintaining borehole stability, as poor quality mudcake reduces the degree of over-equilibration of borehole fluid column pressure to formation pore pressure.

There are a number of experimental methods for evaluating shale hydration with rock chips in the oilfield field, such as linear swelling, cation exchange capacity, and rolling dispersion, but these evaluation methods tend to be qualitative and difficult to compare with each other.

**Time Effect**

When drilling encounters shale formations, time-related wellbore stability problems can often arise. Severe borehole collapse often occurs after a period of drilling fluid immersion, and once the borehole has collapsed, it is difficult to suppress, even if the drilling fluid density is increased. Two main mechanisms control the stress–strain-time characteristics of saturated formations: first, the degree of compaction, which controls the ability to transport pore fluids. For low-permeability shale formations, pore fluid transport is a slow process; the higher the pore pressure, the lower the effective stress, and the more unstable the formation, and it exhibits a time-lag characteristic. Second, creep, under constant stress conditions, the deformation gradually increases with time. For soft mudstone and rock salt formations this characteristic is more prominent, and the borehole shrinkage is serious during drilling, which may lead to the phenomenon of stuck drilling.

**Temperature Effect**
During drilling, the drilling fluid circulation leads to changes in wellbore wall temperature: drilling fluid circulation in the bottom formation leads to a decrease in wellbore wall temperature and drilling fluid circulation in the upper formation leads to an increase in wellbore wall temperature. When drilling is started, circulation stops and the wellbore wall temperature moves back to its original state. Maury and Sauzay (1987) explained the temporal effect of wellbore wall destabilization in terms of temperature changes. The wellbore wall is stable when the borehole is first opened, and as the wall temperature increases, the circumferential and axial stresses in the wellbore wall increase simultaneously according to the following equation

$$\Delta\sigma = \frac{E}{1-\mu}\alpha_T\left(T - T_f\right) \tag{8.28}$$

where $\alpha_T$ is the thermal expansion coefficient of the formation; $E$ is the modulus of elasticity of the formation, $\mu$ is the Poisson's ratio of the formation, and $T_{\mathrm{f}}$ is the original temperature of the formation.

If temperature changes cause significant stress changes, it can lead to borehole collapse. Since thermal stress is proportional to the modulus of elasticity, the harder the formation, the higher the thermal stress. If the formation has a coefficient of thermal expansion of $10^{-5}$/°C, a modulus of elasticity of 10 GPa, and a Poisson's ratio of 0.3, the thermal stress per degree of temperature change is 0.14 MPa/°C.

## 8.4 Model for Predicting Formation Fracture Pressure

### *8.4.1 Leakage Test*

The formation most likely to rupture during drilling is near the casing shoe. Therefore, after casing is cemented, a leakoff test is typically done after 5–10 m of drilling to determine the minimum fracture pressure gradient (Zheng et al. 1988; Zoback 2007; Zoback et al. 1985).

The conventional leakage test procedure is as follows: shut down the well, inject mud into the wellbore with a cement truck through the drill pipe at a constant displacement, and record the injection pressure p and the injection time t or the pumped volume *Q*. At the beginning, the pressure increases linearly as the injection volume increases, and when the pressure reaches a certain value it starts to deviate from a straight line along a downward concave curve that gradually increases and starts to bend downward after reaching a maximum, at which point the pump is stopped.

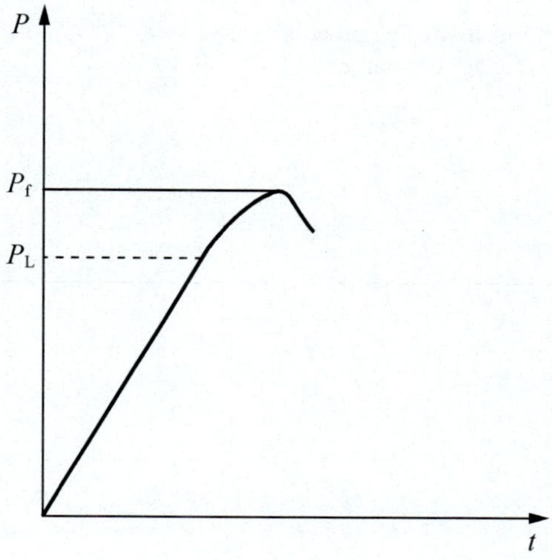


**Fig. 8.17** Conventional leakage test curve

A typical leakoff test curve is shown in Fig. 8.17. The point at which the pressure deviates from the straight line is called the formation leakoff pressure $P_L$, and the maximum pressure is called the formation fracture pressure $P_f$, which is the upper limit allowed by the drilling fluid density. However, for safety reasons in drilling engineering, the leakage pressure is generally used to represent the fracture pressure, but this is not the real physical sense of formation fracture pressure. Studies have shown that: formation leakage pressure is 10% higher than the minimum horizontal in-situ stress.

To obtain more information on the stratigraphy, we recommend a more comprehensive formation leakage test with the typical curve shown in Fig. 8.18. The injection method is the same as the conventional leakage test, with a longer pumping time after reaching the fracture pressure point to allow the fracture to fully expand and the fracture length to exceed the borehole stress concentration zone; then it goes through a transient pumping stop and reopening phase. The following pressure values can be obtained from the full formation leakage test curve.

(1) The formation fracture pressure, $P_f$, is the highest pressure point, reflecting the fact that the liquid column pressure overcomes the tensile strength of the formation to rupture it and form a fracture, and the mud fills the fracture, after which the pressure decreases.
(2) Propagation pressure $P_{pro}$ is the stage of pressure stabilization after the point of formation fracture pressure, when the fracture gradually extends farther out.
(3) The instantaneous stopping pressure $p_s$, when the fracture extends beyond the high stress concentration zone, i.e. 6 times the radius of the borehole (estimated to be about 1 min from the fracture point), stop the pump and record the stopping pressure $p_s$. After stopping the pump, as the fracture is still open, the instantaneous stopping pressure $p_s$ should be balanced with the minimum in-situ stress value of the vertical fracture, i.e., $p_s = \sigma_h$. seepage, the pressure drops, and the fracture closes under the action of the in-situ stress.

Fig. 8.18 Full process leakage test curve

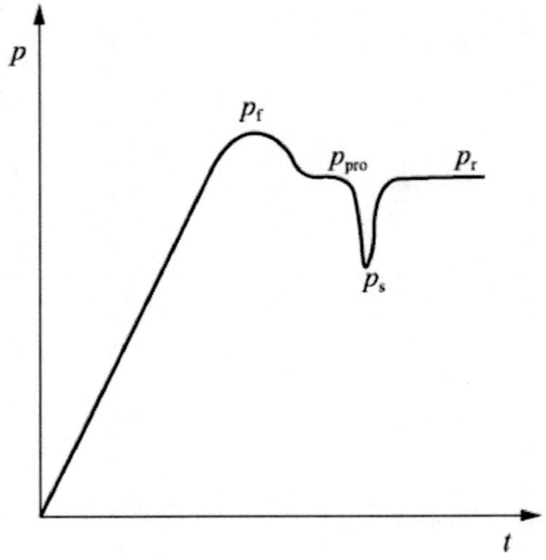


(4) The fracture re-opening pressure $p_r$, which is reopened after the transient pump stop phase, injects drilling fluid into the borehole and re-opens the closed fracture as the pressure increases. Since the pressure $p_r$ required to re-open the fracture does not need to overcome the tensile strength of the formation compared to the fracture pressure $P_f$, the tensile strength of the fractured formation can be approximated as equal to the difference between these two pressures. That is

$$\sigma_\tau = P_f - P_r$$

## 8.4.2 *Model for Predicting Formation Fracture Pressure*

The formation fracture pressure at the casing shoe can only be determined by the leakage test, and it is desirable to give the formation fracture pressure profile of the whole well, i.e., the variation curve of fracture pressure with well depth, in order to do a good job of designing the well structure. So far, many models have been proposed in the research of formation fracture pressure prediction methods in the world. Due to the different factors and assumptions considered, there are significant differences between the models. A few influential methods are described below.

### 8.4.2.1 Eaton Method (1969)

The Eaton method is more widely used in the U.S. Gulf Coast region and the assumptions of the method are as follows

(a) The formation is in a state of uniform horizontal in-situ stress.

(b) The subsurface rock mass is filled with joints, laminations and microfractures (the tensile strength of the formation is zero).
(c) The drilling fluid intrudes along the fracture under pressure and tensions the fracture only to overcome the stress on the vertical fracture face.

The fracture pressure prediction model is

$$P_f = \frac{\mu}{1-\mu}(\sigma_z - P_p) \tag{8.29}$$

The Eaton method is suitable for continuous sedimentary basins with relatively recent stratigraphic deposits and little tectonic influence. In contrast, it is less effective in predicting areas with older stratigraphic age and high tectonic movement influence.

#### 8.4.2.2 Stephon Method (1982)

The basic assumptions of the Stephon method are the same as those of the Eaton method, with the difference that Stephon considers the presence of uniform horizontal tectonic stresses in the formation, which is expressed as follows

$$P_f = \left(\frac{\mu}{1-\mu} + \xi\right)(\sigma_z - P_p) \tag{8.30}$$

The difference between the Stephon method and the Eaton method is that the Stephon method separates the effect of tectonic stresses from the formation Poisson's ratio. This makes it possible to use the actual Poisson's ratio of the formation in the calculation, unlike Eaton, which is back-calculated by the actual formation fracture pressure. Neither the Eaton nor the Stephon method takes into account the stress concentration in the borehole.

#### 8.4.2.3 Huang Rongtun Method (1984)

In the 1980s, Professor Huang Rongtaru of the University of Petroleum proposed a new method for predicting formation fracture pressure with the following basic assumptions.

(a) The formation is under non-uniform tectonic stress.
(b) The presence of stress concentration effects in the vicinity of the borehole.
(c) Wellbore wall rupture is caused by stresses in the wellbore wall exceeding the tensile strength of the rock formation.

Under the above assumptions and considering the stress concentration in the borehole, the expression for the fracture pressure of a vertical well is derived as

$$P_f = 3\sigma_h - \sigma_H - \alpha P_p + \sigma_t \tag{8.31}$$

There is a tectonic stress in the horizontal direction and the values in the two principal stress directions are not equal, then

$$\sigma_H = \left(\frac{\mu}{1-\mu} + \omega_1\right)(\sigma_1 - \alpha P_p) + \alpha P_p$$
$$\sigma_h = \left(\frac{\mu}{1-\mu} + \omega_2\right)(\sigma_v - \alpha P_p) + \alpha P_p \quad (8.32)$$

where $\omega_1$ and $\omega_2$ are the tectonic stress coefficients in the $\sigma_H$ direction and $\sigma_h$ direction, respectively.

Substituting Eq. (8.32) into Eq. (8.31) yields

$$P_f = \alpha P_p + \left(\frac{2\mu}{1-\mu} - k\right)(\sigma_v - \alpha P_p) + \sigma_t \quad (8.33)$$

where $k = \omega_1 - 3\omega_2$.

## 8.5 Example of Borehole Stability Calculation

For deep strata horizontal in-situ stresses are mostly controlled by normal faults or strike slip faults. Examples of calculations of vertical well collapse pressure and fracture pressure for these two combinations of horizontal in-situ stresses are given below.

***Example 1*** A sandstone formation in an oil field with a well depth of 2600 m has a strength that meets the Coulomb-Mohr criterion with a cohesion of 12.0 MPa, internal friction angle of 30°, Poisson's ratio of 0.25, uniaxial tensile strength of 2 MPa, overburden pressure gradient of 22.6 kPa/m, formation pore pressure gradient of 12.2 kPa/m, and Biot coefficient of 1. If the horizontal in-situ stress is non uniform $\sigma_H = 50$ MPa and $\sigma_h = 44.2$ MPa, try to determine a reasonable range of drilling fluid density to open the sandstone formation?

### Solution

(1) Determination of in-situ stress state

Overburden pressure $\sigma_v = 22.6 \times 2600 = 58.76$ MPa.

Horizontal in-situ stress

$$\sigma_H = 50\,\text{MPa}$$
$$\sigma_h = 44.20\,\text{MPa}$$

(2) Formation pore pressure

$$P_P = 12.2 \times 2600/100 = 31.72\,\text{MPa}$$

(3) Formation fracture pressure

$$\begin{aligned} P_f &= 3\sigma_h - \sigma_H + \sigma_t - \alpha P_p \\ &= 3 \times 44.2 - 50 + 2 - 1 \times 31.72 \\ &= 52.88\,\text{MPa} \end{aligned}$$

Equivalent drilling fluid density of fracture pressure: $\rho_f = 100P_f/H = 100 \times 52.88/2600 = 2.03\,\text{g/cm}^3$.

(4) Formation collapse pressure

The stress component on the wellbore wall is given by

$$\begin{aligned} \sigma_r &= P_m - \alpha P_p \\ \sigma_\theta &= \sigma_H + \sigma_h - P_m - 2(\sigma_H - \sigma_h)\cos 2\theta - \alpha P_p \\ \sigma_z &= \sigma_v - 2\mu(\sigma_H - \sigma_h)\cos 2\theta - \alpha P_p \end{aligned}$$

When $\theta = \pm \pi/2$, the stress component on the wellbore wall is:

$$\begin{aligned} \sigma_r &= P_m - \alpha P_p \\ \sigma_\theta &= 3\sigma_H - \sigma_h - P_m - \alpha P_p \\ \sigma_z &= \sigma_v + 2\mu(\sigma_H - \sigma_h) - \alpha P_p \end{aligned}$$

$$f_1 = \left(1 + f^2\right)^{\frac{1}{2}} - f = (1 + \tan^2(30°))^{0.5} - \tan(30°) = 0.57735$$

$$f_2 = \left(1 + f^2\right)^{\frac{1}{2}} + f = (1 + \tan^2(30°))^{0.5} + \tan(30°) = 1.73205$$

The intersection $p_1 = 44.14$ MPa or equivalent density $\rho_1 = 1.7\,\text{g/cm}^3$ from $\sigma_z = \sigma_\theta$;

The intersection $p_2 = 52.9$ MPa or equivalent density $\rho_2 = 2.03\,\text{g/cm}^3$ from $\sigma_r = \sigma_\theta$;

The intersection $p_3 = 61.66$ MPa or equivalent density $\rho_3 = 2.37\,\text{g/cm}^3$ from $\sigma_z = \sigma_r$, as shown in Fig. 8.19.

Examining Region 1 first: $\sigma_1 = \sigma_\theta$, and $\sigma_3 = \sigma_r$, we have

$$\begin{aligned} P_{c1} &= \frac{(3\sigma_H - \sigma_h)f_1 - (f_1 - f_2)\alpha P_p - 2\tau_0}{f_1 + f_2} \\ &= \frac{(3 \times 50 - 44.2) \times 0.57735 + 1.1547 \times 31.72 - 2 \times 12}{2.3094} \\ &= 31.29\,\text{MPa} \end{aligned}$$

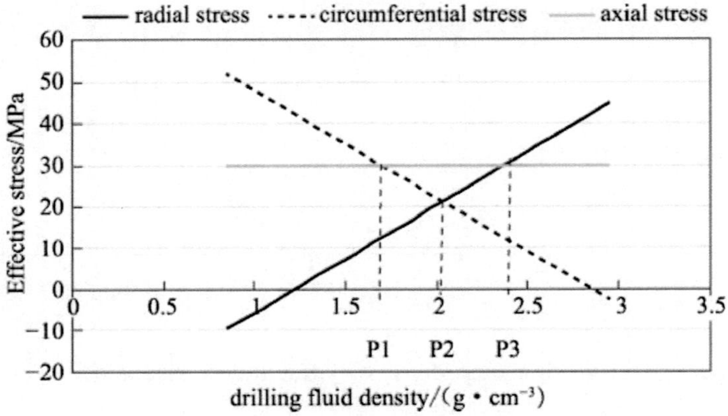


**Fig. 8.19** The effective stress versus mud density for case1

is less than the critical pressure $p_1$ and the calculated value is reasonable. Its equivalent density is 1.203 g/cm$^3$.

Examining Region 4 again: $\sigma_1 = \sigma_r$, $\sigma_3 = \sigma_\theta$, we have

$$P_{c2} = \frac{2\tau_0 + (3\sigma_H - \sigma_h)f_2 + (f_1 - f_2)\alpha P_p}{f_1 + f_2}$$

$$= \frac{2 \times 12 + (150 - 44.2) \times 1.73205 - 1.1547 \times 31.72}{2.3094} = 73.9\,\text{MPa}$$

$P_{c2}$ is greater than the critical pressure $p_3$ and the calculated value is reasonable. Its equivalent drilling fluid density $\rho = 2.842$ g/cm$^2$

Considering the collapse pressure and fracture pressure of the formation together, the safe drilling fluid density window is 1.203–2.03 g/cm$^3$. Then considering the need for balanced pressure drilling, the safety factor is taken as 0.05 g/cm$^3$, and the actual safe density window is 1.27–2.03 g/cm$^3$.

***Example 2*** A mudstone formation in an oil field with a well depth of 3600 m has a strength that meets the Coulomb-Mohr criterion with a cohesion of 10.0 MPa, an internal friction angle of 30°, Poisson's ratio of 0.25, uniaxial tensile strength of 2.5 MPa, overburden pressure gradient of 24.6 kPa/m, formation pore pressure gradient of 12.5 kPa/m, and Biot coefficient of 0.55. If the horizontal in-situ stress is non-uniform $\sigma_H = 92$ MPa and $\sigma_h = 63$ MPa, try to determine a reasonable range of drilling fluid density to open the formation?

**Solution**

(1) Determination of in-situ stress state

Overburden pressure $\sigma_v = 24.6 \times 3600/100 = 88.56\,\text{MPa}$.

Horizontal in-situ stresses

$$\sigma_H = 92\,\text{MPa}$$
$$\sigma_h = 63\,\text{MPa}$$

(2) Formation pore pressure

$$P_P = 12.5 \times 3600/100 = 45\,\text{MPa}$$

(3) Formation fracture pressure

$$\begin{aligned} P_f &= 3\sigma_h - \sigma_H + \sigma_t - \alpha P_p \\ &= 3 \times 63 - 92 + 2.5 - 0.55 \times 45 \\ &= 74.75\,\text{MPa} \end{aligned}$$

Equivalent drilling fluid density of fracture pressure:

$$\rho_f = 100 \times P_f/H = 100 \times 74.75/3600 = 2.08\,\text{g/cm}^3$$

(4) Formation collapse pressure
The stress component on the wellbore wall is given by

$$\begin{aligned} \sigma_r &= P_m - \alpha P_p \\ \sigma_\theta &= \sigma_H + \sigma_h - P_m - 2(\sigma_H - \sigma_h)\cos 2\theta - \alpha P_p \\ \sigma_z &= \sigma_v - 2\mu(\sigma_H - \sigma_h)\cos 2\theta - \alpha P_p \end{aligned}$$

When $\theta = \pm \pi/2$, the stress component on the wellbore wall is:

$$\begin{aligned} \sigma_r &= P_m - \alpha P_p \\ \sigma_\theta &= 3\sigma_H - \sigma_h - P_m - \alpha P_p \\ \sigma_z &= \sigma_v + 2\mu(\sigma_H - \sigma_h) - \alpha P_p \end{aligned}$$

$$f_1 = \left(1 + f^2\right)^{\frac{1}{2}} - f = (1 + \tan^2(30°))^{0.5} - \tan(30°) = 0.57735$$
$$f_2 = \left(1 + f^2\right)^{\frac{1}{2}} + f = (1 + \tan^2(30°))^{0.5} + \tan(30°) = 1.73205$$

The intersection $p_3 = 103.06$ MPa or equivalent density $\rho_3 = 2.86$ g/cm$^3$ from $\sigma_z = \sigma_r$;

The intersection $p_2 = 106.5$ MPa or equivalent density $\rho_2 = 2.96$ g/cm$^3$ from $\sigma_r = \sigma_\theta$;

The intersection $p_1 = 109.94$ MPa or equivalent density $\rho_1 = 3.05$ g/cm$^3$ from $\sigma_z = \sigma_\theta$, as shown in Fig. 8.20.

Examining Region 1 first: $\sigma_1 = \sigma_\theta$, and $\sigma_3 = \sigma_r$, we have

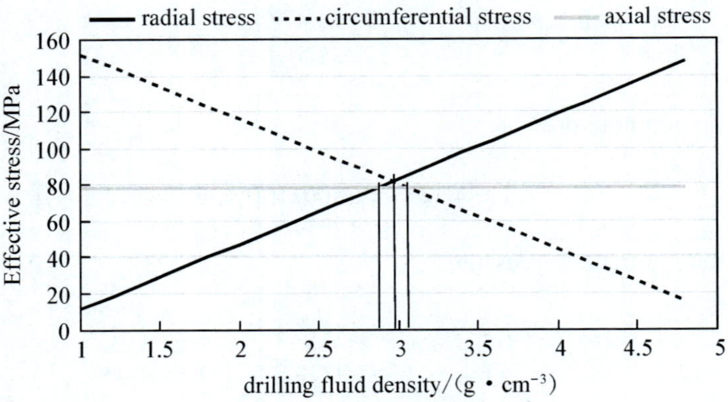


**Fig. 8.20** Effective stress versus mud density for case2

$$P_{c1} = \frac{(3\sigma_H - \sigma_h)f_1 - (f_1 - f_2)\alpha P_p - 2\tau_0}{f_1 + f_2}$$
$$= \frac{(3 \times 92 - 63) \times 0.57735 + 1.1547 \times 24.75 - 2 \times 10}{2.3094} = 56.96\,\text{MPa}$$

$p_{c1}$ is less than the critical pressure $p_1$ and the calculated value is reasonable. Its equivalent density $\rho_{c1} = 1.58$ g/cm$^3$.

Examining Region 4 again: $\sigma_1 = \sigma_r$, $\sigma_3 = \sigma_\theta$, we have

$$P_{c2} = \frac{2\tau_0 + (3\sigma_H - \sigma_h)f_2 + (f_1 - f_2)\alpha P_p}{f_1 + f_2}$$
$$= \frac{2 \times 10 + (92 \times 3 - 63) \times 1.73205 - 1.1547 \times 24.75}{2.3094}$$
$$= 156.03\,\text{MPa}$$

$p_{c2}$ is greater than the critical pressure $p_3$ and the calculated value is reasonable. Its equivalent drilling fluid density $\rho_{c2} = 4.33$ g/cm$^2$.

Considering the slump pressure and fracture pressure of the formation, the safe drilling fluid density window is 1.58–2.08 g/cm$^3$. Considering again the need for balanced pressure drilling, the safety factor is taken as 0.05 g/cm$^3$, and the actual safe density window is still 1.58–2.08 g/cm$^3$.

## Exercises

1. Briefly describe the main mechanisms and hazards of well wall destabilization damage?
2. For the data underlying Example 2, try to analyze the law of different shear damage criteria (Coulomb-Moore and D-P criteria) on the collapse pressure of a straight well?

3. The well depth of an oil field reservoir is 3500 m. The strength meets the Coulomb-Moore criterion with a cohesion of 12.0 MPa, an internal friction angle of 35°, a Poisson's ratio of 0.2, a uniaxial tensile strength of 4. The overlying rock pressure gradient of the formation is 24.5 kPa/m and the formation pore pressure gradient is 12.5 kPa/m. If the horizontal ground stress is non-uniform the maximum horizontal ground stress gradient $\sigma_H/h$ is 19.0 kPa/m, the minimum horizontal ground stress gradient is 16.0 kPa/m, and the effective stress factor is 1. Try to determine a reasonable range of drilling fluid density for a straight well to open this sandstone formation? If the reservoir is developed horizontally, try to determine the variation pattern of collapse pressure with azimuth?

## References


Aadnoy BS, Chenevert ME. Stability of highly inclined boreholes. SPE/IADC16052, New Orleans; 1987.

Bol GM. The effect of various polymers on borehole and cutting stability in water-base shale drilling fluids. SPE14802; 1986.

Bradley WB. Failure of inclined boreholes. J Energy Resour Technol. 1979;101(4):232–9.

Chenevert ME. Shale alteration by water adsorption. J Pet Technol. 1970;22(9):1141–8.

Cheng Y, Huang R. Mechanical analysis of mud shale well wall stability in drilling engineering. J China Univ Pet. 1993;17(4):35–9.

Deng J, Cheng Y, Chen M, Wei B. Well wall stability prediction technology. Beijing: Petroleum Industry Press; 2008.

Erling F, Rune M, Holt PH, Arne MR. Petroleum related rock mechanics. Amsterdam: Elsevier; 1992.

Ewy RT, Cook NGW. Deformation and fracture around cylindrical openings in rock—I. Observations and analysis of deformations. II. Initiation, growth and interaction of fractures. Int J Rock Mech Min Sci Geomech Abstr. 1990;27:387–407, 409.

Guenot A. Borehole breakouts and stress fields. Int J Rock Mech Min Sci Geomech Abstr. 1990;26:185–95.

Maury VM, Sauzay JM. Borehole stability: case histories, rock mechanics approach and results. SPE16051; 1987.

Mclean MR, Addis MA. Wellbore stability: the effect of strength criteria on mud weight recommendations. SPE20405, New Orleans; 1990.

Santarelli FJ. Theoretical and experimental investigation of the stability of the axisymmetric wellbore. Ph.D. thesis, University of London; 1987.

Santarelli FJ, Brown ET, Maury V. Analysis of borehole stresses using pressure-dependent linear elasticity. Int J Rock Mech Sci Geomech Abstr. 1986;23:445–9.

Xu T, et al. New technologies for well wall stabilization in drilling engineering. Beijing: Petroleum Industry Press; 1999.

Zheng ZQ, Cook NGW, Myer LR. Borehole breakout and stress measurement. Key questions in rock mechanics. Proceedings of 29th US Symposium. Minneapolis, June 1988;471–8.

Zoback MD. Reservoir geomechanics. London: Cambridge University Press; 2007.

Zoback MD, Moos D, Mastin L, Anderson RN. Wellbore breakouts and in situ stress. J Geophys Res. 1985;18:5523–30.

# Chapter 9
# Mechanics of Hydraulic Fracturing

Since its inception in 1946, hydraulic fracturing technology has been increasingly used in the petroleum industry and has become a major technical tool for solving various production enhancement challenges. These challenges relate to increasing production from oil wells, increasing injection of water wells, energy storage and environmental protection.

Hydraulic fracturing mechanics is a description of the important processes and mechanisms in hydraulic fracturing technology. Hydraulic fracturing is very complex and it involves different mechanical components such as fluid mechanics, rock mechanics, fracture mechanics and thermodynamics. In hydraulic fracturing, hydrodynamics describes the flow of single-, two-, or three-phase fluids in the tubing, wellbore, and fractures; rock mechanics describes the deformation or fracturing of the rock due to changes in fluid pressure; fracture mechanics describes the damage and cracking that occurs near the end of a hydraulic fracture; and thermodynamics describes the heat exchange between the fracturing fluid and the formation. Adding to the complexity is the fact that the responses used are all coupled and interact with each other. The key to hydraulic fracturing design is an understanding of the mechanical behavior of the rock, fractures, and fluids. In order to develop hydraulic fracturing design and analysis tools, it is necessary to build models that adequately describe the various responses.

## 9.1 The Role of Fracturing

### *9.1.1 Flow Characteristics of Fractured Wells*

The role of fracturing in increasing production in oil and gas wells is examined below by analyzing the flow characteristics of the wellbore. The reservoir is a typical porous medium with typical pore channels having a diameter of $10^{-7}$–$10^{-4}$ m. Typical



Y. Cheng et al., *Foundations of Rock Mechanics in Oil and Gas Engineering*,
https://doi.org/10.1007/978-981-99-1417-3_9

**Table 9.1** Pressure gradients and dimensionless pressure functions for the three forms (Dake 1978)

| | $\Delta p$ | $p_D$ |
|---|---|---|
| Transient state flow | $p_i$-$p_{wf}$ | $p_D = -\frac{1}{2}\mathrm{Ei}\left(-\frac{1}{4t_D}\right)$; $t_D = \frac{kt}{\phi\mu c_i r_w^2}$ |
| Steady state flow | $P_e$-$p_{wf}$ | $p_D = \ln\frac{r_e}{r_w}$ |
| Quasi steady flow | $\overline{p}$-$p_{wf}$ | $p_D = \ln\frac{0.472r_e}{r_w}$ |

where $E_i$ is the exponential function, $p_i$ is the initial reservoir pressure, $p_e$ is the outer reservoir boundary pressure, $\overline{p}$ is the average reservoir pressure, $p_{wf}$ is the bottom well flow pressure, $r_e$ is the outer reservoir radius, and $r_w$ is the borehole radius

sandstone porosity is in the range of 0.10–0.25 and carbonate formation porosity is in the range of 0.1–0.4.

Combining the continuity Eq., the Eq. of state, and Darcy's law yields the diffusion Eq. as follows:

$$\frac{\partial^2 p}{\partial^2 r} + \frac{1}{r}\frac{\partial p}{\partial r} = \frac{\phi\mu C_t}{k}\frac{\partial p}{\partial t} \tag{9.1}$$

$C_t$ is the total compression factor of the reservoir and t is time.

The solution for its constant production rate (steady-state flow) is

$$q = \frac{2\pi kh\Delta p}{\mu p_D} \tag{9.2}$$

$p_D$ is a dimensionless pressure function.

The flow of fluids in a reservoir can be classified into three types: transient flow, steady-state flow (constant outer boundary pressure), and pseudo steady-state flow (closed outer boundary). The driving pressure gradient $\Delta p$ and the dimensionless pressure function $P_D$ for these three regimes is given in Table 9.1.

Drilling, completion, and downhole operations processes can cause reservoir damage, which is generally characterized by a dimensionless skin factor s. The skin factor affects the pressure drop in steady-state flow. Equation (9.2) introduces the effect of the skin coefficient and yields.

$$q = \frac{2\pi kh\Delta p}{\mu(p_D + s)} \tag{9.3}$$

The above Eq. is further transformed to obtain the expression for the oil recovery index.

$$J = \frac{q}{\Delta p} = \frac{2\pi kh}{\mu(p_D + s)} \tag{9.4}$$

The skin coefficient s is generally positive and the larger the value the more severe the reservoir contamination (e.g., $s = 20$); whereas $s = 0$ indicates no contamination in vertical wells, while for directional wells zero skin coefficient also exists. Negative skin coefficients indicate a production-enhancing treatment of the reservoir matrix, i.e., the permeability of the near-well zone is greater than the original permeability. This state of the skin coefficient can be achieved by applying various chemical treatments to the reservoir skeleton. If the chemical treatments do not achieve satisfactory results, hydraulic fracturing is performed.

Here gives an example of borehole flow characterization of an unfractured well. Assume a well with radius $r_w$ of 0.1 m, reservoir radius $r_e = 300$ m, reservoir thickness $h$ of 20 m, and $p_e = 35$ MPa. Fluid viscosity 1 mPa·s and skin factor s of 10. The skin factor decreases from 10 to 0 and permeability increases from 0.01 to 100 mD before and after chemical treatment of the reservoir skeleton. Borehole pressure $p_{wf} = 20$ MPa and calculate the production of the well before and after chemical treatment.

According to the steady-state flow Eq. (9.3), it is obtained that

$$q = \frac{(2)\pi\left(9.87 \times 10^{-14}\right)(20)\left(3.5 \times 10^7 - 2 \times 10^7\right)}{\left(1 \times 10^{-3}\right)\left[\ln \frac{300}{0.1} + s\right]} = \frac{0.186}{8 + s}$$

Before chemical treatment s = 10 and from the above Eq. we get.

$$q_1 = 1.03 \times 10^{-2}\, \mathrm{m}^3/\mathrm{s} = 889.92\, \mathrm{m}^3/\mathrm{d}$$

After chemical treatment s = 0, and the above Eq. gives.

$$q_2 = 2.32 \times 10^{-2}\, \mathrm{m}^3/\mathrm{s} = 2004.48\, \mathrm{m}^3/\mathrm{d}$$

The increments in yield after chemical treatment are.

$$q_2 - q_1 = 1.29 \times 10^{-2}\, \mathrm{m}^3/\mathrm{s} = 1114.56\, \mathrm{m}^3/\mathrm{d}$$

Considering the economics of field development, a minimum well production of 7.95 $m^3$/d is required, and the minimum permeability of the reservoir is obtained as 0.4 mD according to Eq. (9.3), by making $s = 0$. If the permeability of the reservoir is below 0.4md, hydraulic fracturing is required.

Once hydraulically fractured, the borehole is connected to the fracture and reservoir fluids flow first from the reservoir to the fracture and then from the fracture to the borehole. For the vast majority of vertical wells, the fracture is a large vertical seam. Flow characteristics for infinite inflow capacity conditions are given by Gringarden and Ramey (1974) and for finite inflow capacity conditions by Cinco-Ley and Samaniego (1981). For the infinite inflow case, the fluid flow is linear, i.e., it flows from the reservoir into the fracture. Once the fluid enters the fracture, it is assumed that it then enters the wellbore with infinite fracture infusion capacity. Whereas for

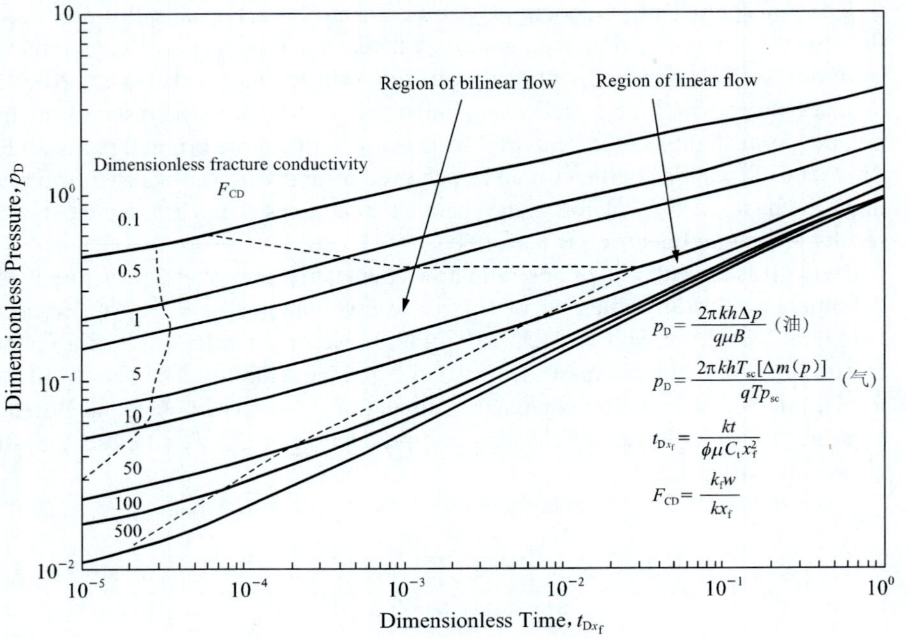


**Fig. 9.1** Relationship between causeless pressure and causeless time (Agarwal et al. 1979)

the finite inflow case, the flow is described by bilinear flow, i.e., from the reservoir into the fracture and then from the fracture into the borehole.

The solution for transient flow in the finite inflow case is given in Fig. 9.1. The vertical coordinate is the dimensionless pressure $p_{\mathrm{D}}$, and the horizontal coordinate is the dimensionless time $t_{\mathrm{DXf}}$. The parameter is the dimensionless inflow capacity $F_{\mathrm{CD}}$.

Define the dimensionless conductivity of a fracture.

$$F_{\mathrm{CD}} = \frac{k_{\mathrm{f}} w}{k x_{\mathrm{f}}} \tag{9.5}$$

After fracturing, the fracture passes through the contamination zone, resulting in a significant reduction in the skin factor, resulting in a significant increase in production. If the reservoir fluid is a liquid, then.

$$p_{\mathrm{D}} = \frac{2\pi kh(p_{\mathrm{i}} - p_{\mathrm{wf}})}{qB\mu} \tag{9.6}$$

$$t_{\mathrm{DX_f}} = \frac{kt}{\phi\mu c_{\mathrm{t}} x_{\mathrm{f}}^2} \tag{9.7}$$

$$F_{\mathrm{CD}} = \frac{k_{\mathrm{f}} w}{k x_{\mathrm{f}}} \tag{9.8}$$

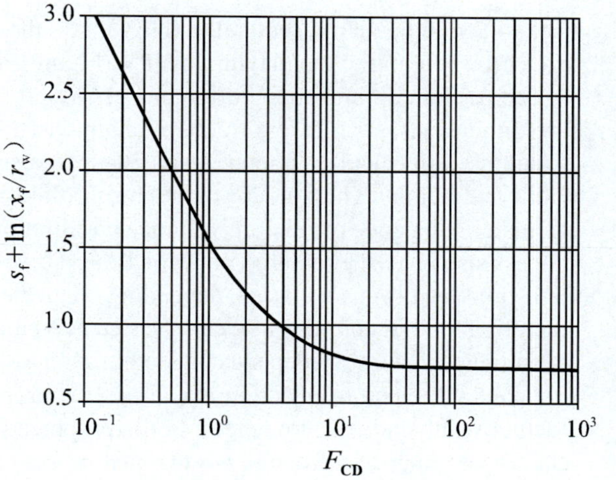


**Fig. 9.2** The role of skin factors (Cinco-Ley and Samaniego 1981; Cinco-Ley et al. 1978)

$x_f$ is the half-fracture length; $k_f$ is the fracture permeability; $w$ is the support fracture width.

Half-fracture length and fracture conductivity are the key parameters that determine the flow characteristics of fractured wells. When $F_{CD} > 300$, there is no difference between Cinco-Ley's solution and Gringarten's solution. In practice, when $F_{CD} > 70$, the two solutions can be considered consistent. After longer production times, the fluid flow in the fractured well is a proposed radial flow, with half-slit lengths and conductivity expressed by equivalent skin coefficients, as in Fig. 9.2.

In order to compare the flow characteristics of fractured and unfractured wells, Assume that the permeability of the well in Example 1 is $k = 0.39$ mD, chemical modification is ineffective, and hydraulic fracturing is required. The fracture permeability $k_f$ is 100,000 mD and the support fracture is $5 \times 10^{-3}$ m wide, try to compare the production dynamics of the well before and after fracturing.

According to Eq. (9.8), $F_{CD} = 4.2$, and according to Fig. 9.2, we find $s_f + \ln(x_f/r_w) = 0.96$, bringing in the values of $x_f$ and $r_w$ to obtain: $s_f = -7$.

Again, according to Eq. (9.3), we get $q = 7.35 \times 10^{-4}$ m$^3$/s = 63.5 m$^3$/d, which is eight times the yield of an uncontaminated unfractured vertical well.

It is important to note that once a well is fractured, the vast majority of the production comes from the fracture, which becomes a conduit through the contaminated zone. Therefore, the role of the pre-frac radial skin factor can be ignored.

### 9.1.2 *Optimal Design Process for Hydraulic Fracturing*

The objective of optimal hydraulic fracturing design is to exchange the minimum operational cost for the maximum economic benefit. The economic criterion often used today is the net present value (NPV), i.e., the design objective is the fracture size corresponding to the maximum NPV (Meng 1989; Meng and Brown 1987).

The fracture optimization design process usually starts with optimizing the fracture dimensions, such as the fracture length. A suitable fracture expansion model is then used to calculate fracture geometry parameters in detail, including fracture width and height. The amount of fracturing fluid required is estimated based on the material balance of fracture volume and filtration loss volume. This calculation process also gives the required injection time. There are several calculation methods for estimating proppant dosage, depending on the flow pattern of the proppant in the fracturing fluid. A common model is the step pumping procedure (Nolts 1986), where the initial concentration depends on the filtration loss characteristics. The amount of proppant at the end of injection is equal to the proppant dose corresponding to the fracture width and fracture length, or the proppant dose calculated by some empirical models, such as a fracture width equal to three times the proppant diameter. In addition, the choice of proppant type is critical, as fracture permeability is related to proppant strength under in situ stress conditions (Brown 1992).

Therefore, the support fracture width, fracture permeability, half fracture length, and reservoir permeability are known, and Eq. (9.3) is used to predict the flow characteristics of the well after compression, and a prediction of future returns can be made to determine the NPV.

Central to this design process is determining the amount of fracturing fluid, the amount of proppant and the pumping time. These are also central to construction costs.

This process is used repeatedly to obtain the NPV for different half-fracture lengths, and the optimized half-fracture length $x_f$, which gives the maximum NPV, i.e., the optimal construction solution. Typical design results are that the half-fracture length is less than 100 m for high permeability reservoirs and up to 500 m for low permeability reservoirs. For high permeability loose reservoirs, short wide fractures are produced using Fra-Pack techniques. The optimization method of fracture geometry parameters for certain proppant dosage conditions is given here.

The parameters used are the same as the above ($k = 0.39$ mD, $h = 20$ m, $r_e = 300$ m, $\mu = 1$ mPa·s, $p_e = 35$ MPa, $p_{wf} = 20$ MPa) to determine the optimal half-fracture length, proppant fracture width, and steady-state yield. Using the actual permeability of the fracture, consider proppant breakage with $k_f = 1 \times 10^{11}$ m$^2$. Assume that the fracture height is equal to the reservoir thickness. Using the Cinco-Samaniego diagram (Fig. 9.2), a radial flow is assumed to be proposed.

The same volume of proppant can produce either a long narrow fracture or a short wide fracture. According to Eq. (9.3), the yield Eq. is

$$q = \frac{\frac{2\pi k h \Delta p}{\mu}}{\ln \frac{r_e}{r_w} + s_f}$$

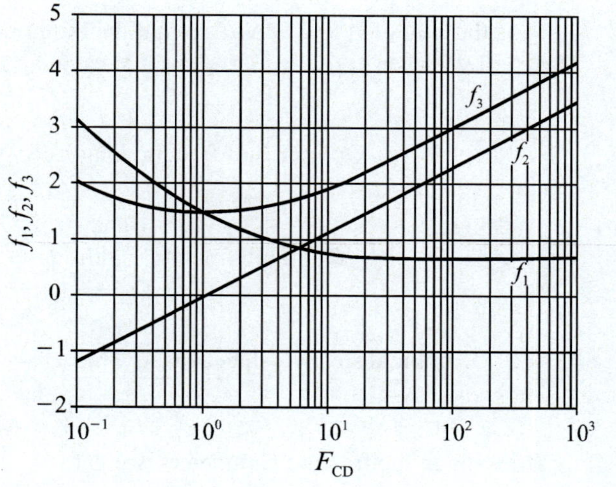


**Fig. 9.3** Function to optimize fractured well capacity

The proposed radial steady-state flow indicates relatively high reservoir permeability and generally performs Fra & Pack treatment. Obviously, our aim is to minimize the denominator. Let $f_1 = \lg F_{CD}$, , and according to the Cinco- Samaniego diagram (Fig. 9.2), the function f$_1$ is defined as

$$f_1 = s_f + \ln \frac{x_f}{r_w}$$

It is easy to draw the relationship curve of $f_1$ with $F_{CD}$ (as in Fig. 9.3). For a given $f_1$, the denominator of the yield Eq. is expressed as follows.

$$\ln \frac{r_e}{r_w} - \ln \frac{x_f}{r_w} + f_1$$

Further simplifying, $\ln r_e - \ln x_f + f_1$.

Setting $V_f = 2whx_f$, and using Eq. (9.8), eliminating the half fracture length for the above equation gives

$$y = \ln r_e - \ln \sqrt{\frac{V_f k_f}{2hk}} + \frac{2.303}{2} \log_{10} F_{CD} + f_1(\lg F_{CD})$$

Le t $\frac{dy}{dF_{CD}} = 0$, and obtain the extreme of $y$.

The only unknown quantity is $F_{CD}$. The first two are constants and therefore do not affect the location of the minimum, and the last two contain no problem-specific data. Thus, the optimal $F_{CD}$ is a given constant for any reservoir, well, and proppant. In turn, the same $F_{CD}$, yielding the same proposed steady-state capacity.

For simplicity, two new functions are introduced.

$f_2(\lg F_{CD}) = \frac{2.303}{2} \lg F_{CD}$, as in the straight line in Fig. 9.3.

$f_3$ is the sum of $f_1$ and $f_2$ used to find the minimum value. As shown in Fig. 9.3, $f3_{.opt} = 1.45$ when $f_{CD.opt} = 1.2$. Thus, the calculation is as follows.

$$\text{Optimal half fracture length} \quad x_f = \sqrt{\frac{V_f k_f}{2.4hk}}$$

$$\text{Optimum fracture width} \quad w = \sqrt{\frac{0.6V_f k}{hk_f}}$$

$$\text{optimal steady - state capacity} \quad q = \frac{\frac{2\pi kh\Delta p}{\mu}}{\ln r_e - \ln\sqrt{\frac{V_f k_f}{2hk}} + 1.45}$$

Bringing in the fracture parameters, we get.

$$q = 4.54 \times 10^{-3}\,\text{m}^3/\text{s} = 392.3\,\text{m}^3/\text{d}$$

It is usually important to check whether $x_f$ exceeds $r_e$, and if it exceeds $r_e$, $x_f = r_e$. Similarly, check whether the fracture width is greater than three times the particle diameter. Both conditions are satisfied for this example.

### 9.1.3 Introduction to Fracturing Fluids

Hydraulic fracturing is a large-scale construction project, with fracturing fluid typically exceeding 2000 m$^3$, proppant exceeding $5 \times 10^5$ kg, bottomhole pressure exceeding 50 MPa, and wellhead pressure exceeding 20 MPa. At least two fracturing pumps are used, each with 1500–2000 hp. Offshore operations also require specially designed fracturing vessels with similar power.

Figure 9.4 gives a flow chart of fracturing treatment, where fracturing fluid is mixed with proppant and injected into the fractured formation through a certain type of pipe column.

A brief description of fracturing fluid types, desirable properties and admixtures is given below.

#### 9.1.3.1 Fracturing Fluid Selection

Fracturing fluids should be of a nature to accommodate fracture initiation, extension and proppant transport with minimal filtration loss and long-term residual damage to the proppant fractures.

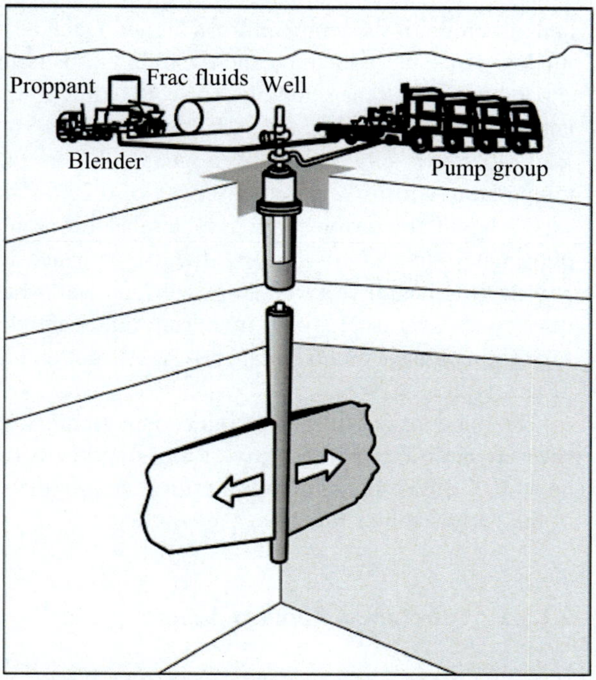


**Fig. 9.4** Schematic diagram of fracturing construction

Viscosity is the most important parameter of the fracturing fluid and is increased with admixtures during treatment and reduced with additional admixtures at the end of treatment.

The ideal fracturing fluid should have a low viscosity in the pipe sink and a high viscosity when it enters the fracture so that it has a high sand-carrying capacity and creates a fracture with a large fracture width. In addition, the role of viscosity builder is to form mud cake on the wall of the fracture and reduce the filtration loss. After the treatment is completed, high viscosity is detrimental to the flow of formation fluids or the flow of injected water and therefore needs to be reduced. This conflicting functional need is a fundamental consideration in fracturing fluid design.

Fracturing fluids include water-based fracturing fluids, oil-based fracturing fluids, oil–water mixed fracturing fluids, gas–water mixed fracturing fluids and oil–gas mixed fracturing fluids.

Common viscosity builders used in water-based fracturing fluids are hydroxyethyl cellulose HEC and hydroxypropyl guar gum HPG, added at 2.4–9.6 kg/m$^3$. Under ambient conditions, the viscosity of these polymer solutions can reach 0.1 pa·s, but at reservoir temperatures ($T = 65$–115 °C) the viscosity drops to 0.02 Pa·s, which is not sufficient to suspend the proppant. The minimum viscosity required within the fracture is 0.1 Pa·s (Brown and Economides 1992), with the maximum shear rate at the fracture tip and the minimum viscosity. To increase the viscosity significantly, a polymer cross-linker is required, either a boric acid cross-linker when the temperature is less than 115 °C, or an organometallic cross-linker such as a mixture of titanium and zirconium if the temperature is higher. Delayed crosslinking is required to meet the low viscosity in the tube and the high viscosity in the fracture. Activated at high shear rates, e.g., at the shot hole. To avoid oxide degradation, deaerators are generally added.

40-lb boric acid cross-linked guar solution (4.8 kg/m$^3$) at 90 °C reservoir temperature with an apparent viscosity of 0.2 pa·s after 4 h of shear injection.

Oil-based fracturing fluids are suitable for water-sensitive formations and use phosphate esters as crosslinkers, the use of which is being phased down for safety and environmental reasons. Non-injurious water-based fracturing fluids are being developed. $CO_2$ or $N_2$ foam fracturing fluids are also frequently used to minimize formation damage and facilitate post-construction rejection, with a 50–90% fraction of gas, typically 70%.

The main task at the end of injection is to break the glue; unbroken polymer can severely harm the permeability of the support fractures, so a glue breaker needs to be added. Breaking the glue is a critical step in the success of fracturing and is one of the current topics for further research.

#### 9.1.3.2 Selection of Support Agent

If the bottomhole pressure is less than the closure pressure, the fracture width will drop to zero unless proppant material is placed inside the fracture. The strength and size of the proppant are the two most important parameters concerning the proppant. Proppants are classified into three main categories: low strength, medium strength and high strength. The strength selection is related to the level of in-situ stress.

Low-strength proppant is generally selected from natural sand (particle size 12/20–20/40 mesh, average particle size 0.1–0.2 mm) and is generally used at well depths less than 2000 m; if the sand is wrapped with a layer of resin on the outside, it can improve the strength of the sand and can be used in higher stress environments. Synthetic high-strength proppant can be used for fracturing at well depths of 3000–5000 m.

## 9.2 Fracture Stress Field Analysis

The classical hydraulic fracturing model is generally a plane strain model, where the plane in which plane strain occurs in the model should be perpendicular to the direction of maximum fracture extension, which is the vertical plane for the PKN model and the horizontal plane for the GDK model. To better understand the hydraulic fracturing model, let's first analyze the elastic solution of the linear fracture.

### 9.2.1 Solution of Internally Compressed Linear Fractures

A hollow two-dimensional fracture exists in an infinitely large plane. Pressure exists inside the fracture, which causes the fracture to open. This is typical of the Griffith fracture problem.

Let the *xy*-plane be an infinite plane and a linear fracture of length 2c is located in the center. For the convenience of analysis, assume that the linear fracture is elliptical and the width of the fracture is much smaller than the length of the fracture, as shown in Fig. 9.5.

The boundary conditions are as follows.

$$\begin{aligned} &\sigma_{yy}(x,0) = -p(x),\ 0 \le x \le c \\ &u_y(x,0) = 0 \qquad x > c \\ &\tau_{xy}(x,0) = 0 \qquad x \ge 0 \end{aligned} \tag{9.9}$$

The first condition is that there is a constant normal distributed force $p(x)$ acting on the inner surface of the linear fracture and that $p(x)$ is an even function.

At infinity the stress is 0. The solution to this problem includes both the stress state and the displacement state. We are interested in the displacement at the surface of the fracture and the stress at the tip.

Based on Muskhelishvili's work (Muskhelishvili 1953; England and Green 1963; Green and Zerna 1968; Sneddon 1973), first construct a pressure function.

$$g(\xi) = \int_0^{\xi} \frac{p(x)dx}{(\xi^2 - x^2)^{1/2}} \quad 0 < \xi < c \tag{9.10}$$

Note that the dummy parameter ξ has the same magnitude as x and $g(\xi)$ has the magnitude of pressure.

If the $g(\xi)$ function is known, the displacement at a point on the upper surface of the fracture (half-length 0– $c$) can be solved for as

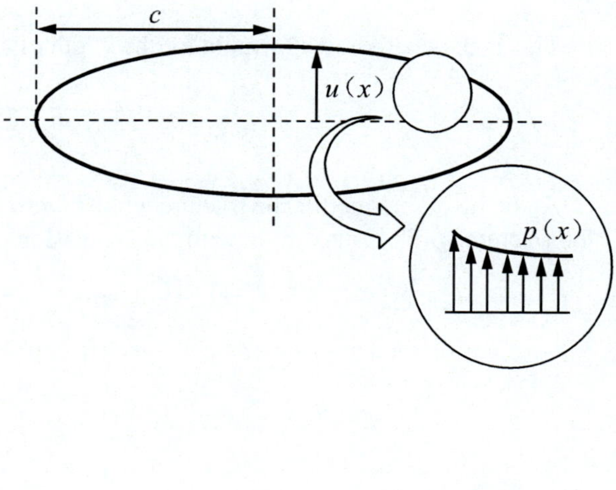


**Fig. 9.5** Schematic diagram of an internal pressure linear fracture

$$u_y(x,0) = -\frac{4}{\pi E'}\int_x^c \frac{\xi g(\xi)\mathrm{d}\xi}{(x^2-\xi^2)^{1/2}} \quad 0 \le x \le c \tag{9.11}$$

where *E*' is the plane strain modulus of elasticity.

According to the second boundary condition, the displacement beyond the fracture is zero. Clearly, an expression for the displacement at the lower surface of the fracture can be obtained. Thus the fracture width is twice the width of Eq. (9.11).

If the normal force at the fracture boundary is known, the positive y-directional stress on the *x*-axis beyond the fracture tip (Sneddon 1973) is

$$\sigma_{yy}(x,0) = \frac{2}{\pi}\left[\frac{xg(c)}{(x^2-c^2)^{1/2}} - g(0) - x\int_0^c \frac{g'(\xi)d\xi}{(x^2-\xi^2)^{1/2}}\right] \quad x > c \tag{9.12}$$

For several specific functions $p(x)$, the above integrals have analytic solutions.

### 9.2.2 Constant Pressure Distribution in the Seam

If the pressure in the seam is constant at $p_0$, then the expression for $g(\xi)$ is

$$g(\xi) = \int_0^\xi \frac{p_0 dx}{(\xi^2-x^2)^{1/2}} \approx \frac{p_0\pi}{2} \quad 0 < \xi < c \tag{9.13}$$

It can be seen that $g(\xi)$ is also a constant.

The displacement at the upper boundary of the fracture is

$$u_y(x,0) = -\frac{4}{\pi E'}\int_x^c \frac{\xi g(\xi)d\xi}{(x^2-\xi^2)^{1/2}} = \frac{2p_0}{E'}\sqrt{c^2-x^2}, 0 \le x \le c \tag{9.14}$$

This is the well-known formula for the width of an elliptical fracture.

$$w(x) = \frac{4p_0}{E'}\sqrt{c^2-x^2} \tag{9.15}$$

At the tip of the fracture the fracture width is zero and at the middle of the fracture the fracture width is maximum with the expression

$$w_0 = \frac{4cp_0}{E'} \tag{9.16}$$

If $g(\xi)$ is a constant, then the last term of Eq. (9.12) is zero. The remaining term is.

$$\sigma_{yy}(x,0) = p_0\left[\frac{x}{(x^2 - c^2)^{1/2}} - 1\right], x > c$$

The singularity of the stress at x = c can be seen from the expressions of $\sigma_{yy}$. The stress and displacement solutions for a linear fracture at constant pressure are shown in Fig. 9.6, with the pressure distribution in the fracture given in Fig. 9.6a, the $g(\xi)$ distribution in Fig. 9.6b, the displacement distribution in Fig. 9.6c, and the stress distribution in Fig. 9.6d. The stress due to constant internal pressure is infinitely amplified at the fracture tip, but decreases rapidly after leaving the tip, and the stress is nearly zero after a few times the fracture length.

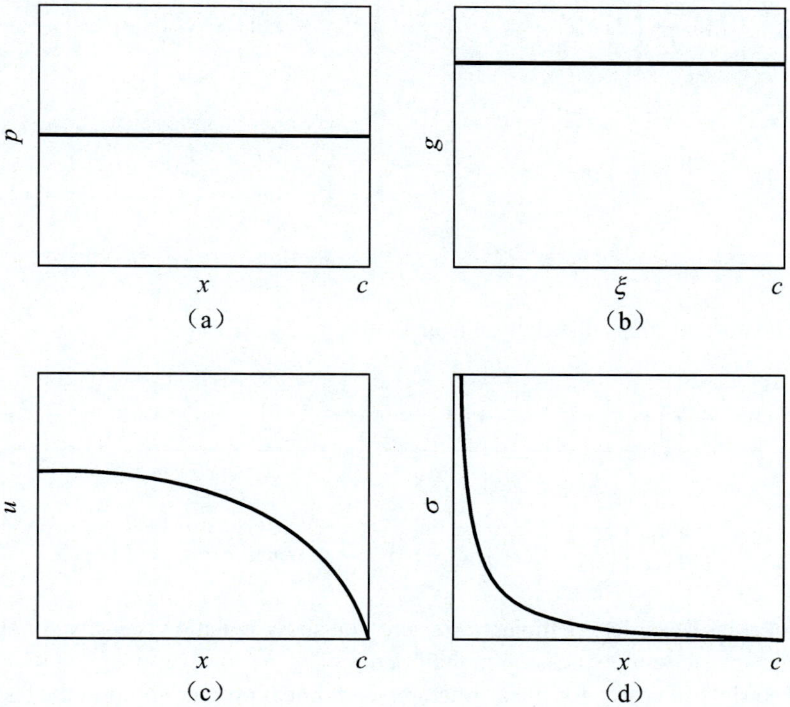


**Fig. 9.6** Solution of constant pressure for linear cracks

### 9.2.3 The Pressure Distribution in the Fracture as a Polynomial

If the pressure distribution in the fractur is polynomial.

$$p(x) = p_0 + p_1 x + p_2 x^2 + \cdots \tag{9.17}$$

Can be expanded into a taylor series, the $g$-function is

$$g(\xi) = \frac{\pi p_0}{2} + p_1 \xi + \frac{\pi p_2 \xi^2}{2} + \frac{2 p_3 \xi^3}{3} + \cdots \tag{9.18}$$

The normal displacement of the upper surface of the fracture is

$$\begin{aligned} u_y(x,0) = \frac{1}{E'} \Bigg\{ & 2p_0\left(c^2 - x^2\right)^{\frac{1}{2}} + \frac{2p_1}{\pi}\left[c\left(c^2 - x^2\right)^{\frac{1}{2}} + x^2 \ln \frac{c + \left(c^2 - x^2\right)^{\frac{1}{2}}}{x}\right] \\ & + \frac{p_2}{3}\left(c^2 + 2x^2\right)\left(c^2 - x^2\right)^{\frac{1}{2}} \\ & + \frac{p_3}{\pi}\left[\left(\frac{2}{3}c^3 + cx^2\right)\left(c^2 - x^2\right)^{\frac{1}{2}} + x^4 \ln \frac{c + \left(c^2 - x^2\right)^{\frac{1}{2}}}{x}\right] + \cdots \Bigg\} \end{aligned} \tag{9.19}$$

The normal stress distribution is given by

$$\begin{aligned} \sigma_{yy}(x,0) = {} & p_0\left[x\left(x^2 - c^2\right)^{\frac{-1}{2}} - 1\right] + \frac{2p_1}{\pi}\left[cx\left(x^2 - c^2\right)^{\frac{-1}{2}} - x \arctan\left(c\left(x^2 - c^2\right)^{\frac{-1}{2}}\right)\right] \\ & + p_2\left[\left(\frac{c^2 x}{2} - x^3\right)\left(x^2 - c^2\right)^{\frac{-1}{2}} - x^2\right] \\ & + \frac{2p_3}{\pi}\left[\left(\frac{c^3 x}{3} - cx^3\right)\left(x^2 - c^2\right)^{\frac{-1}{2}} - x^3 \arctan\left(c\left(x^2 - c^2\right)^{\frac{-1}{2}}\right)\right] + \cdots \end{aligned} \tag{9.20}$$

If the higher order coefficients are zero, the above equation is easily simplified to a constant pressure expression in the fracture.

The stress analysis for linear fractures with linear pressure drop in the fracture is as follows. Assume that the half-fracture length $x_f = 10$ m and that the pressure in the fracture is 0.5 MPa at the center of the crack and decreases linearly to zero at the tip of the fracture. Assuming a plane strain state in the horizontal direction, $E = 1$ GPa, $\mu = 0.25$, determine the maximum fracture width.

For the horizontal plane strain problem, $c = x_f = 10$ m. The pressure gradient in the fracture $p_1 = -p_0/x_f = 50$ kPa/m, while $p_2 = p_3 = 0$.

The fracture width is two times the normal displacement and the expression is as follows.

$$w(x) = \frac{4}{E'\pi}\left[(p_0\pi + p_1 c)\left(c^2 - x^2\right)^{\frac{1}{2}} + p_1 x^2 \ln\left(\frac{c + \left(c^2 - x^2\right)^{\frac{1}{2}}}{x}\right)\right]$$

The maximum fracture width is located at the center of the fracture, at $x = 0$.

$$w_0 = \frac{4}{E'\pi}(p_0\pi + p_1 c)c = \frac{4(1 - 0.25^2)}{\pi \times 10^9}(5 \times 10^5 \times \pi - 5 \times 10^4 \times 10)$$
$$= 1.28 \times 10^{-2}\ \text{m} = 12.8\ \text{mm}$$

Figure 9.7 shows the characteristic diagram of the fracture, Fig. 9.6a gives the pressure distribution, Fig. 9.6b gives the $g(\xi)$ distribution, Fig. 9.6c gives the displacement distribution, and Fig. 9.6d gives the stress distribution. Comparing Fig. 9.6 and Fig. 9.7, the displacement and stress distributions are similar.

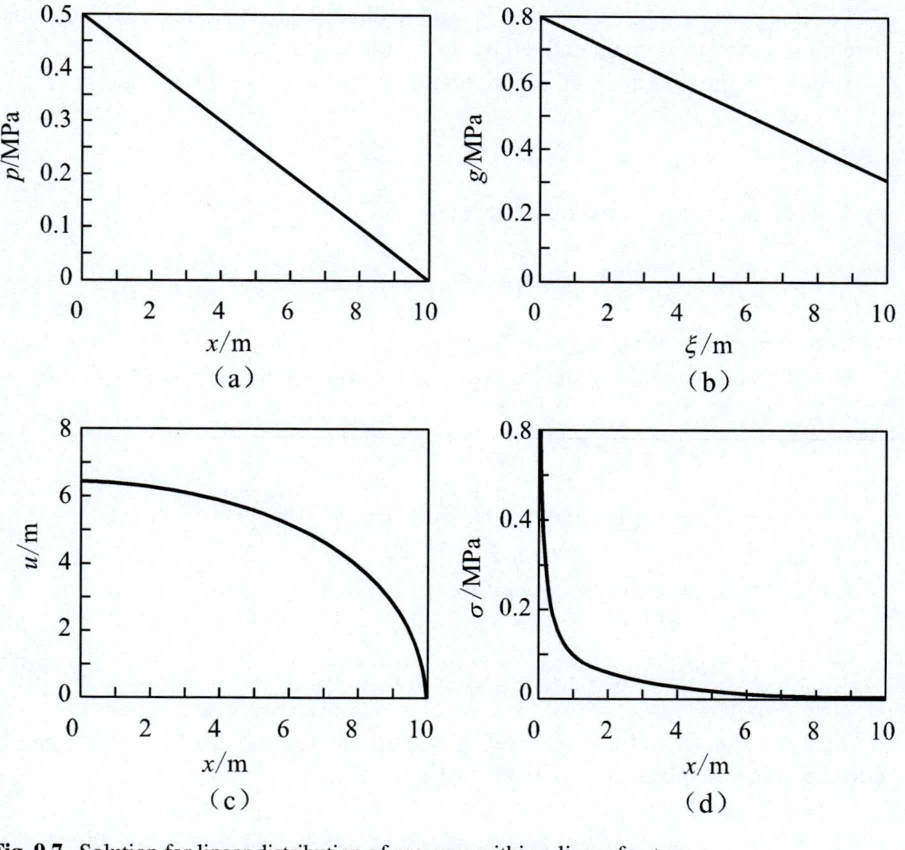


**Fig. 9.7** Solution for linear distribution of pressure within a linear fracture

### 9.2.4 Smooth Closure of Fractures

Assume that from somewhere near the fracture end, $p(x)$ starts to be negative, which corresponds to a tensile force that closes the fracture. This can lead to a smooth closure of the fracture at this point, thus eliminating the stress singularity at the fracture tip (Barenblatt 1962). It can be seen from Eq. (9.12) that in order to avoid the singularity of stress at the crack tip, when $\xi = c$, let $g(\xi) = 0$; then according to Eq. (9.11), the derivative of the displacement with respect to x is 0 and the crack closes smoothly. Thus, for a fracture with only pressure acting, smooth closure is not possible. But if we allow the pressure to be distributed in two segments, one in the middle of the fracture, as pressure, and the other at the tip of the fracture, as tension. The fracture can be closed smoothly. This particular pressure distribution plays a key role in the study of the classical fracture stress field (Barenblatt 1962; Khristianovitch and Zheltov 1955; Zheltov and Khristianovitch 1955; Geertsma and Klerk 1969).

For this purpose a jump function is used to characterize the pressure distribution. The jump function is characterized by three constants.

$s_1$ denotes the pressure in the middle of the fracture,

$$p(\mathrm{x}) = \mathrm{s}_1 \quad x \le x_0$$

$s_2$ denotes the tension at the tip of the crack,

$$p(x) = -s_2 \quad x_0 < x \le x_c$$

The third parameter is the jump position $\mathrm{x}_0$.

Note that all parameters are positive and the corresponding g functions are

$$\begin{aligned} g(\xi) &= \frac{s_1}{2}\pi \qquad x \le x_0 \\ g(\xi) &= \frac{-s_2}{2}\pi + (s_1 + s_2)\arctan\left[x_0\left(\xi^2 - x_0^2\right)^{\frac{-1}{2}}\right] \quad x > x_0 \end{aligned} \tag{9.21}$$

Let the position of the jump point for $g(c) = 0$ be

$$x_0 = c\,\sin\frac{\pi s_2}{2(s_1 + s_2)} \tag{9.22}$$

This condition was first proposed by Khristianovich and Zheltov (1955) and the corresponding displacement is given by Eq. (9.11).

$$u_y(x,0) = \frac{1}{E'\pi}(s_1 + s_2)\left[4x_0\ln\frac{q_1 + q_2}{q_3} + x\ln\frac{c^2x_0^2 - 2q_1q_2x_0x + c^2x^2 - 2x_0^2x^2}{c^2x_0^2 + 2q_1q_2x_0x + c^2x^2 - 2x_0^2x^2}\right]$$

Where $\mathrm{q}_1 = \left(c^2 - x_0^2\right)^{\frac{1}{2}} q_2 = \left(c^2 - x^2\right)^{\frac{1}{2}}$

$$q_3 = \begin{cases} \left(x_0^2 - x^2\right)^{\frac{1}{2}} \; if \; x \le x_0 \\ \left(x^2 - x_0^2\right)^{\frac{1}{2}} \; if \; x > x_0 \end{cases} \tag{9.23}$$

The maximum fracture width at $x = 0$ is two times the displacement, i.e.,

$$\begin{aligned} w_0 &= \frac{8}{E'\pi}(s_1 + s_2)x_0 \ln\left(\frac{c + q_1}{x_0}\right) \\ &= \frac{8(s_1 + s_2)c}{E'\pi} \sin\left[\frac{\pi s_2}{2(s_1 + s_2)}\right] \times \ln \frac{1 + \cos\left(\frac{\pi s_2}{2(s_1+s_2)}\right)}{\sin\left(\frac{\pi s_2}{2(s_1+s_2)}\right)} \end{aligned} \tag{9.24}$$

Finally, the stress at the tip of the fracture is

$$\sigma_{yy}(x, 0) = -s_1 + \frac{2}{\pi}\left[(s_1 + s_2)\arctan \frac{x\left(c^2 - x_0^2\right)^{\frac{1}{2}}}{x_0\left(x^2 - x_0^2\right)^{\frac{1}{2}}}\right] \tag{9.25}$$

The fracture width of smooth closed fracture under the action of jump function is calculated below. Assume that $s_1 = 1$ MPa, half- fracture length $c = 10$ m, $s_2 = 2$ MPa, $E = 9.1$ GPa, and $\mu = 0.3$.

According to Eq. (9.22), $x_0 = 10. \sin(2\pi/2(1 + 2)) = 8.66$ m

$$E' = \frac{1}{1 - \mu^2}E = 10\,\text{Pa}$$

According to Eq. (9.23), we get $w_0 = 3.63$ mm.

The pressure distribution and the crack size and stress distribution inside the crack are shown in Fig. 9.8, from which it can be seen that the assumptions made are a good solution to the problem of stress singularity at the crack tip.

### 9.2.5 *Shape of Fractures and Net Pressure Concept Under In-Situ Stress Conditions*

The stress distribution and displacement distribution of internally pressure fractures in the absence of far-field stress conditions were analyzed in the previous section. How can these results be applied to the actual fracturing analysis in the presence of in situ stress conditions? The combination of fractures and far-field stresses will be discussed below. Since this problem is quite complex, only one linear fracture parallel to the maximum principal stress direction is considered here, as shown in Fig. 9.9. Similar to the analysis above, since the material is linearly elastic, consistent with the stress superposition principle, the fracture boundary coincidence is a linear superposition of the in-fracture pressure and the far-field stress. The difference

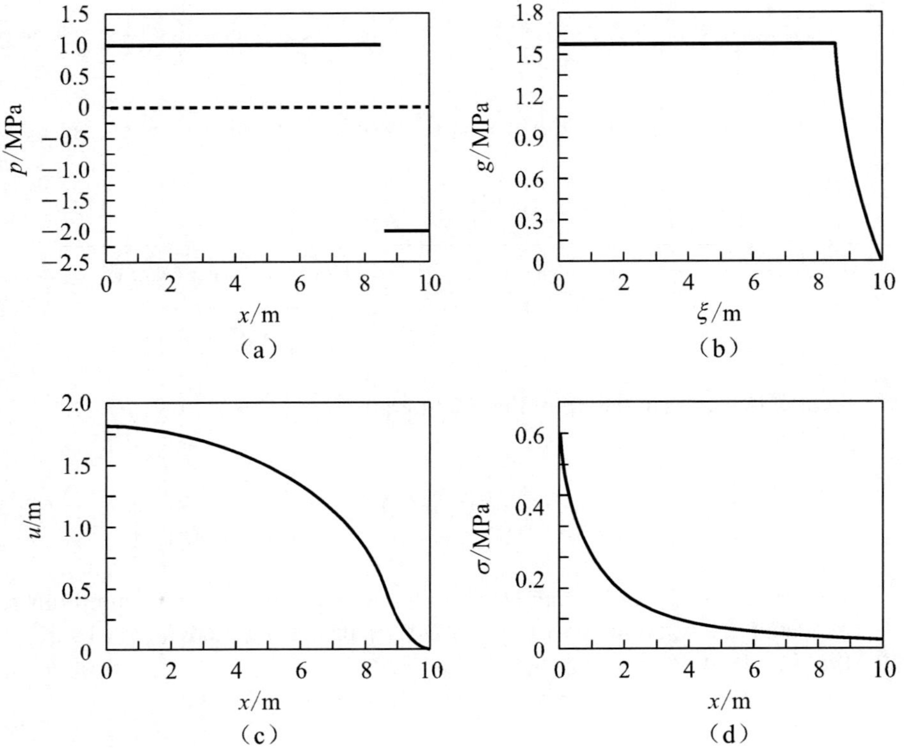


**Fig. 9.8** Solution of smooth closed fractures

between these two forces is important in hydraulic fracturing analysis and is referred to as the net pressure $p_n = p - \sigma_h$. The corresponding solution in the presence of in situ stress can be obtained using $p_n$ in place of the in-fracture pressure P.

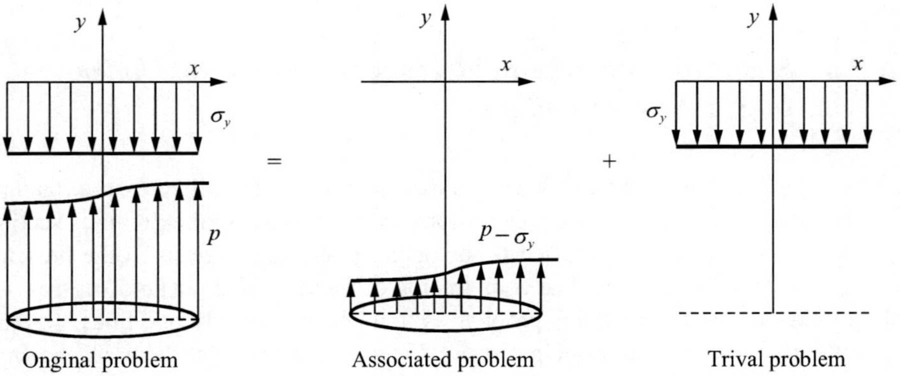


**Fig. 9.9** Internal pressure cracks under far-field stress

Below, the fracture width of smooth closed crack under original in-situ stress is calculated. Assume average internal pressure of the fracture $\overline{p} = 3$ MPa, minimum horizontal in-situ stress in the formation $S_{min} = \sigma_y = 2$ MPa, and fracture half-fracture length $x_f = 10$ m. At $x_0 < |x| \le x_L$, the pressure is zero. $e = 9.1$ GPa, $\mu = 0.3$.

Using the solution to smooth closed fracture under the action of jump function according to the superposition principle, the $s_1 = \overline{p} - \sigma_{yy} = 1$ MPa, $s_2 = \sigma_{yy} = 2$ MPa, $c = x_f = 10$ m, and the jump point $x_0$ is given by Example 9.2.2, the

$$x_0 = 10 \times \sin\left[\frac{\pi \times s_2}{2(s_1 + s_2)}\right] = 10 \times \sin\left[\frac{\pi \times 2}{2 \times (1 + 2)}\right] = 8.66\,\text{m}$$

The crack width is determined by (9.23).

$$w(x) = \frac{2}{E'\pi}\overline{p}\left[4x_0 \ln\frac{q_1 + q_2}{q_3} + x \ln\frac{x_f^2 x_0^2 - 2q_1 q_2 x_0 x + x_f^2 x^2 - 2x_0^2 x^2}{x_f^2 x_0^2 + 2q_1 q_2 x_0 x + x_f^2 x^2 - x_0^2 x^2}\right]$$

Of which

$$q_1 = \left(x_f^2 - x_0^2\right)^{\frac{1}{2}}$$

$$q_2 = \left(x_f^2 - x^2\right)^{\frac{1}{2}}$$

$$q_3 = \begin{cases} \left(x_0^2 - x^2\right)^{\frac{1}{2}}, \; if \;\; x \le x_0 \\ \left(x^2 - x_0^2\right)^{\frac{1}{2}}, \; if \; x > x_0 \end{cases}$$

The fracture width can be found by taking the parameters into account:

$$w(x) = \frac{6 \times 10^{-4}}{\pi}\left[4 \times 8.66 \ln\frac{5 + \sqrt{100 - x^2}}{\sqrt{75 - x^2}} + x \ln\frac{150 - x^2 - x\sqrt{3(100 - x^2)}}{150 - x^2 + x\sqrt{3(100 - x^2)}}\right]$$

$0 \le x \le x_0$

$$w(x) = \frac{6 \times 10^{-4}}{\pi}\left[4 \times 8.66 \ln\frac{5 + \sqrt{100 - x^2}}{\sqrt{x^2 - 75}} + x \ln\frac{150 - x^2 - x\sqrt{3(100 - x^2)}}{150 - x^2 + x\sqrt{3(100 - x^2)}}\right]$$

$x_0 \le x \le x_f$

The stress distribution is the same as in solution of smooth closed fracture, which considers the in situ stress as a net pressure.

The fracture width at $x = 0$ is the maximum fracture width, i.e.:

$$w_0 = \frac{8}{E\prime\pi}(s_1 + s_2)x_0 \ln\left(\frac{c + q_1}{x_0}\right)$$

$$= \frac{8(s_1 + s_2)c}{E'\pi} \sin\left[\frac{\pi s_2}{2(s_1 + s_2)}\right] \times \ln \frac{1 + \cos\left(\frac{\pi s_2}{2(s_1+s_2)}\right)}{\sin\left(\frac{\pi s_2}{2(s_1+s_2)}\right)} = 3.63 \text{ mm}$$

If the real effect of in situ stress is considered, Geertsma and de Klerk give the following equation. for the seam width.

$$w(x) = \frac{4}{E'\pi}\overline{p}\left[x \ln \frac{\left|1 - \frac{x_0 q_4}{x}\right|}{1 + \frac{x_0 q_4}{x}} - x_0 \ln \frac{|1 - q_4|}{1 + q_4}\right] \tag{9.26}$$

Of which

$$q_4 = \left(\frac{x_f^2 - x^2}{x_f^2 - x_0^2}\right)^{1/2}$$

As $x_0$ converges to $x_f$, Geertsma and de Klerk give the maximum seam width formula

$$w_0 = \frac{8}{\pi E'}\overline{p}\frac{x_0}{x_f\sqrt{x_f^2 - x_0^2}} \tag{9.27}$$

### 9.2.6 Circular Cracks

The mathematical treatment of a circular crack is similar to that of a linear crack, assuming that the fracture lies in the xy-plane and is open along the z-direction, with the boundary conditions and schematic shown in Fig. 9.10. The surface of the fracture is circular and the boundary conditions at $z = 0$ are as follows.

$$\begin{cases} \sigma_{zz} = -p(r), \ 0 \le r \le R \\ u_z = 0, \qquad r > R \\ \sigma_{xz} = 0, \qquad r \ge 0 \end{cases} \tag{9.28}$$

Condition 1 indicates that the internal pressure acts on the crack surface with equal stresses and that the internal pressure is a function of position $r$.

The first step in the solution is to construct the $g$-function. Similar to the linear fracture, the pressure function:

$$G(\xi) = \frac{2}{\pi}\int_0^{\xi} \frac{rp(\xi)dr}{(\xi^2 - r^2)^{1/2}} \quad 0 < \xi < R \tag{9.29}$$

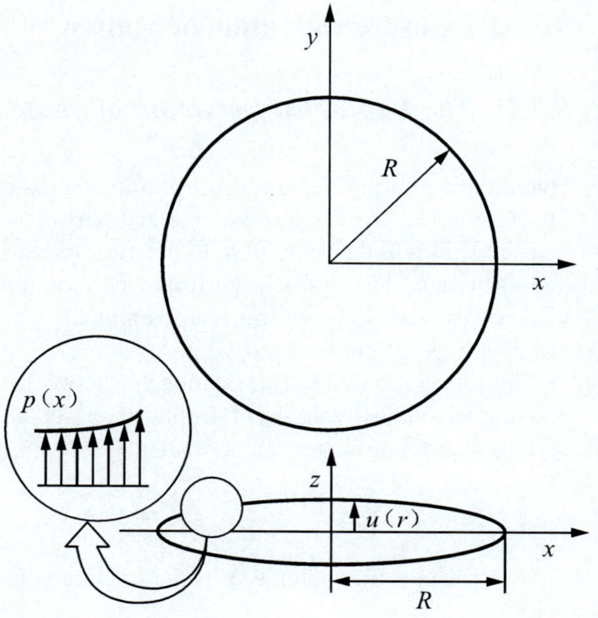


**Fig. 9.10** Internal pressure circular cracks

The G function is known and the corresponding displacement function is

$$u_z(r, 0) = \frac{2}{E'} \int_c^R \frac{\xi g(\xi) d\xi}{(\xi^2 - r^2)^{1/2}} \quad 0 \le r \le R \tag{9.30}$$

$G(R) = 0$ for a smooth closed fracture.

The fracture width is twice the displacement, and for the constant internal pressure case (Green and Zerna 1968), there are.

$$G(\xi) = p_0 \xi \tag{9.31}$$

$$w(r) = \frac{8}{\pi E'} p_0 \sqrt{R^2 - r^2} \tag{9.32}$$

Its maximum fracture width is

$$w_0 = \frac{8 R p_0}{\pi E'} \tag{9.33}$$

## 9.3 Law of Conservation of Matter

### 9.3.1 The Law of Conservation of Matter

Hydraulic fracturing requires the injection of a large volume of fracturing fluid, part of which forms the fracture volume and part of which is filtered out through the fracture wall into the formation. In practice, fracturing fluids are injected in stages, and each stage has different performance requirements. For proppant fracturing, there are two categories of fronts and sand-carrying fluids, with sand-carrying fluids carrying large amounts of proppant.

This section establishes the relationship between injected volume, fracture volume and filtrate volume by means of the law of conservation of matter, the most basic law of conservation of matter can be written as.

$$V_\mathrm{i} = V + V_\mathrm{L} \tag{9.34}$$

$V_\mathrm{i}$ is the injection volume, $V$ is the fracture volume, and $V_\mathrm{L}$ is the filtration loss volume.

The above equation ignores the compressibility of the fluid and the difference between the fracture volume and the fluid volume inside the fracture (there may be non-wetting regions at the fracture tip). For convenience, the parameters associated with a single winged fracture are used below.

The fracture-making efficiency can be expressed as.

$$\eta = \frac{V}{V_i} \tag{9.35}$$

The volume of the fracture is.

$$V = A\overline{w} \tag{9.36}$$

$A$ is the area of the single wing; $\overline{w}$ is the average fracture width, defined as the crack volume divided by the fracture face area, or the single wing volume divided by the single wing area.

Let the constant injection rate be $2q$ (the monoflanking fracture injection rate is $q$), then the amount of monoflanking fracture injected after injection time $t$ is $V_i = qt$.

Thus Eq. (9.34) is rewritten as

$$qt = A\overline{w} + V_L \tag{9.37}$$

The fracture-making efficiency is.

$$\eta = \frac{A\overline{w}}{qt} \tag{9.38}$$

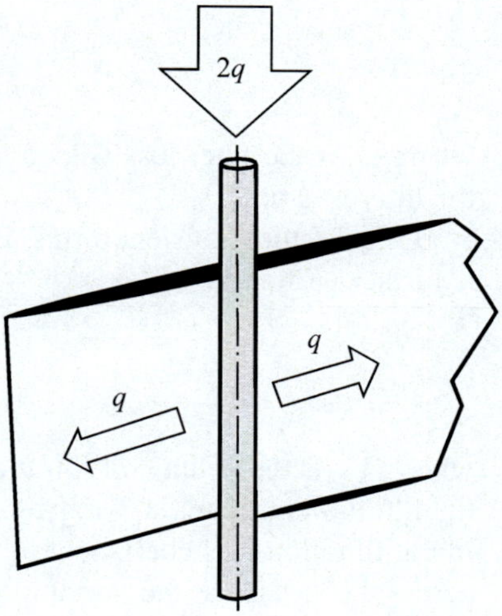


**Fig. 9.11** Fracture morphology and injection rate distribution

Assuming a constant fracture height, the half fracture length is expressed as

$$x_f = \frac{A}{h_f} \tag{9.39}$$

This allows a convenient representation of the half fracture length $x_f$, fracture height $h_f$ and average fracture width $\overline{w}$ at any pumping moment t or end of pumping moment $t_e$. These parameters and their interrelationships are central to the hydraulic fracturing effort, and Eqs. (9.34)–(9.39) also makes it easy to discern the reasonableness of the numerical results when designing for fracturing (Fig. 9.11).

## 9.3.2 Fluid Filtration Loss and Initial Filtration Loss

With respect to fracturing filtration loss, there are two analytical approaches, one that considers the material properties of the fluid/rock system and the other that considers the flow of fluid into the porous medium.

### 9.3.2.1 Carter Filtration Test

During the flow, the polymer in the fracturing fluid gradually forms a mud cake at the fracture wall, which hinders further fluid inflow and the filtration loss rate gradually decreases. Filtration loss is a complex process determined by a multi-factor coupled system and is analyzed based on experimental phenomena using the Carter equation (Howard and Fast 1957). The rate of filtration loss is defined as follows.

$$v_L = \frac{C_L}{\sqrt{t}} \tag{9.40}$$

where $v_L$ is the filter loss velocity, m/s; $C_L$ is the filter loss coefficient, m/s$^{1/2}$; $t$ is the filter loss time, s.

If the fracture transient forms with a constant rate of filtration loss, the above Eq. is further written as

$$\frac{V_L}{A_L} = 2C_L\sqrt{t} + S_P \tag{9.41}$$

where, $V_L$ is the volume of filtration loss generated from 0 to $t$ time period through the filtration loss area $A_L$ (regarded as the area of the fracture monoflap). $S_p$ is the initial filtration loss coefficient in m, and is the volume of fluid flowing transiently across the surface at the initial moment of filtration loss at the wall. $c_L$, $S_p$ both of which can be determined from indoor experiments.

Typical experiments for the determination of filtration coefficient are given below. The fracturing fluid is applied to the core cross section at typical pressure, and the experimental data are shown in the Table 9.2.

First find the square root of time, then find the filter loss velocity ($V_L/A_L$), and draw Fig. 9.12. The slope m of Fig. 9.12 is 6.9 × $10^{-5}$ m/s$^{1/2}$, and the intercept $b$ = 2.4 × $10^{-3}$ m.

Comparing this with Eq. (9.41) gives

$$C_L = \frac{m}{2} = 3.5 \times 10^{-5} \text{m/s}^{\frac{1}{2}}$$

$$Sp = b = 2.4 \times 10^{-3}\,\text{m}$$

**Table 9.2** Results of filter loss experiments

| Filter loss time t/min | Filter loss volume $V_L$/cm$^3$ |
|---|---|
| 1 | 5.2 |
| 2 | 6.7 |
| 4 | 7.3 |
| 10 | 8.6 |
| 20 | 9.7 |
| 30 | 10.6 |
| 40 | 11.4 |
| 50 | 12.5 |
| 60 | 13.2 |

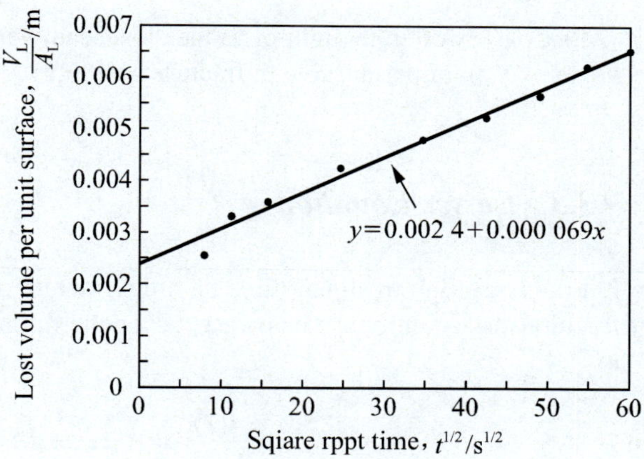


**Fig. 9.12** Analysis of the results of the filter loss experiment

#### 9.3.2.2 Cracking Time Allocation Factor

The treatment time for hydraulic fracturing varies from a few minutes to several hours. The fracture surfaces near the borehole were in tension throughout the treatment time, while the fracture tip surfaces were in tension for a short period of time. When applying Eq. (9.40) and Eq. (9.41), it is necessary to record the time of opening at different locations on the fracture wall surface. Neglecting the initial filtration loss and rewriting Eq. (9.34) for conservation of matter using Eq. (9.41), we obtain

$$V_i = V + \kappa\left(2AC_L\sqrt{t}\right) \tag{9.42}$$

$\kappa$ is the time allocation factor, and the expression is:

$$\kappa = \frac{V_L}{2AC_L\sqrt{t}} \tag{9.43}$$

Its value depends on the fracture evolution process or the moment of fracture opening. In particular, the value of $\kappa$ is maximum when there is no initial filtration loss, and the transient opening of the fracture wall (full opening at the moment of initiation), $\kappa$ is equal to 2.

Based on the previous definition, the time allocation factor $\kappa$, seam making efficiency and average seam width are related as follows.

$$\kappa = \frac{(1-\eta)\overline{W}}{2\eta C_L\sqrt{t}} \tag{9.44}$$

tantamount to

$$\eta = \frac{\overline{W}}{2\kappa C_L\sqrt{t} + \overline{W}} \tag{9.45}$$

$2\kappa C_L\sqrt{t}$ is often thought of as the "lost seam width". The time allocation factor $\kappa$ can play an important role in fracturing design.

### 9.3.3 Carter Equation

Let the fracturing treatment time be $t$, the fracture opening moment be $\tau$, and each fracture surface unit have its own $\tau$. Given the filtration loss rate of the surface unit as

$$\frac{\partial V_L}{\partial t} = \frac{C_L}{\sqrt{t-\tau}}\partial A \tag{9.46}$$

Suppose that the fracture surface is known to have an open history $A(\tau)$, or $\tau$ $(A)$. Then the rate of filtration loss through the two wall surfaces is

$$2\sum_{0}^{A(t)} \frac{C_L}{\sqrt{t-\tau(A)}}\mathrm{d}A = 2\int_{0}^{t} \frac{C_L}{\sqrt{t-\tau}}\left(\frac{\mathrm{d}A}{\mathrm{d}\tau}\right)\mathrm{d}\tau \tag{9.47}$$

The fracture volume growth rate is expressed as.

$$w\frac{\mathrm{d}A}{\mathrm{d}t} + A\frac{\mathrm{d}w}{\mathrm{d}t} \tag{9.48}$$

From the initial filtration loss rate of

$$2S_\mathrm{p}\frac{\mathrm{d}A}{\mathrm{d}t} \tag{9.49}$$

Carter gave an equation for the material balance determined from the flow rate. He argues that at moment $t$, the monoflank injection rate is $q$, which should be equal to the rate of each filtration loss plus the rate of fracture volume growth, i.e.,

$$q = 2\int_{0}^{\tau} \frac{C_L}{\sqrt{t-\tau}}\frac{\mathrm{d}A}{\mathrm{d}\tau}\mathrm{d}\tau + (W + 2S_p)\frac{\mathrm{d}A}{\mathrm{d}t} + A\frac{\mathrm{d}W}{\mathrm{d}t} \tag{9.50}$$

For the constant injection rate condition, Carter simplified the material balance equation by neglecting the initial filtration loss and considering that the fracture width at the start of fracturing fluid injection is equal to the width at the last moment of the end of treatment, i.e., $\frac{\mathrm{d}w}{\mathrm{d}t} = 0$. An approximate solution to the above equation can be obtained as follows.

$$A(t) = \frac{(W + 2S_P)q}{4C_L^2\pi}\left[\exp(\beta^2)\mathrm{erfc}(\beta) + \frac{2\beta}{\sqrt{\pi}} - 1\right] \tag{9.51}$$

Of which

$$\beta = \frac{2C_L\sqrt{\pi t}}{W + 2S_p} \tag{9.52}$$

Equation (9.51) gives the surface pattern of fractures for a given fracture width and treatment time, and is called Carter's Equation.

### 9.3.4 Approximation of the Power-Law Growth of the Fracture Surface Area with Treatment Time

If, in a double logarithmic coordinate system, the plot from Carter's equation. the curve of the fracture area calculated as a function of time is shown in Fig. 9.13. It can be seen that at the beginning the slope of the curve is 1 and later becomes 1/2.

Based on this relationship, Nolte (1986) assumes that for constant displacement injection, fracture area growth conforms to the power law equation.

$$\frac{A}{A_e} = \left(\frac{t}{t_e}\right)^{\alpha} \tag{9.53}$$

where the index $\alpha$ is kept constant.

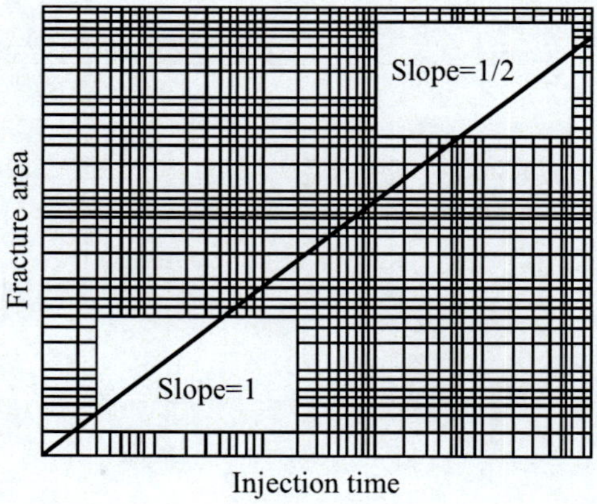


**Fig. 9.13** Double logarithmic relationship of crack surface area with construction time

#### 9.3.4.1 Results Derived from a Power-Law Increase in Fracture Surface Area with Treatment Time

Nolte (1986, 1989) further assumes that the initial filtration loss is 0. Introduce a new function.

$$g_0(\alpha) = \frac{V_{Le}}{2C_L A_e \sqrt{t_e}} \tag{9.54}$$

The subscript e indicates the moment when the pump is stopped.

To obtain a closed solution of the function $g_0(\alpha)$, consider a cell surface d$A$ with the moment of opening $\tau$, the moment of end of treatment $t_\mathrm{e}$ and the filtered volume of the cell d$A$ during treatment as

$$\mathrm{d}V_\mathrm{L} = \mathrm{d}A \int_\tau^{t_e} \frac{C_\mathrm{L}}{\sqrt{t-\tau}} \mathrm{d}t \tag{9.55}$$

The surface area of the fracture increases from 0 to $A_\mathrm{e}$ and the filtered volume of the fluid is the integral of Eq. (9.55) along the entire surface.

$$V_{Le} = 2 \int_0^{A_e} \int_\tau^{t_e} \frac{C_L}{\sqrt{t-\tau}} dt dA \tag{9.56}$$

Bring Eq. (9.56) into Eq. (9.54) yields.

$$g_0(\alpha) = \frac{1}{A_e} \int_0^{A_e} \frac{1}{\sqrt{t_e}} \left( \int_\tau^{t_e} \frac{C_L}{\sqrt{t-\tau}} dt \right) dA \tag{9.57}$$

According to Eq. (9.53), the moment of fracture opening is

$$\tau = t_e \left( \frac{A}{A_e} \right)^{\frac{1}{\alpha}} \tag{9.58}$$

Dimensionless each variable, $t_\mathrm{D} = \frac{t}{t_e}$, $A_\mathrm{D} = \frac{A}{A_e}$, $\tau_\mathrm{D} = \frac{\tau}{t_e}$, Eq. (9.58) becomes,

$$\tau_D = A_D^{\frac{1}{\alpha}} \tag{9.59}$$

Equation (9.57) transforms to

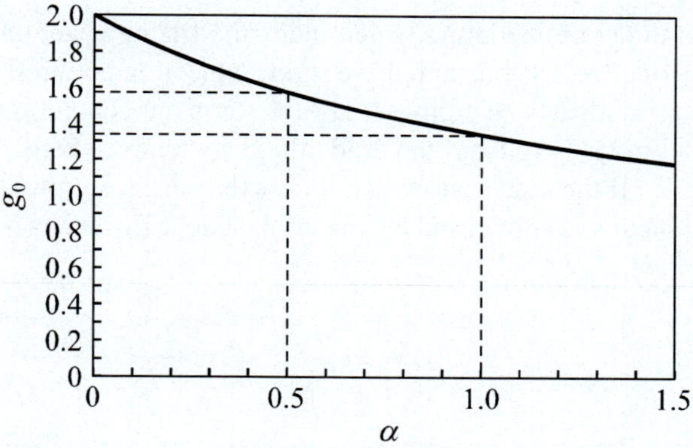


**Fig. 9.14** $g_0$ ($\alpha$) function curve

$$g_0(\alpha) = \int_0^1 \left( \int_{A_D^{1/\alpha}}^1 \frac{1}{\sqrt{t_D - A_D^{1/\alpha}}} \mathrm{d}t_D \right) \mathrm{d}A_D \tag{9.60}$$

The points give.

$$g_0(\alpha) = \frac{\alpha\sqrt{\pi}\Gamma(\alpha)}{\Gamma(3/2 + \alpha)} \tag{9.61}$$

where $\Gamma(\alpha)$ is the Eulerian gamma function, and the curve for the function $g_0$ ($\alpha$) is given in Fig. 9.14. Surprisingly, the function at $\alpha = 0$ is equal to 2, i.e., the function $g_0$ ($\alpha$) is the maximum of the coefficient $\kappa$. Indeed, if the exponent $\alpha$ is 0, the entire fracture surface is fully opened at the beginning of fracturing and the volume of filtration loss is maximum.

Due to the special nature of the Eulerian gamma function, its alternative form (Valko and Economides 1993) is

$$g_0(\alpha) = \frac{2^{2\alpha}\alpha[\Gamma(\alpha)]^2}{(1 + 2\alpha)\Gamma(2\alpha)} \tag{9.62}$$

Two points should be emphasized.

- The function $g_0$ ($\alpha$) is simply a function of the index $\alpha$.
- $g_0\left(\frac{1}{2}\right) = \frac{\pi}{2} \cong 1.57$ and $g_0(1) = \frac{4}{3} \cong 1.33$ (9.63)

It can be seen that $\alpha = 0.5$ and 1, the calculated change in the opening distribution factor is less than 20%. Equation (9.63) is commonly used as the upper and lower limit values for the calculation. Because Nolte and many other researchers have concluded that there is a strong correlation between $\alpha$ and $\eta$,, and maximum filtration loss ($\eta \cong 0$), $\alpha = 0.5$; no filtration loss ($\eta \cong 1$), $\alpha = 1$. In practice, it is difficult to determine along which index the fractures are tensioned. However, we note that once $\alpha$ is a constant, there is no connection between $\alpha$ and $\eta$, since $\alpha$ is constant and $\eta$ is always changing. It always starts out equal to 1 and then drops. If the pumping time is very long, the fluid efficiency tends to zero.

If the assumption that the fracture surface grows by a power law holds, the exponent $\alpha$ is known and the material balance Eq. at any point during pumping is rewritten as

$$\overline{W} = \frac{qt_e^{\alpha}}{A_e} t^{1-\alpha} = \frac{2C_L\alpha\sqrt{\pi}\Gamma(\alpha)}{\Gamma(3/2+\alpha)}\sqrt{t} - 2S_p \tag{9.64}$$

The average fracture width at the end of treatment was.

$$\overline{w}_e = \frac{qt_e}{A_e} - \frac{2C_L\alpha\sqrt{\pi}\Gamma(\alpha)}{\Gamma\left(\frac{3}{2}+\alpha\right)} - 2S_p \tag{9.65}$$

Many researchers have noted that the exponent $\alpha$ is related to the rheology of the fluid. Despite this, formula (9.64) has never been used in design, probably because its analytical form (Eq. 9.62) is unknown. In contrast, the difference technique given below is widely used.

#### 9.3.4.2 A Combination of the Power Law Growth Hypothesis and the Difference Method

Returning to the formal material balance Eq. (9.42), at constant displacement injection, the Eq. is rewritten as follows.

$$\frac{qt}{A} = \overline{w} + C_l\sqrt{t}(2\kappa) \tag{9.66}$$

For the last term, a different approximation can be used for the treatment.

Assumption (1):

$$\begin{cases} \kappa = \frac{\pi}{2}, \eta \to 1 \\ \kappa = \frac{4}{3}, \eta \to 0 \end{cases} \tag{9.67}$$

Assumption (2): the $\kappa$ value varies linearly within the upper and lower bounds and passes the final $\eta$ value.

Then the overall material balance equation is.

$$\frac{qt}{A} = \overline{W} + C_L\sqrt{t}\left[\frac{8}{3}\eta + (1-\eta)\pi\right] \tag{9.68}$$

Again, we note: it is difficult to accept this assumption because the exponent α is considered to be constant during fracture extention, while the fluid efficiency decreases monotonically. Indeed, if the exponent $\alpha$ is determined by the final value of fluid efficiency, which is related to the length of treatment time, how can the surface exponent be determined? However Eq. (9.68) is widely used in fracturing design (Meng 1989). In software design mode it is a useful Eq. when the fracture surface is specified and the fracture width is calculated to exceed the material balance. A quadratic Eq. will be obtained, from which the pumping time is derived. Because a reasonable value for fluid efficiency is included, Eq. (9.68) is combined with Eq. (9.38) and embedded into the difference cycle.

### 9.3.5 *Numerical Methods for Equilibrium Equations of Matter*

If the filtering loss is consistent with the Carter test result, the fracture width reaches its final value at the outset and Eq. (9.51) is strictly valid. Because the assumption that the fracture width reaches its final value at the beginning is incorrect, it is important to ask at this point, can Eq. (9.51) still be used as a constraint equation for the surface expansion of the fracture?

Assuming a constant injection rate, constant step size discrete construction time, current time denoted by the subscript n ($t_n = n\Delta t$), and allowing the fracture surface and average fracture width to vary with treatment time, the finite difference form of Eq. (9.50) is

$$q\Delta t = 2C_L\sqrt{\Delta t}\sum_{j=1}^{n}\frac{A_j - A_{j-1}}{\sqrt{n-j+\xi}} + 2S_p(A_n - A_{n-1}) + A_n w_n - A_{n-1}w_{n-1} \tag{9.69}$$

The constant $\xi$ is introduced with a value between 0 and 1. If $\xi = 1$, which means that the calculation is done using the filter loss rate at the end of the time step, the result may be low; if $\xi = 0.5$, which means that the calculation is done using the filter loss rate at the middle of the time step, the result is just right; the optimal choice is $\xi = 0.25$, because the actual filter loss rate is equal to the average filter loss rate of the first time step. The basic equation used in the numerical calculation is as follows.

$$A_n = \frac{q\Delta t + (2S_p + 4C_L\sqrt{\Delta t} - W_{n-1})A_{n-1} - 2C_L\sqrt{\Delta t}\sum_{j=1}^{n-1}\frac{A_j - A_{j-1}}{\sqrt{n-j+0.25}}}{W_n + 2S_p + 4C_L\sqrt{\Delta t}} \tag{9.70}$$

Here, $A_0 = 0$ and the law of growth of the fracture width must be known, it may be assumed that

$$W_j = W_e \left( \frac{j \Delta t}{t_e} \right)^{\varphi} = W_e \left( \frac{j}{n_e} \right)^{\varphi} \tag{9.71}$$

The solution is deterministic if $\phi = 0$, which is the constant fracture width case.

## 9.4 PKN Model and KGD Model

Models for hydraulic fracturing engineering applications require joint elastic mechanics, fluid mechanics and material balance equations, and other extended criteria. Under the conditions of a given fluid pumping procedure, fracturing models can predict the fracture geometry and the pattern of borehole pressure changes with treatment time. From a design perspective, a rough estimate of fracture geometry is sufficient, so those simplified geometric models that can predict fracture length and average width at the end of pumping are very useful. From a production perspective, the fracture length is a very important parameter; followed by the seam width to determine the inflow capacity. The model has a fixed fracture height in predicting these two parameters and is called a 2D model. For circular fractures, the fracture height is not fixed and is also referred to as a 2D model.

The PKN model and the KGD model are two classical 2D models of hydraulic fracturing design with constant fracture height. Both 2D models implicitly assume plane strain. For the PKN model it is the vertical plane strain model (Geertsma and Klerk 1969) and for the KGD model it is the horizontal plane strain model (Geertsma 1989). For vertical planes, parallel vertical planes are allowed to slide against each other if the seam length is greater than the seam height. This model was originally proposed by Perkins and Kern in 1961, and was later developed and refined by Nordgren, hence the name PKN model (1972). For the horizontal plane, the plane strain assumption represents an infinite number of horizontal planes perpendicular to the seam height and is used to describe short and tall cracks. This model was originally proposed by Khristianovic and Zheltov (1955), and was further investigated by Geertsma and de Klerk, hence the name KGD model.

Figure 9.15 depicts the PKN model for a fracture that is elliptical in both the horizontal and vertical directions. In both directions the fracture width is much smaller than the dimensions in the other direction, the fracture height is constant and the fracture length is large. This may differ somewhat from the actual fracture geometry.

An important characteristic parameter for fracture design is the average fracture width, defined as the fracture volume divided by the fracture face area, or the single flank volume divided by the single flank area. The maximum fracture width at the borehole is related to the average fracture width as follows.

$$\overline{w} = \left( \frac{\pi}{4} \right)^2 w_0 \tag{9.72}$$

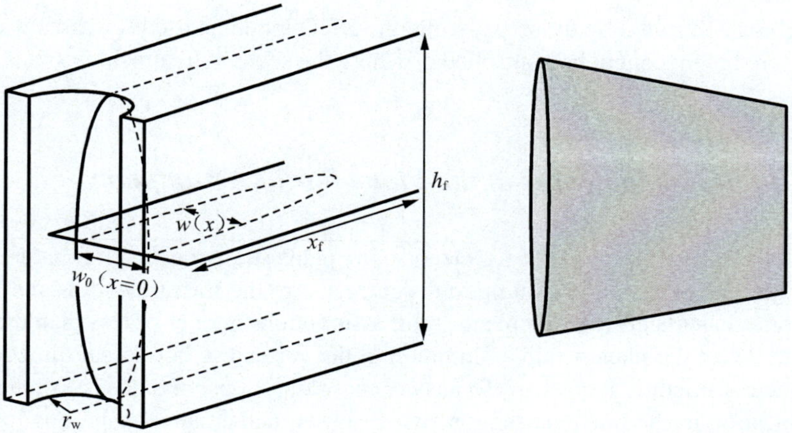


**Fig. 9.15** PKN model

where, $\left(\frac{\pi}{4}\right)^2$ is defined as a shape factor to characterize the relationship between the average fracture width and the maximum fracture width, usually taken as $3\pi/16$.

For the KGD model, as shown in Fig. 9.16, the cross section is rectangular and the fracture height is constant. The maximum fracture width on the well wall is related to the average fracture width as

$$\overline{w} = \frac{\pi}{4} w_0 \tag{9.73}$$

The PKN and KGD models are widely used in engineering design and are consistent with the assumptions of the Gringarten & ramey and Cinco-Ley & Samaniego capacity models. The PKN model is generally used for fracturing where the fracture

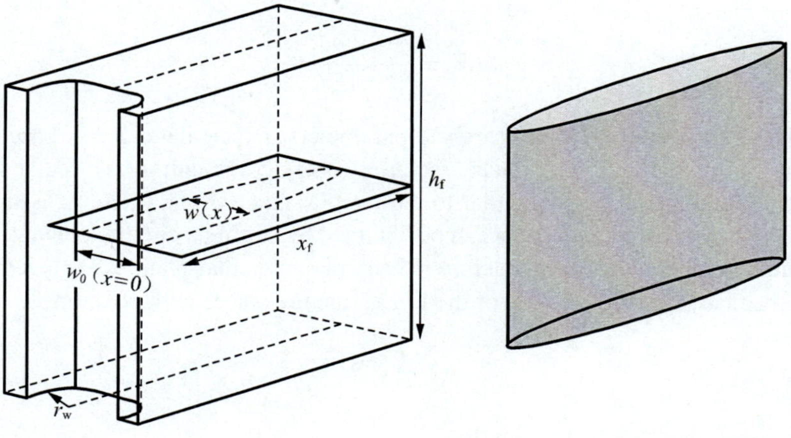


**Fig. 9.16** KGD model

height is controlled by lithology, while the KGD model is used for cases where the fracture height cannot be controlled or where the scale of treatment is small.

### 9.4.1 Reasonableness of the Plane Strain Assumption

Both the PKN model and the KGD model are plane strain models. It is natural to ask whether the plane strain assumption is correct. Are the fracture widths and volume energies calculated with the plane strain assumptions greater or less than the actual values? Can the plane strain assumption in the vertical or horizontal direction give the actual fracture geometry? To answer these questions, one needs to refer to the description in the previous section (which gives mathematical solutions for linear and circular fractures). The following analytical approach is theoretical in nature, but its conclusions are important for practical purposes.

We consider a circular fracture of radius R located at the center of a vertical plane with constant internal pressure $p_{n0}$. According to the previous theory, the theoretical solutions for the fracture width, volume, average crack width, and elastic energy are

$$w(r) = \frac{8}{\pi E'} p_{n0} \sqrt{R^2 - r^2} \tag{9.74}$$

$$w_0 = \frac{8 R p_{n0}}{\pi E'} \tag{9.75}$$

$$V = \frac{16 R^3 p_{n0}}{3E'} \tag{9.76}$$

$$\overline{w} = \frac{16 R p_{n0}}{3\pi E'} \tag{9.77}$$

$$W_0 = \frac{16 p^2 p_{n0} R^3}{3E'} \tag{9.78}$$

The approximate solutions for these parameters are calculated below based on the plane strain assumption. A linear fracture is obtained by cutting a circular fracture with a plumb plane perpendicular to the circular plane of the fracture, as shown in Fig. 9.17. A set of linear cracks can be obtained by cutting a circular plane with a set of plumb planes. The linear fracture in each perpendicular plane is analyzed below, and the characteristic length $c$ of the linear fracture varies with position $x$.

$$c(x) = \sqrt{R^2 - x^2} \tag{9.79}$$

According to Eq. (9.15), it follows that

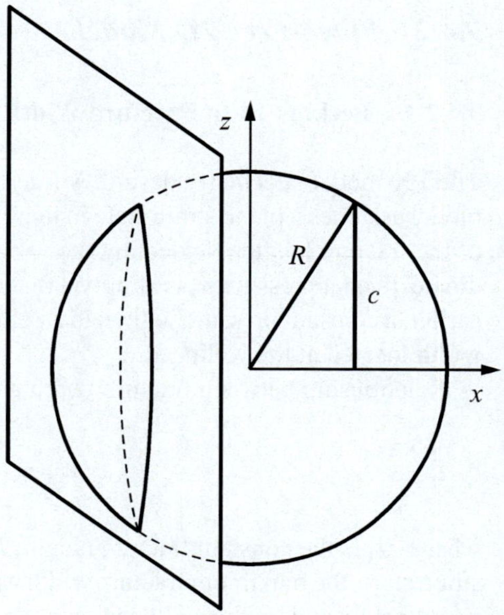


**Fig. 9.17** Plane strain assumption for circular cracks

$$w(x, z) = \frac{4}{E'} p_{n0} \sqrt{R^2 - x^2 - z^2} \tag{9.80}$$

$$r^2 = x^2 + z^2$$

Comparing the two fracture width Eqs. (9.74) and (9.80) above shows that the latter is $\frac{\pi}{2}$ times larger than the former, with an error of about 57% in the plane strain assumption results. The approximate solutions for the maximum fracture width and volume as well as the average fracture width are also larger than the exact solution. The reason for this is that the plane strain model only considers the limiting effect of the crack tip in the $z$-direction and ignores the limiting effect in the $x$-direction, resulting in large calculated fracture widths, while the linear elastic theory solution considers the limiting effect in all directions and has small calculated values.

It is clear that the plane strain model will overestimate the fracture width. If the length perpendicular to the plane strain plane direction is larger, the calculation error is smaller. In particular, for the KGD model, the fractures of short length and high height are small errors, and for the PKN model the fracture of long length and short height are small errors.

## 9.4.2 *Filter-Free 2D Model*

### 9.4.2.1 Perkins-Kern Fracture Width Formula

The PK model succinctly describes fracture extension, as shown in Fig. 9.15. The model assumes a plane strain state in the vertical plane perpendicular to the direction of the fracture length. Neglecting the variation of fluid pressure in the vertical coordinate, the net pressure $p_n$ is simply a function of the fracture length $x$. As mentioned earlier, a constant pressure will yield an elliptical fracture with the maximum fracture width located at the wellhead.

Relationship between fracture width and net pressure is.

$$W_0 = \frac{2h_f P_n}{E'} \tag{9.81}$$

where, $H_f$ is the constant fracture height; $E'$ is the modulus of elasticity in the plane strain state; the maximum fracture width $w_0$ for any section is related to the transverse coordinate $x$ and, at the wellhead, is $w_{w,0}$.

For a Newtonian fluid, the pressure drop in an elliptical fracture flowing in the direction of the fracture length is

$$\frac{\Delta p}{L} = \frac{64\mu q}{\pi W_0^3 h_f} \tag{9.82}$$

where, $\mu$ is the fluid viscosity and $q$ is the flow rate of the single wing fracture.

The PK model assumes a constant single-wing injection rate of $q$ (in practice, $q$ decreases as the fluid is filtered out and the fracture width increases), and taking Eq. (9.81) into Eq. (9.82) yields

$$\frac{\mathrm{d}p}{\mathrm{d}x} = -\frac{8\mu_i q E'^3}{\pi h_f^4 P_n^3} \tag{9.83}$$

To obtain a solution to Eq. (9.83), one needs to specify the net pressure at any location x. The PK model assumes that the net pressure is zero at the crack tip. Integrating Eq. (9.83) gives

$$p_{n,w}^4 - 0^4 = \frac{32E'^3 \mu q x_f}{\pi h_f^4} \tag{9.84}$$

Thus, Eqs. (9.81) and (9.84) are jointly derived as

$$w_{w,0}^4 = \frac{512\mu q x_f}{\pi E'} \tag{9.85}$$

Opening the above equation four times, the maximum seam width at the borehole is

$$w_{w,0} = \left(\frac{512}{\pi}\right)^{\frac{1}{4}} \left(\frac{\mu q x_f}{E'}\right)^{\frac{1}{4}} = 3.57\left(\frac{\mu q x_f}{E'}\right)^{\frac{1}{4}} \tag{9.86}$$

Define the maximum fracture width at the borehole multiplied by the fracture shape factor equal to the average fracture width of the fracture, i.e.,

$$\overline{W} = \gamma W_{w,0} \tag{9.87}$$

Since the fracture vertical shape is elliptical, the shape factor contains $\frac{\pi}{4}$, and then considering the variation of the fracture lateral dimensions, the lateral component of the shape factor can also be found.

For Eq. (9.81), this can be further transformed as

$$\frac{w_0}{w_{w,0}} = \frac{p_n}{p_{n,w}} = \sqrt[4]{1 - \frac{x}{x_f}} \tag{9.88}$$

The transverse component of the shape factor is

$$\int_0^1 \sqrt[4]{1-\xi}\,\mathrm{d}\xi = \frac{4}{5}$$

Thus, the shape factor is

$$\gamma = \pi \frac{1}{4}\frac{4}{5} = \frac{\pi}{5} = 0.628 \tag{9.89}$$

Combining Eqs. (9.86), (9.87) and (9.89) yields the average fracture width equation as

$$\overline{w} = \left(\frac{512\pi^3}{625}\right)^{\frac{1}{4}} \left(\frac{\mu q x_f}{E'}\right)^{\frac{1}{4}} = 2.24\left(\frac{\mu q x_f}{E'}\right)^{\frac{1}{4}} \tag{9.90}$$

For a constant injection rate and no filtration loss, the material balance equation is

$$qt = \overline{W} x_f h_f = 2.24\left(\frac{\mu q x_f}{E'}\right)^{\frac{1}{4}} x_f h_f \tag{9.91}$$

Therefore, the fracture lengths at different treatment moments for constant displacement conditions are

$$x_f = \left(\frac{625}{512\pi^3}\right)^{\frac{1}{5}}\left(\frac{q^3E'}{\mu h_f^4}\right)^{\frac{1}{5}} t^{\frac{4}{5}} = 0.524\left(\frac{q^3E'}{\mu h_f^4}\right)^{\frac{1}{5}} t^{\frac{4}{5}} \tag{9.92}$$

According to Eqs. (9.86) and (9.92), the maximum fracture width at the well wall is obtained as

$$w_{w,0} = \left(\frac{2560}{\pi^2}\right)^{\frac{1}{5}}\left(\frac{q^2\mu}{E'h_f}\right)^{\frac{1}{5}} t^{\frac{1}{5}} = 3.04\left(\frac{q^2\mu}{E'h_f}\right)^{\frac{1}{5}} t^{\frac{1}{5}} \tag{9.93}$$

Combining terms Eqs. (9.92) and (9.81) gives the net pressure at the borehole as

$$p_{n,w} = \left(\frac{80}{\pi^2}\right)^{\frac{1}{4}}\left(\frac{E'^4\mu q^2}{h_f^6}\right)^{\frac{1}{5}} t^{\frac{1}{5}} = 1.52\left(\frac{E'^4\mu q^2}{h_f^6}\right)^{\frac{1}{5}} t^{\frac{1}{5}} \tag{9.94}$$

Thus, for the no-filter loss PK model, the fracture length varies with increasing treatment time by a power exponent of 4/5, and the maximum fracture width at the borehole, the average fracture width, and the net pressure vary with treatment time by a power exponent of 1/5.

#### 9.4.2.2 Geertsma-De Klerk's Eq. for Fracture Width

The first model for hydraulic fracturing design was first proposed by Geertsma-de klerk and is shown in Fig. 9.15. The authors assumed that the fracture openings have the same fracture width and a fixed fracture height in any horizontal section. The physical implication is that the fracture surfaces at the top and bottom of the reservoir are free to move so that their vertical cross-sections are rectangular. The fracture width is simply a function of the fracture length $x$. The analysis is carried out for arbitrary horizontal cross sections according to the plane strain assumption. Khristianovitch and Zheltov (1955), Zheltov and Khristianovitch (1955) made another interesting assumption: the existence of a non-wetting zone at the fracture tip. Geertsma-de klerk accepted the hypothesis proposed by Khristianovitch and Zheltov (1955), Zheltov and Khristianovitch (1955) and, using a new mathematical approach, derived the explicit formula.

For the horizontal plane strain assumption, the constant net pressure $\overline{p}_n$ produces a fracture width at the borehole of

$$W_w = \frac{4x_f\overline{p_n}}{E'} \tag{9.95}$$

Assuming that at any location the flow rate of the fluid is equal to the injection rate $q$, since the flow cross section is rectangular, the total pressure drop from the borehole to the tip of the fracture is

$$p_{n,w} - p_{n,tip} = \frac{12\mu q x_f}{h_f}\left(\frac{1}{x_f}\int_0^{x_f}\frac{1}{w^3}dx\right) \tag{9.96}$$

Assuming that the pressure at any point within the fracture is equal to the borehole pressure, the average pressure of Eq. (9.95) can be replaced by the borehole pressure, giving

$$w_w = \frac{4x_f p_{n,w}}{E'} \tag{9.97}$$

The net pressure at the tip of the crack is 0.

$$p_{n,tip} = 0 \tag{9.98}$$

The average value of $1/w^3$ can be obtained by multiplying the value at the borehole by a constant $\frac{7}{\pi}$, so that

$$\left(\frac{1}{x_f}\int_0^{x_f}\frac{1}{w^3}dx\right) = \frac{7}{\pi w_w^3} \tag{9.99}$$

Jointing Eqs. (9.96)–(9.99) yields.

$$p_{n,w} = \frac{84\mu q x_f}{\pi h_f w_w^3} \tag{9.100}$$

Based on Eqs. (9.97) and (9.100), it follows that.

$$w_w = \frac{336\mu q x_f^2}{\pi E' h_f} \tag{9.101}$$

Therefore, the width of the fracture at the borehole is

$$w_w = \left(\frac{336}{\pi}\right)^{\frac{1}{4}}\left(\frac{\mu q x_f^2}{E' h_f}\right)^{\frac{1}{4}} = 3.22\left(\frac{\mu q x_f^2}{E' h_f}\right)^{\frac{1}{4}} \tag{9.102}$$

Equation (9.102) is the KGD model. At this point, the shape factor has no vertical component and the horizontal component is $\pi/4$ (constant pressure elliptical fracture). Thus, the shape factor is.

$$\gamma = \frac{\pi}{4} = 0.785 \tag{9.103}$$

The average seam width is

$$\overline{w} = \left(\frac{21\pi^3}{16}\right)^{\frac{1}{4}} \left(\frac{\mu q x_f^2}{E' h_f}\right)^{\frac{1}{4}} = 2.53 \left(\frac{\mu q x_f^2}{E' h_f}\right)^{\frac{1}{4}} \tag{9.104}$$

Comparing Eqs. (9.104) and (9.90), the ratio of the average Geertsma-de Klerk fracture width to the average Perkins-kern fracture width is

$$\frac{\overline{w}_{GK}}{\overline{w}_{PK}} = \left(\frac{21 \times 625}{32 \times 512}\right)^{\frac{1}{4}} \left(\frac{2x_f}{h_f}\right)^{\frac{1}{4}} = 0.95 \left(\frac{2x_f}{h_f}\right)^{\frac{1}{4}} \tag{9.105}$$

Thus, both equations give approximately the same average fracture width for the same fracture height and horizontal dimensions. As mentioned earlier, the horizontal plane strain assumption (KGD) is more appropriate for $2x_f < h_f$, while the vertical plane strain assumption (PKN) is more appropriate for $2x_f > h_f$. Equation (9.105) shows that the conversion between these two models is much "smoother".

We have derived the Perkins-kern fracture width Eq., and below we combine the Geertsma-de Klerk fracture width equation with the material balance equation for the constant injection rate, no filtration loss case, to obtain

$$qt = \overline{w} x_f h_f = \left(\frac{21\pi^3}{16}\right)^{\frac{1}{4}} \left(\frac{\mu q x_f^6 h_f^3}{E'}\right)^{\frac{1}{4}} = 2.43 \left(\frac{\mu q x_f^6 h_f^3}{E'}\right)^{\frac{1}{4}} \tag{9.106}$$

This gives the expression for the fracture length

$$x_f = \left(\frac{16}{21\pi^3}\right)^{\frac{1}{6}} \left(\frac{q^3 E'}{\mu h_f^3}\right)^{\frac{1}{6}} t^{\frac{2}{3}} = 0.539 \left(\frac{q^3 E'}{\mu h_f^3}\right)^{\frac{1}{6}} t^{\frac{2}{3}} \tag{9.107}$$

Based on Eqs. (9.101) and (9.107), the fracture width at the borehole can be calculated

$$w_w = \left(\frac{5376}{\pi^3}\right)^{\frac{1}{6}} \left(\frac{q^3 \mu}{E' h_f^3}\right)^{\frac{1}{6}} t^{\frac{1}{3}} = 2.36 \left(\frac{q^3 \mu}{E' h_f^3}\right)^{\frac{1}{6}} t^{\frac{1}{3}} \tag{9.108}$$

Bringing Eqs. (9.107) and (9.108) into Eq. (9.100) gives

$$p_{n,w} = \left(\frac{21}{16}\right)^{\frac{1}{3}} \left(E'^2 \mu\right)^{\frac{1}{3}} t^{-\frac{1}{3}} = 1.09 \left(E'^2 \mu\right)^{\frac{1}{3}} t^{-\frac{1}{3}} \tag{9.109}$$

Thus, for the no-filter Geertsma-de Klerk model, the fracture length is proportional to the 2/3rd power of t, the fracture width at the borehole is proportional to the 1/3rd power of *t*, and the net pressure at the borehole is proportional to the − 1/3rd power of *t*.

Equation (9.109) shows that the net bottomhole pressure decreases instead with increasing treatment time, whereas in practice it generally increases with time; more surprisingly, Eq. (9.109) assumes that the net bottomhole pressure is independent of the rate of fracturing fluid injection, which also contradicts everyday experience in fracturing treatment.

Turning to Eq. (9.102), and to emphasize the contrast with the Perkins-kern model, whose derivation is given here, the Geertsma-de Klerk model does not use the three assumptions of Eqs. (9.51)–(9.53). Instead, the internal pressure at the fracture tip is assumed to be negative and equal to the minimum horizontal in-situ stress. That is, the fracture tip is a vacuum. They consider a segmented constant pressure distribution for a fracture with constant pressure $\overline{p}$ at $x = 0$ to $x_0$ and 0 at $x_0$ to the crack tip $x_{\text{f}}$. The fracture width is a function of location, as shown in Sect. 9.2.4.

$$w(x) = \frac{2}{E^{'}\pi}\overline{p}\left(4x_0 \ln \frac{q_1 + q_2}{q_3} + x \ln \frac{x_f^2 x_0^2 - 2q_1 q_2 x_0 x + x_f^2 x^2 - 2x_0^2 x^2}{x_f^2 x_0^2 + 2q_1 q_2 x_0 x + x_f^2 x^2 - 2x_0^2 x^2}\right) \tag{9.110}$$

Of which.

$$\begin{aligned} q_1 &= \left(x_f^2 - x_0^2\right)^{\frac{1}{2}} \\ q_2 &= \left(x_f^2 - x^2\right)^{\frac{1}{2}} \\ q_3 &= \begin{cases} \left(x_0^2 - x^2\right)^{\frac{1}{2}} \ if\ x \le x_0 \\ \left(x^2 - x_0^2\right)^{\frac{1}{2}} \ if\ x > x_0 \end{cases} \end{aligned} \tag{9.111}$$

$$x_0 = x_f \sin \frac{\pi p_{n,w}}{2\left(p_{n,w} - p_{n,tip}\right)} = x_f \sin \frac{\pi\left(\overline{p} - \sigma_{\min}\right)}{2\overline{p}} \tag{9.112}$$

According to Eqs. (9.110) and (9.111), Geertsma-de Klerk gives some approximate values for the column. The approximate expression for the fracture width at the borehole is

$$w_w \approx \frac{8}{\pi E^{'}}\overline{p}\frac{x_0}{x_f}\sqrt{x_f^2 - x_0^2} \tag{9.113}$$

Equation (9.113) is the approximate solution to Eq. (9.110) when $x_0 \to x_f$. The approximate solution to Eq. (9.112) is

$$\overline{p} - \sigma_{\min} \approx \frac{2}{\pi}\overline{p}\frac{x_0}{x_f}\sqrt{1 - \frac{x_0^2}{x_f^2}} \tag{9.114}$$

#### 9.4.2.3 Formula for Circular Crack Widths

The fracture width Eq. can be derived from the PKN and KGD models, only differing in the selection of the shape factor, which can be obtained by taking $x_f = R$ from Eq. (9.91): $\overline{W} = 2.24\left(\frac{\mu q R}{E'}\right)^{1/4}$. For the same average seam width, the volume of the fracture obtained is smaller than the previous two models because the surface area of the fracture is smaller.

The derivation process is not described in detail, and the results are as follows: the R varies with 4/9 of time, the width varies with 1/9 of time, and the net pressure varies with − 1/3 of time; here the net pressure variation is logical, since the growth of circular cracks implies the absence of height growth hindrance.

### *9.4.3 2D Model When Considering Filtering Loss*

Consider the effect of filtration loss on fracture size, assuming that $h_f$, $E'$, $q$, $\mu$, $C_L$ and $S_p$ are known, and consider two problems. During design stage, the half-slit length $x_f$, treatment time $t_e$ is determined, also given the final net pressure at the borehole, the fracture width at the borehole, the average fracture width and the fluid efficiency. During the simulation process, the pumping time is set and the fracture length is found by the software and other relevant parameters are obtained.

The main idea of the model calculations below is the correlation of the geometric control equations with the material balance equations. It is important to remember that the derivation of the fracture width equation includes the assumption of equal flow velocities at any location within the fracture. The assumption of constant flow velocity in the $x$-direction does not hold because the fracture width is varying and the filtration loss on the fracture surface is also varying. This computational procedure is therefore only approximate, but the results are valid.

#### 9.4.3.1 PKN-C Model

In the literature, Eq. (9.86) is considered less accurate and is replaced by the following equation.

$$w_{w,0} = 3.27\left(\frac{\mu q x_f}{E'}\right)^{\frac{1}{4}} \tag{9.115}$$

The equation factor at this point is 3.27, given by Nordgren (1972). Using the shape factor given by Eq. (9.90), the average fracture width is

$$\overline{W} = \frac{\pi}{5} w_{w,0} = 2.05\left(\frac{\mu q x_f}{E'}\right)^{1/4} \tag{9.116}$$

The coefficient in Eq. (9.92) becomes 2.05 instead of 2.24.

Several other shape factors such as $\frac{3\pi}{16}, \frac{\pi^2}{16}$, are also used by other authors (Nordgren 1972; Economides et al. 1994).

The symbol C indicates that we use the Carter formula for the material balance equation. According to Eq. (9.51), the resulting equation for the fracture length is

$$x_{\mathrm{f}} = \frac{(\overline{W} + 2S_{\mathrm{P}})q}{4C_{\mathrm{L}}^2 \pi h_{\mathrm{f}}} \left[ \exp(\beta^2)\mathrm{erfc}(\beta) + \frac{2\beta}{\sqrt{\pi}} - 1 \right] \tag{9.117}$$

Here $\beta = \frac{2C_L\sqrt{\pi t}}{\overline{W}+2S_p}$

Associating the fracture length and the fracture width dichotomously into a closed body can be solved numerically to obtain $x_{\mathrm{f}}$ or $t_{\mathrm{e}}$ (corresponding to the simulation model or the design model, respectively).

The corresponding net wellbore pressure is:

$$p_{n,w} = \frac{E'}{2h_f} W_{w,0} \tag{9.118}$$

An example of fracturing simulation and fracturing design using the PKN-C model is given below. The input parameters are given in Table 9.3, the pumping time is 200 min, note that the injection rate q is the single flank discharge, E' is the plane strain modulus of elasticity, determine the fracture length, maximum fracture width at the borehole, average fracture width, and fluid efficiency in the simulation mode.

Equations (9.116) and (9.117) are combined into one equation to obtain an expression for $x_{\mathrm{f}}$, which, when brought into the data in Table 9.4, gives

$$x_f = \frac{2.05\left(\frac{0.0662 \times 0.2 \times x_f}{6.13 \times 10^{10}}\right)^{\frac{1}{4}} 0.0662}{4(9.84 \times 10^{-6})^2 \times \pi \times 51.8} \left[ \exp(\beta^2) erfc(\beta) + \frac{2\beta}{\sqrt{\pi}} - 1 \right]$$

Of which

**Table 9.3** Input parameters

| | |
|---|---|
| $C_L$/ (m·s$^{-1/2}$) | $9.84 \times 10^{-5}$ |
| $S_p$/m | 0 |
| $h_f$/m | 51.8 m |
| $E'$/Pa | $6.13 \times 10^{10}$ |
| $\mu$/( Pa·s) | 0.2 |
| $q$/(m$^3$·s$^{-1}$) | 0.0662 |
| $t_e$/s | 12,000 |

**Table 9.4** PKN-C simulation results

| | |
|---|---|
| $x_f = 1340\,\mathrm{m}$ | $\eta = 74.1\%$ |
| $w_{w,0} = 1.35 \times 10^{-2}\,\mathrm{m}$ | $P_{n,w} = 7.98 \times 10^6\,\mathrm{Pa}$ |
| $\overline{w} = 8.48x10^{-3}\,\mathrm{m}$ | |

$$\beta = \frac{2 \times 9.84 \times 10^6 \sqrt{\pi \times 12,000}}{2.05\left(\frac{0.062 \times 0.2 \times x_f}{6.13 \times 10^{10}}\right)^{\frac{1}{4}}}$$

The numerical algorithm, with three decimal places retained, gives $x_f = 1340$ m. The other parameters are then obtained from Eqs. (9.115), (9.116), (9.38), (9.118). The results are shown in Table 9.2.

Similar to the simulation model, assuming the fracture length is 1000 m, the results of the calculations in the design model are obtained as follows.

Bringing $x_f = 1000$ m into Eq. (9.116) and calculating the average fracture width, the average width

$$\overline{w} = 2.05\left(\frac{0.0662 \times 0.2 \times 1000}{6.13 \times 10^{10}}\right)^{1/4} = 7.88 \times 10^{-3}\,\mathrm{m}$$

Then Eq. (9.117) gives

$$1000 = \frac{7.88 \times 10^{-3} \times 0.0662}{4\left(9.84 \times 10^{-6}\right)^2 \times \pi \times 51.8}\left[\exp(\beta^2)erfc(\beta) + \frac{2\beta}{\sqrt{\pi}} - 1\right]$$

Of which

$$\beta = \frac{2 \times 9.84 \times 10^{-6}\sqrt{\pi \times t}}{7.88 \times 10^{-3}}$$

The results of the calculations are shown in Table 9.3.

#### 9.4.3.2 KGD-C Model

Similar to the calculation steps of the PKN-C model, the Geertsma-de Klerk Eq. associates the Carter formula that can be computed for both the simulation and design models. Instead of using Eqs. (9.115), (9.116), (9.102) and (9.104) are used. Equation (9.118) is replaced by the following Eq., the net pressure in the wellbore

$$P_{n,w} = \frac{E'}{4x_f} W_w \tag{9.119}$$

**Table 9.5** Results of PKN-C design model calculations

| | |
|---|---|
| $t_e = 8060\,\mathrm{s}$ | $\eta = 76.5\%$ |
| $w_{w,0} = 1.25 \times 10^{-2}$ m | $p_{n,w} = 7.42 \times 10^6$ pa |
| $\overline{w} = 7.88 \times 10^{-3}$ m | |

An example of fracture simulation and fracture design using KGD-C model is given below. Recalculate using the data in the Table 9.1. Equations (9.104) and (9.117) are combined to give

$$x_f = \frac{2.53\left(\frac{0.0662\times0.2\times x_f^2}{6.13\times10^{10}\times51.8}\right)^{\frac{1}{4}}0.0662}{4(9.84\times10^6)^2\times\pi\times51.8}\left[\exp(\beta^2)erfc(\beta)+\frac{2\beta}{\sqrt{\pi}}-1\right]$$

Of which

$$\beta = \frac{2\times9.84\times10^6\sqrt{\pi\times12000}}{2.53\left(\frac{0.0662\times0.2\times x_f^2}{6.13\times10^{10}\times51.8}\right)^{\frac{1}{4}}}$$

The results are shown in Table 9.4.

Let $x_f = 1000$ m, the results of the design mode calculations are shown in Table 9.5.

As can be seen from the equations, the fracture width of the KGD model is larger than the PKN model fracture width, and the KGD model has a shorter fracture length and longer pump-in time than the PKN model. The fracture width produced in engineering practice is intuitively larger than the PKN model width.

#### 9.4.3.3 PKN-N and KGD-N Models

Replace Carter formula with Eq. (9.68), i.e.

$$\frac{qt}{h_f x_f} = \overline{W} + C_L\sqrt{t}\left[\frac{8}{3}\eta + (1-\eta)\pi\right] \tag{9.120}$$

Among others $\eta = h_f x_f \overline{w}/(qt)$.

In the calculation procedure, the average fracture width is calculated first, and the corresponding fracture length can be found by assuming a certain time, and similarly, the corresponding fracture width can be found by assuming a certain fracture length. That gives the PKN-N model and the KGD-N model.

The calculation example of PKN-N model is given below. According to Eq. (9.116), obtained the average fracture width,

**Table 9.6** Simulation results

| | |
|---|---|
| $x_f = 748\,\text{m}$ | $\eta = 85.7\%$ |
| $w_w = 2.24 \times 10^{-2}\,\text{m}$ | $p_{n,w} = 4.58 \times 10^5\,\text{pa}$ |
| $\overline{w} = 1.76 \times 10^{-2}\,\text{m}$ | |

$$\overline{w} = 2.05\left(\frac{0.0662 \times 0.2 \times 1000}{6.13 \times 10^{10}}\right)^{1/4} = 7.88 \times 10^{-3}\,\text{m}$$

Equation (9.120) becomes.

$$\frac{0.0662 \times t}{1000 \times 51.8} = 7.88 \times 10^{-3} + 6.13 \times 10^{10}\sqrt{t}\left[\frac{8}{3}\frac{1000 \times 51.8 \times 7.88 \times 10^{-3}}{0.0662 \times t} + \left(1 - \frac{1000 \times 51.8 \times 7.88 \times 10^{-3}}{0.0662 \times t}\right)\pi\right]$$

The results are shown in Table 9.6.

#### 9.4.3.4 PKN-$\alpha$ Versus KGD-$\alpha$ Models

Parameter $\alpha$ indicates a power-law growth in fracture surface area. For fracture length growth consistent with the power law, the matter balance equation (Eq. 9.65) is coupled to the arbitrary fracture width Eq., obtain the length of facture.

$$x_f = \frac{qt/h_f}{\overline{W} + 2S_P + 2C_L\sqrt{t}[\frac{\alpha\sqrt{\pi}\Gamma(\alpha)}{\Gamma(3/2+\alpha)}]} \tag{9.121}$$

From Sect. 9.1.1, it is assumed that $\alpha = \frac{4}{5}$ when there is no filtering; in the presence of filtering, the coefficients remain constant.

Applying Eq. (9.46), we cannot find the end of pumping time. Equation (9.121) satisfies the principle that the future does not affect the past and can be used to calculate the fracture length at any time without regard to the fracture width or fluid efficiency at the end of pumping. But, the Carter- or Nolt-type material balance equations, i.e., Eqs. (9.117) and (9.120), can't satisfy this basic requirement.

The calculation results of PKN-α model are given below, using the data in Table 9.1. Equations (9.116) and (9.121) are combined into one equation, giving

$$x_f = \frac{0.0662\frac{12000}{51.8}}{2.05\left(\frac{0.0662 \times 0.2 \times x_f}{6.13 \times 10^{10}}\right)^{\frac{1}{4}} + 2 \times 9.84 \times 10^{-6}\sqrt{12000} \times g_0\left(\frac{4}{5}\right)}$$

among others

**Table 9.7** Results of design mode calculations

| | |
|---|---|
| $t_e = 1.88x10^4 s(313\text{ min})$ | $\eta = 84.7\%$ |
| $w_w = 2.59 \times 10^{-2}$ m | $p_{n,w} = 5.30 \times 10^5$ pa |
| $\overline{w} = 2.03 \times 10^{-2}$ m | |

$$g_0\left(\frac{4}{5}\right) = \frac{0.8\sqrt{\pi}\Gamma(0.8)}{\Gamma\left(\frac{3}{2} + 0.8\right)} = 1.415$$

The results of the calculations are shown in Table 9.7, and the results are almost identical to PKN-C model. The results obtained from the arithmetic examples show that the PKN-$\alpha$, PKN-C and PKN-N models give almost identical results.

#### 9.4.3.5 Circular Fractures

**R-C model**

Using the constant fracture width formula, the average width

$$\overline{w} = 2.24\left(\frac{\mu q R}{E'}\right)^{\frac{1}{4}} \tag{9.122}$$

Radius equation,

$$R^2 = \frac{(\overline{W} + 2S_P)q}{4C_L^2\pi^2}\left[\exp(\beta^2)erfc(\beta) + \frac{2\beta}{\sqrt{\pi}} - 1\right] \; where \; \beta = \frac{2C_L\sqrt{\pi t}}{\overline{w} + 2S_p} \tag{9.123}$$

The net pressure at the borehole is:

$$P_{n,w} = \frac{\pi E'}{8R} W_{w,0} \tag{9.124}$$

The shape factor (Nolte 1986) that relates the average fracture width to the maximum fracture width is taken as $\gamma = 8/15$.

**R-N model**

The radius formula uses the following equation.

$$\frac{qt}{R^2\pi} = \overline{W} + C_L\sqrt{t}\left[\frac{8}{3}\eta + (1 - \eta)\pi\right] \text{ Of which:} \eta = \frac{R^2\pi\overline{W}}{qt} \tag{9.125}$$

### 9.4.3.6 Results for Non-newtonian Fluids

For the fracture width equation, there are several ways to consider the non-Newtonian properties of the fluid. The most convenient method is to add an additional equation that establishes the relationship between apparent viscosity and flow rate, assuming that the power-law is satisfied and that the apparent viscosity is as follows.

$$\mu_e = K\left[\frac{1 = (\pi - 1)n}{\pi n}\right]^n \left(\frac{2\pi u_{avg}}{w_0 h_f}\right)^{n-1} \tag{9.126}$$

$u_{avg}$ is the mean linear velocity and $w_0$ is the maximum fracture width of the elliptical fracture. For the PK model, the flow rate is assumed to be constant and equal to the injection rate. Since the average cross-sectional area is $\overline{w}h_f$, the average velocity of the elliptical cross-section is $u_{avg} = q/(\overline{w}h_f)$. Denote the maximum value of the average elliptical transect by $w_0 = 4\overline{w}/\pi$ (note that $w_0$ is not the seam width at the borehole, but the fracture width at the average location).

Equation (9.51) becomes

$$\begin{aligned}\mu_e &= K[\frac{1 + (\pi - 1)n}{\pi n}]^n \left(\frac{\pi^2}{1}\right)^{n-1} \left(\frac{q}{\overline{w}^2 h_f}\right)^{n-1} \\ &= K\left[\frac{1 + (\pi - 1)n}{\pi n}\right]^n (\frac{25}{2})^{n-1} \left(\frac{q}{W_{w,0}^2 h_f}\right)^{n-1}\end{aligned} \tag{9.127}$$

The shape factor $\gamma = \pi/5$ is used here to represent the maximum fracture width at the borehole in relation to the average fracture width.

Bringing the equivalent apparent viscosity Eq. (9.127) into the PKN-C model (9.115) yields the maximum fracture width equation for the PKN model.

$$\begin{aligned}w_{w,0} &= 9.15^{1/(2n+2)} \times 3.98^{n(2n+2)} \left[\frac{1 + 2.14n}{n}\right]^{n(2n+2)} \\ &\times K^{1/(2n+2)} \left(\frac{q^n h_f^{1-n} x_f}{E'}\right)^{1/(2n+2)}\end{aligned} \tag{9.128}$$

For the KGN model,

$$\mu_e = \frac{2^{n-1}}{3} K \left(\frac{1 + 2n}{n}\right)^n \left(\frac{q}{\overline{w}^2 h_f}\right)^{n-1} \tag{9.129}$$

The equation for the width of the KGD fracture is

**Table 9.8** Results of PKN-N design model calculations

| | |
|---|---|
| $t_e = 8080\text{s}(135\,\text{min})$ | $\eta = 76.2\%$ |
| $\overline{w} = 7.88 \times 10^{-3}\,\text{m}$ | $p_{n,w} = 7.42 \times 10^6\,\text{pa}$ |
| $w_{w,0} = 1.25 \times 10^{-2}\,\text{m}$ | |

**Table 9.9** PKN-α Simulation results

| | |
|---|---|
| $x_f = 1330m$ | $\eta = 73.5\%$ |
| $w_{w,o} = 1.35 \times 10^{-2}m$ | $p_{n,w} = 7.97 \times 10^6 pa$ |
| $\overline{w} = 8.46 \times 10^{-3}$ | |

**Table 9.10** Parameters of the power law

| Parameters | Values |
|---|---|
| $n$ | 0.5 |
| $K$ | 2.87 Pa.s$^{0.5}$ |

$$w_w = 11.1^{1/(2n+2)} \times 3.24^{n/(2n+2)} \left[\frac{1+2n}{n}\right]^{n/(2n+2)} K^{1/(2n+2)} \left(\frac{q^n x_f^2}{h_f^n E'}\right)^{1/(2n+2)} \tag{9.130}$$

Considering non-Newtonian fluids (see Table 9.8 for parameters), the PKN-C and KGD-C models and the data in Table 9.1 were used to recalculate the fracturing simulation (Tables 9.9 and 9.10).

The joint cubic Eqs. (9.128), (9.116) and (9.117) give the expression for $x_f$ as

$$x_f = \frac{\overline{w} \times 0.0662}{4(9.84 \times 10^{-6})^2 \times \pi \times 51.8}\left[\exp(\beta^2)erfc(\beta) + \frac{2\beta}{\sqrt{\pi}} - 1\right]$$

Of which

$$\beta = \frac{2 \times 9.84 \times 10^{-6}\sqrt{\pi \times 12000}}{\overline{w}}$$

$$\overline{w} = \frac{\pi}{5} 9.15^{1/(2\times0.5+2)} \times 3.98^{0.5\times(2\times0.5+2)} \left[\frac{1+2.14\times0.5}{0.5}\right]^{0.5\times(2\times0.5+2)}$$
$$\times 2.87^{1/(2\times0.5-2)} \left(\frac{0.0662^{0.5} \times 51.8^{(1-0.5)}}{6.13\times10^{10}\times51.8} x_f\right)^{\frac{1}{(2\times0.5+2)}}$$

The calculation results of PKN-C and KGD-C fracturing simulation modes are shown in Tables 9.11 and 9.12 respectively.

**Table 9.11** PKN-C simulation results (power-law fluids)

| | |
|---|---|
| $x_f = 1200\,\text{m}$ | $\overline{w} = 9.86 \times 10^{-3}\,\text{m}$ |
| $\mu_e = 0.409\,\text{pa.s}$ | $\eta = 76.9\%$ |
| $w_{w,0} = 1.57 \times 10^{-2}\,\text{m}$ | $p_{n,w} = 7.40 \times 10^5\,\text{pa}$ |

**Table 9.12** KGD-C simulation results (power-law fluid)

| | |
|---|---|
| $x_f = 600\,\text{m}$ | $\overline{w} = 2.27 \times 10^{-2}\,\text{m}$ |
| $\mu_e = 0.858\,\text{pa.s}$ | $\eta = 88.6\%$ |
| $w_w = 2.88 \times 10^{-2}\,\text{m}$ | $p_{n,w} = 7.40 \times 10^5\,\text{pa}$ |

## 9.5 Factors Influencing Fracture Extension

### *9.5.1 Fracture Extensions in Vertical Wells*

The idea that hydraulic fractures extend along a direction perpendicular to the minimum horizontal in-situ stress is borne out by a large amount of practical production data. However, for shallow formations, the hydraulic fractures may be horizontal and such reservoirs generally have high permeability and are not suitable for fracturing.

The stress distribution in the near wellbore zone is different from the far-field stress, so that the fracture initiation direction may be different from the propagation direction behind, and the fracture initiation is also closely related to the injection method.

Figure 9.18 depicts the relationship between fracture initiation and propagation. Several fractures in the near-well zone may merge and eventually develop into one main fracture, and near-well effects (twisting, etc.) may result in additional frictional drag.

The pressure distribution in the fracture is closely related to the energy dissipation generated by the various resistances. For a zero-phase penetration, the first single-winged fracture is fractured first, and the second fracture starts when the resistance generated by the expansion of the first fracture is greater than the fracture initiation pressure of the second fracture. The generation of the second fracture depends on factors such as cementation between the casing and the formation, the injection point, and the spacing between the two fractures. If the fracture initiation pressure of the second fracture is too high, it is possible that only a single flanking fracture will be produced because the pressure increment required for fracture expansion is small. To avoid this, penetration phasing is important and is generally considered to be at least 120°, while current penetration guns are usually deployed at 30°.

Fractures are prone to near-well zone distortion when fracturing slanting and horizontal wells and are generally not recommended for fracturing directional wells. Near-well fracture distortion can result in a reduction in the width of the main fracture, which can result in less than expected production after fracturing. If a directional

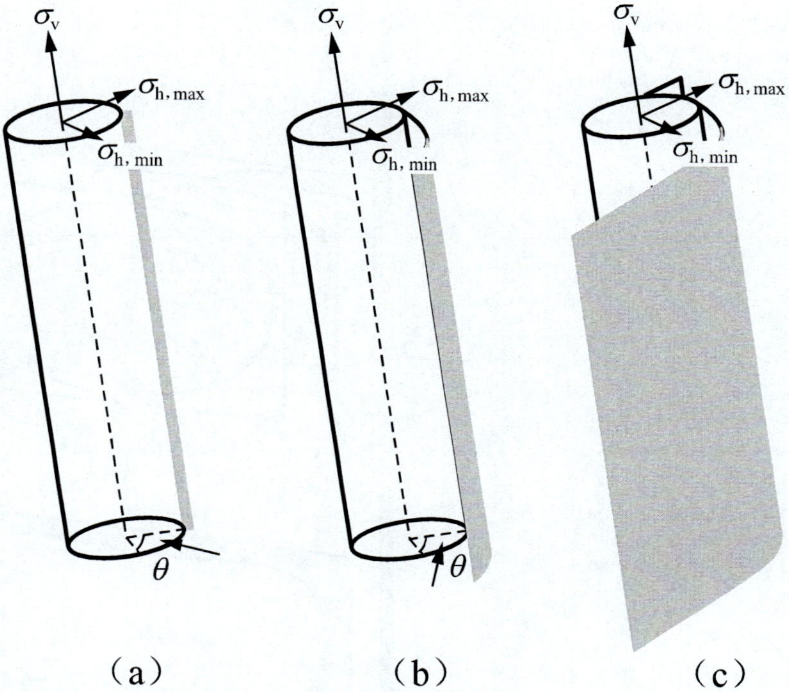


**Fig. 9.18** Fracture initiation process for a straight well in any orientation

well needs to be fractured, it is recommended that the well section become a vertical borehole as it is drilled to the target formation.

### *9.5.2 The Fracture Extension in Horizontal Well*

Horizontal wells emerged in the 1980s and in many cases fracturing of horizontal wells was desired in order to increase production per well. Fracturing of horizontal wells may produce axial vertical fractures or axial transverse fractures due to different borehole orientations. The optimal borehole orientation depends on reservoir characteristics and production requirements.

The difference between horizontal well transverse fractures and vertical fractures of vertical well is that for vertical well and vertical fractures, flow from reservoir to fracture and from fracture to borehole is linear, whereas for horizontal well transverse fractures, flow from reservoir to fracture is linear and flow from near-well fracture to borehole is radial, as shown in Fig. 9.19 (Dake 1978). To form multiple transverse fractures, the segment length of each penetration cluster should be short, otherwise it will produce fracture steering as shown in Fig. 9.20, which is not desired (Brown and Economides 1992).

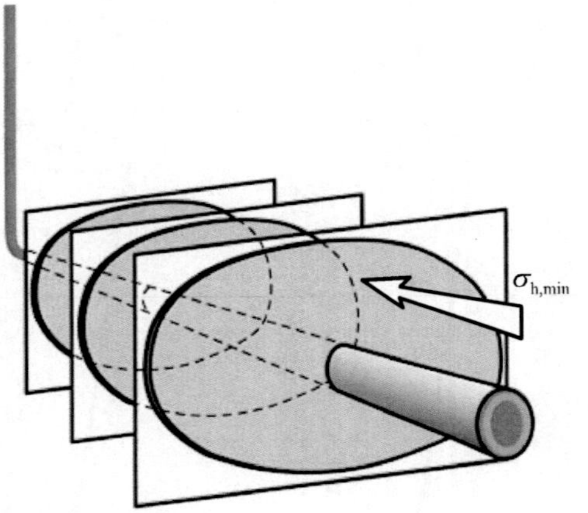


**Fig. 9.19** Fracturing of horizontal wells with multiple transverse fractures

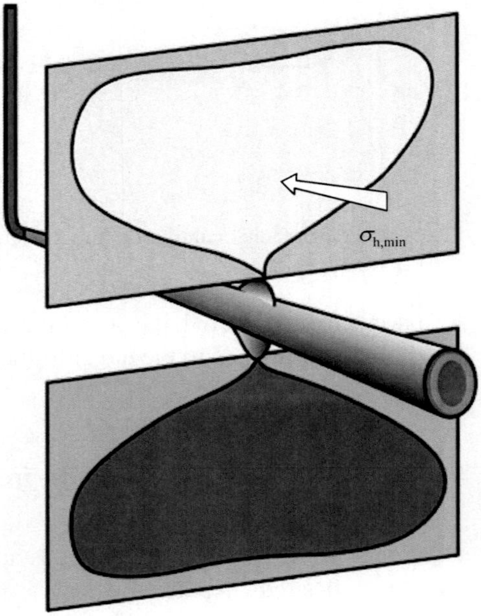


**Fig. 9.20** Transformation of horizontal well axial seam into a cross-sectional seam

Deinbacher investigated the difference between the fracture width after the axial fracture was turned to the transverse fracture and the ideal transverse fracture width, with the following results.

$$\frac{w_l}{w_t} = \frac{\pi}{2}\frac{D}{L} \tag{9.131}$$

where, $w_l$ is the fracture width after the axial fracture is turned to the transverse fracture, $w_t$ is the fracture width of the ideal transverse fracture, $L$ is the length of the penetration section, and $D$ is the borehole diameter.

From Eq. (9.131), it can be seen that if the penetration section length $L$ is greater than $1.5D$, an axial fracture may be created and turned to the transverse fracture. Therefore, the transverse fracture penetration section length is required to be small.

Horizontal wells are prone to axial fractures, but their yields are not higher than those of vertical well fracturing, especially for low permeability reservoirs. However, for relatively high permeability reservoirs, horizontal wells with axial fractures are still attractive and have much higher yields than vertical well fracturing.

### 9.5.3 Multi-layered Fracture Profiles

The fracture height of a vertical well may migrate, and Fig. 9.21 gives a schematic diagram of a vertical fracture with the reservoir in the middle portion and the compartment above and below. Differences in elastic parameters between the formations result in different fracture widths.

High fracture extension is undesirable. The risk of desanding may arise when fracturing fractures penetrate into adjacent impermeable formations. Figure 9.22 depicts the case where the reservoir is separated by a thin layer of high elastic modulus, and when the fracture height penetrates the compartment, the fracture width is narrow enough for fracturing fluid to enter, but not proppant. This can lead

**Fig. 9.21** Three-layer model

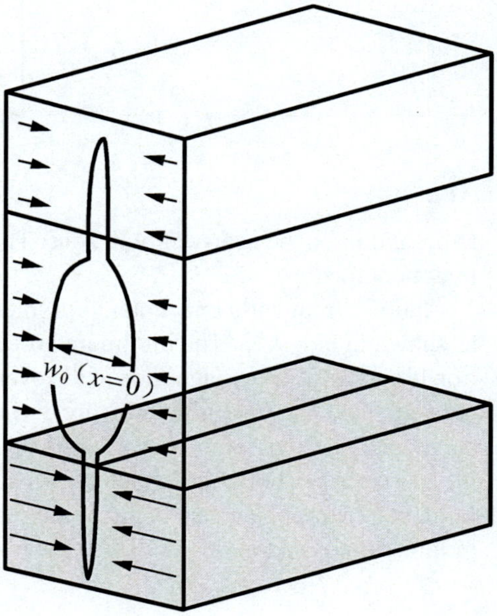

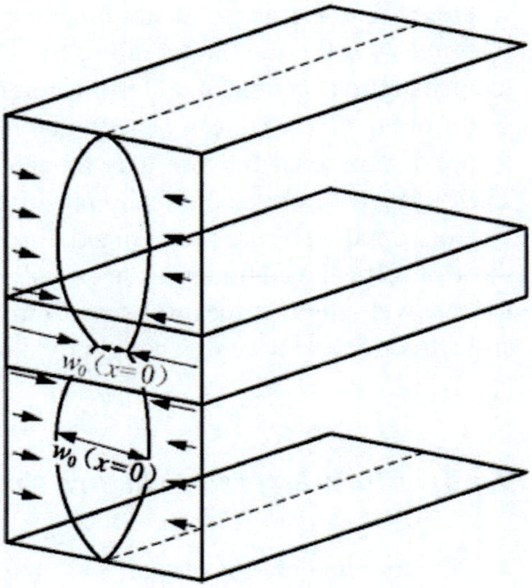


**Fig. 9.22** Variation of seam width along seam height

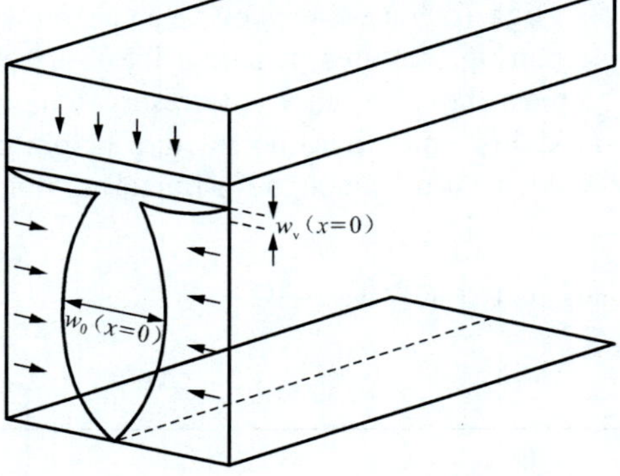


**Fig. 9.23** T-seam

to desanding in the near-wellbore zone. The treatment pressure rises rapidly, causing treatment to stop.

Another frequently encountered problem is the generation of T-shaped fractures, as shown in Fig. 9.23. The minimum horizontal in-situ stress at the fractured formation is comparable in magnitude to the overburden stress, and when vertical fractures are generated, the fracturing pressure during treatment is greater than the overburden stress, but the overlying rock is difficult to fracture and prone to T-shaped fractures. The upper horizontal fracture can accept a considerable amount of fracturing fluid, but the proppant cannot enter, thus easily causing the lower sand blockage and termination of operation.

## Exercises

1. What are the differences in flow characteristics of fractured straight wells with infinite and finite inflow capacity?
2. What is the physical meaning of dimensionless pressure function and dimensionless inflow capacity?
3. Briefly describe the process for optimizing the design of hydraulic fracturing in straight wells?
4. derive the equation for the seam width of a two-dimensional Griffith crack (seam length $2c$) for constant internal pressure $p_0$?
5. Assuming an elliptical crack with a seam length of $2c$ ($c = 10$ m), try to compare the difference in maximum seam width, crack tip stress for two internal pressure conditions: constant internal pressure ($p_0 = 5$ MPa) and smooth closed crack ($s = 5$ MPa, $s_2 = -10$ MPa)?
6. What is the net pressure of a crack?
7. A low-permeability sandstone reservoir is known to be 30 m thick with an elastic modulus of 15,000 MPa and a Poisson's ratio of 0.2. Developed by fracturing a straight well with a fracturing fluid filtration loss coefficient of $9.84 \times 10^{-5}$ m/s$^{1/2}$, an initial filtration loss $S_p$ of 0, a fracturing fluid viscosity of 0.2 Pa.s, a fracturing fluid injection rate of 6 m$^3$/min. and a pumping time of 7200 s. The fracturing fluid is simulated using the PKN model and the KGD model, respectively. PKN model and KGD model for fracturing simulations to determine the seam length at the end of construction, the empirical maximum seam width, the average seam width and the fluid efficiency. What are the simulation results if the filtration loss is not considered?
8. For horizontal well fracturing, how to consider the shot-hole cluster length to create a simple fracture in the near-well zone?

## References

Agarwal RG, Carter RD, Pollock CB. Evaluation and prediction of performance of low permeability gas wells stimulated by massive hydraulic fracturing. JPT. 1979;31(3):362–72.

Barenblatt GI. Mathematical theory of equilibrium cracks. Adv Appl Mech. 1962;7(01):55–129.

Brown JE, Economides MJ. An analysis of hydraulically fractured horizontal wells. In: SPE; 1992. p. 24322.

Brown JE, Economides MJ. Practical considerations in fracture treatment design. In: Economides MJ. Practical companion to reservoir stimulation. Amsterdam: Elsevier; 1992.

Cinco-Ley H, Samaniego F. Transient pressure analysis for fractured wells. JPT. 1981;33(9):1749–977.

Cinco-Ley H, Samaniego F, Domingue AN. Transient pressure behavior for a well with a finite-conductivity vertical fracture. SPEJ; 1978: Aug. 253–264.

Dake LP. Fundamentals of reservoir engineering. Amsterdam: Elsevier; 1978.

Economides MJ, Hill AD, Ehlig-Economides CA. Petroleum production systems. Englewood Cliffs: Prentice Hall; 1994.

England AH, Green AE. Some two dimensional pinch and crack problems in classical elasticity. In: Proceeding Cambridge Philosophical Society London; 1963. (Vol. 59, pp. 489–500).
Geertsma J. Two-dimensional fracture-propagation models, in Recent Advances in Hydraulic fracturing. In: Gidley JL, et al., editors. SPE monograph vol. 12. SPE, Richardson, TX; 1989.
Geertsma J, de Klerk F. A rapid method of predicting the width and extent of hydraulically induced fractures. JPT. 1969;21(12):1571–81.
Green AE, Zerna W. Theoretical elacticity. London: Oxford University Press; 1968.
Gringarden AC, Ramey AJ. Unsteady state pressure distributions created by a well with a single-infinite conductivity vertical fracture. SPEJ. 1974;14(4):247–360.
Howard GC, Fast CR. Optimum fluid characteristics for fracture extension. Drilling and Produ Prac API 1957;261–270.
Khristianovitch SA, Zheltov YP. Formation of vertical fractures by means of highly viscous fluids. In: Proceedings of world petroleum congress, Rome, 1955. (Vol. 2, p. 579).
Meng H-Z. The optimization of propped fracture treatments. In: Economides MJ, Nolte KG, editors. Reservoir stimulation 2nd ed. Englewood Cliffs: Prentice Hall; 1989.
Meng HZ, Brown KE. Coupling of production forecasting fracture geometry requirements and treatment schedules in the optimum hydraulic fracturing design. In: SPE; 1987. p. 16435.
Muskhelishvili NI. Some basic problems of the mathematical theory of elasticity. Nordhoof, Holland; 1953.
Nolte KG. Fracturing pressure analysis. In: Recent advances in hydraulic fracturing. In: Gildly JL, et al. editors. Monograph Series, SPE, Richardson, TX; 1989.
Nolts KG. Determination of proppant and fluid schedules from fracturing pressure declines. SPEPE. 1986;1(4):255–65.
Nordgren RP. Propagation of a vertical hydraulic gracture. SPEJ. 1972;12(4):306–14.
Sneddon IN. Integral transform method, in mechanics of fracture 1, method of analysis and solutions of crack problems. In: Sih GC, editors, Leyden : Nordhoff International; 1973.
Valko P. Economides M.J. Fracture height containment with continuum damage mechanics. In: SPE; 1993. p. 26598.
Zheltov YP, Khristianovitch SA. On the mechanism of hydraulic fracture of an oil-bearing stratum. Izv AN SSSR, OTN. 1955;5:3–41.

# Chapter 10
# Mechanics of Oil Well Sand Production

In the oil industry, sand production from oil and gas wells is a widespread problem worldwide, and hundreds of millions of dollars are spent annually on this research. The main reasons for the difficulty of such research are as follows:

- Oil field development takes place deep in the ground and cannot be directly observed at the surface.
- The deformation of loose sandstone formations often exhibits large deformations, the mechanical properties of which may vary over a wide range, and the coring of deep formations is not only expensive, but also subject to certain uncertainties, such as the difficulty of maintaining the water content, temperature and pressure conditions of the formation at the surface.
- As production proceeds and various production enhancement measures are implemented, the reservoir becomes more complex, which also makes it difficult to study the mechanism of sand emergence.
- Well sand production is influenced by many complex factors such as: geological conditions, production parameters, etc.

Regardless of the cost of sand control measures, it is the only way to make a well recoverable for wells that have been abandoned or cannot be developed due to sand production. Excessive use of sand control measures, however, not only increases production costs, but also contaminates the formation and reduces well production. Therefore, it is critical to accurately predict whether a well will be sanded out before it is finalized and during its production.



Y. Cheng et al., *Foundations of Rock Mechanics in Oil and Gas Engineering*,
https://doi.org/10.1007/978-981-99-1417-3_10

## 10.1 Basic Processes and Hazards of Oil Well Sanding

### *10.1.1 Basic Process of Sand Emergence from Oil Wells*

Theoretical studies have found that formation sands can be divided into two types of filling sands and skeletal sands. When the flow rate of fluid reaches a certain value, it first makes the uncemented sand particles filled in the pore channel of the formation move and the well starts to produce sand, the outflow of such filled sand is inevitable and plays the role of unblocking the pore channel of the formation; on the contrary, if these filled sands stay in the formation, they may block the pore of the formation, causing the permeability to drop and the production to decrease. Therefore filling sand is not a target of control. When the flow rate or production pressure difference reaches a certain value, the stress on the rock will reach or exceed its strength, causing damage to the rock structure and turning the skeletal sand into loose sand, which will be carried away by the fluid and cause a large amount of sand in the well, a situation that should be avoided in the production process. For the skeletal sand to become loose sand, from the macroscopic point of view, it is caused by tensile damage or shear damage of the rock; from the microscopic point of view, it is the individual sand grains flaking off from the rock skeleton due to the erosion of the produced fluid.

Well sand production can be broadly divided into two stages: the first stage is the change from skeletal sand to loose sand, which is a necessary condition leading to sand production; the second stage is the transport of loose sand. Only after these two conditions are met will large-scale sand production be caused. For the first stage of sand emergence, stress factors, such as borehole pressure, in situ stress state, and rock strength, are the main factors affecting sand emergence; the role of hydraulic factors, such as flow rate, permeability, viscosity, and relative permeability of two-phase or three-phase flow, is mainly manifested in the second stage of sand emergence, which is the transport of loose sand formed due to shear damage. The above-mentioned view can explain the following phenomena in the field production process: for example, in a block of Dongxin Oil Production Zone of Shengli Oilfield, with a medium-strength formation, the production output is high at the beginning, at which time the formation is not damaged and no sand emerges; as the production proceeds, the formation is damaged due to pressure decay, water intrusion, etc., and a damage zone is produced around the borehole, which does not produce high output but causes sand emergence.

Therefore, to achieve effective sand control for formations with a certain cementation strength, the first step is to prevent damage to the formation, i.e., to prevent the necessary conditions for sand production from being met, which is mainly achieved by controlling stress factors, such as maintaining reservoir pressure and reducing the production pressure difference. However, as production proceeds, the reservoir pressure decays, the rock strength decreases, and rock damage inevitably occurs, so that the transition from the first stage of sand production to the second stage is achieved by controlling the flow rate to prevent the transport of loose sand. For weakly cemented and uncemented reservoirs, the conditions of the first stage of sand emergence are easily met, so the key to sand control is to prevent the conditions required for the second stage of sand emergence from being met, and the purpose of sand control can be achieved by controlling the flow rate and producing differential pressure.

### 10.1.2 Factors Influencing Sand Emergence from Oil Wells

Oil and gas well sand emergence is one of the major problems encountered in oil recovery, and a lot of manpower and resources are spent on prevention and research every year. Sand emergence not only leads to production reduction or shutdown operations and abrasion of surface and downhole equipment, but also causes casing damage and well scrapping. Through the analysis of the data of a sparse sandstone block in Shengli Oilfield, it is found that; with the intensification of sand emergence and the reduction of formation pressure, the number of casing damage wells increases year by year, and the casing damage of oil wells in sparse sandstone reservoirs is serious, which has accounted for more than 10%, and individual blocks have reached more than 20%.

There are many factors that affect well sand emergence, which can be summarized into three broad categories.

- Geomechanical factors, including in situ stress state, pore pressure, in situ stratigraphic temperature, geological formations, etc.
- Comprehensive properties of the sandstone reservoir, including well depth, strength and deformation characteristics of the sandstone, porosity, permeability, radius of drainage, composition of fluids (oil, gas, water content and distribution, etc.), clay content and composition, particle size and shape, compaction, etc.
- Engineering factors, including completion type, well structure parameters (well depth, well inclination, orientation, well diameter), performance of completion fluids, production enhancement measures (fracturing, acidizing, etc.), production process parameters (flow rate, production pressure differential and flow rate), extent of formation damage (increased skin factor), well switch scheme, artificial lift technology, extent of reservoir depletion, cumulative sand production, etc. These factors and parameters interact and influence each other, making the sand emergence problem very complex. For a particular field, it is difficult to determine which factors are the main factors for sand emergence from the formation based on field experience alone.

The effects of geomechanical factors and formation properties on sand emergence will be described in later sections in relation to specific borehole conditions. The following section focuses on the effect of some engineering factors on sand emergence.

#### 10.1.2.1 Perforation Size

If tensile damage occurs, the perforation hole size will become larger. The stable perforation hole size depends on the depth and density of the perforation.

#### 10.1.2.2 Selective Perforation

The strength of the rock may vary considerably with depth within the same reservoir. Homogeneous layers are typically a few centimeters thick. If the fluids between these small layers are interconnected, the following issues should be considered.

- Choose to perforation holes at high intensity layers so that higher production rates can be achieved at higher production pressure differentials.
- The perforation density may be increased in the low strength layer, which reduces the flow resistance and shear forces in order to achieve the same production capacity at a lower production pressure differential.
- Consider the bubble point of the fluid. If the oil is extracted below the bubble point, two-phase flow and high resistance will occur; if the oil is extracted above the bubble point, the resistance will be reduced.

#### 10.1.2.3 Cyclic Loading

When an oil well is shut in for a period of time and then produced, the formation will be cyclically loaded and a decrease in formation strength will occur. It is often seen that an increase in sand production will occur after a period of well shut-in and then production. Therefore, attention should be paid to well shut-in operating procedures to reduce the cyclic loads they may generate.

#### 10.1.2.4 Impact of Water Intrusion

In order to maintain the balance of reservoir pressure, water injection has become a common development method during oilfield development. On the one hand, as the injection water advances from the water well to the oil well, the water content of the reservoir near the oil well gradually increases; on the other hand, as the reservoir pressure decreases during the extraction process, the edge and bottom water of the reservoir also gradually migrates to the producing well, causing water intrusion. The effect of water intrusion on sand production can be summarized as follows.

**Water intrusion reduces the strength of the rock**

Water intrusion can dissolve some of the cement between the sand grains, reducing the cement strength of the formation. For example, the strength of some muddy cemented rocks can be reduced by up to 30% when the clay minerals undergo hydration and swelling after water intrusion.

**Water intrusion disrupts the continuity of oil flow in the pore space**

Studies have shown that oil sand particles are generally surrounded by extremely thin clay films, with more microporous channels between the sand layers and many very thin clay interlayers inside the oil layer, and the difference in permeability between the small layers separated by these clay interlayers is quite large. When water intrusion occurs, the clay around the sand grains swells, narrowing the oil flow channels and reducing the relative permeability of the oil, greatly increasing the resistance to oil flow and the dragging force of the produced fluid on the sand grains, creating conditions for the second stage of sand emergence.

As shown in Fig. 10.1a, when the oil-bearing saturation of the formation is high, the oil flow is continuous inside the pore space. This is when a small amount of bound water is at the periphery of the pore space and holds the very small free particles in place so that they are not washed away even at considerable oil flow velocities. When the water intrusion is large, it breaks the continuity of the oil flow into droplets of unequal size, thus changing the single-phase flow of crude oil into two-phase flow of oil and water and increasing the oil flow resistance. In addition, when water becomes the continuous phase of flow, the shear surface of the flow is the surface of the sand grain, and as long as the flow rate increases slightly, the loose particles originally stabilized on the surface of the sand grain will be washed away, and accumulation will occur in the appropriate parts, blocking the flow pores and seriously reducing the permeability, as shown in Fig. 10.1b. With the flow rate remaining constant, the production pressure differential increases due to the decrease in permeability, thus providing both phases of sand discharge conditions are created.

**Produces a water-lock effect and increases resistance to oil flow**

This effect can be illustrated by the curve shown in Fig. 10.2. The curve is drawn from experimental data obtained by using specific core specimens for the experiment. The curves shown in Fig. 10.2 represent the effect of varying the water saturation within the pore on the relative permeability of oil and water. Position 1–1' is the original condition with 20% bound water in the core. At this point the relative permeability

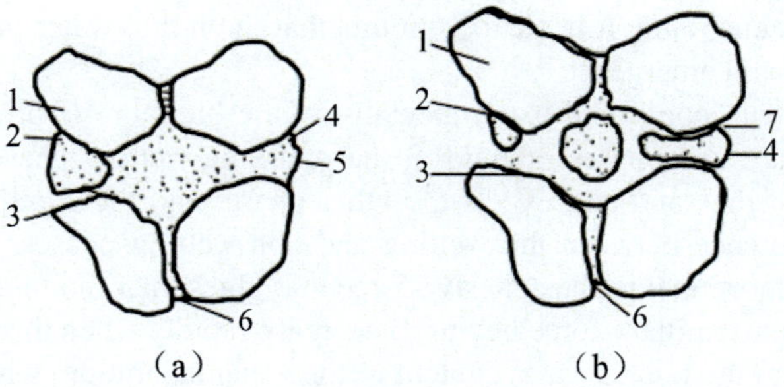


**Fig. 10.1** Effect of water intrusion on the flow of crude oil within the reservoir. **a** 1—Oil sand particles; 2—Loose particles; 3—Shear surface; 4—Oil flow; 5—Cementation, **b** 1—Oil sand particles; 2—Loose sand particles; 3—Shear surface; 4—Water; 5—Small oil streams; 6—Cements

Fig. 10.2 Relative permeability curve

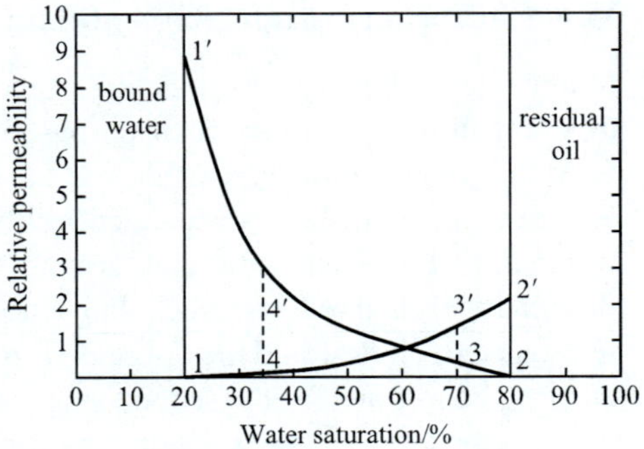


of the oil phase is equal to 0.9 and the relative permeability of the water phase is zero. If the absolute permeability is $10 \times 10^{-3}$ $\mu m^2$, the effective permeability of oil phase is $9 \times 10^{-3}$ $\mu m^2$. With the increase of water content saturation, the relative permeability of oil will keep decreasing, while the relative permeability of water increases. At position 4–4', water content saturation is 34%, but the effective reference permeability of oil phase is only $3 \times 10^{-3}$ $\mu m^2$, which is 1/3 of the original condition. Position 4 indicates that the effective water phase permeability is very low. It can be seen that the water content saturation only increased by 14%, while the effective oil permeability and oil flow rate both decreased significantly.

There is a difference between capillary force and cohesive between particles. From field practice, for example, in the North Sea oil field, a well drilled in 1981 started with a reservoir pressure of 44.13 MPa and a production rate of 795 $m^3$/d, and after the supplemental perforation the production rate reached 1908 $m^3$/d, and even at such a high production rate, no sand came out. As the reservoir pressure decayed, as shown in Fig. 10.3, water intrusion began by September 1985, but still no sand was produced. Subsequently, as water injection increased reservoir pressure, sudden and severe sand production began in January 1986, with sand production as high as 570 g/$m^3$. The well did not produce sand when the reservoir pressure dropped and water intrusion began to occur. Instead, significant sand production began only after water injection, when reservoir pressure increased and water-bearing saturation reached a certain value. It is clear from this that the initial water intrusion was not the cause of sand emergence.

This has been confirmed experimentally in the literature (Durrett et al. 1977). When there are two phases of fluid in the pore, the wetting phase spreads along the surface of the sand and expels the other phase, and the capillary force is the pressure difference between the wetting and non-wetting phases, and the surface tension is proportional to the adhesive force that glues the sand together. As shown in Fig. 10.4, the capillary force begins to decrease rapidly when the water content is increased from the bound water content to the water saturation, while the adhesive force increases, and is greatest when the water saturation is about 80%. When the pore medium is completely filled with a liquid, there is no cohesive between the sand grains.

This is briefly analyzed below, assuming a spherical pore of radius R with a sand grain of radius r, as shown in Fig. 10.5, the relationship between the oil–water

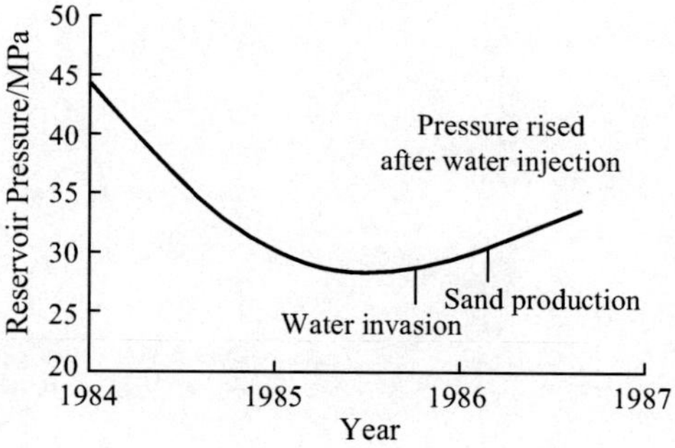


**Fig. 10.3** Example of sand emergence from a well

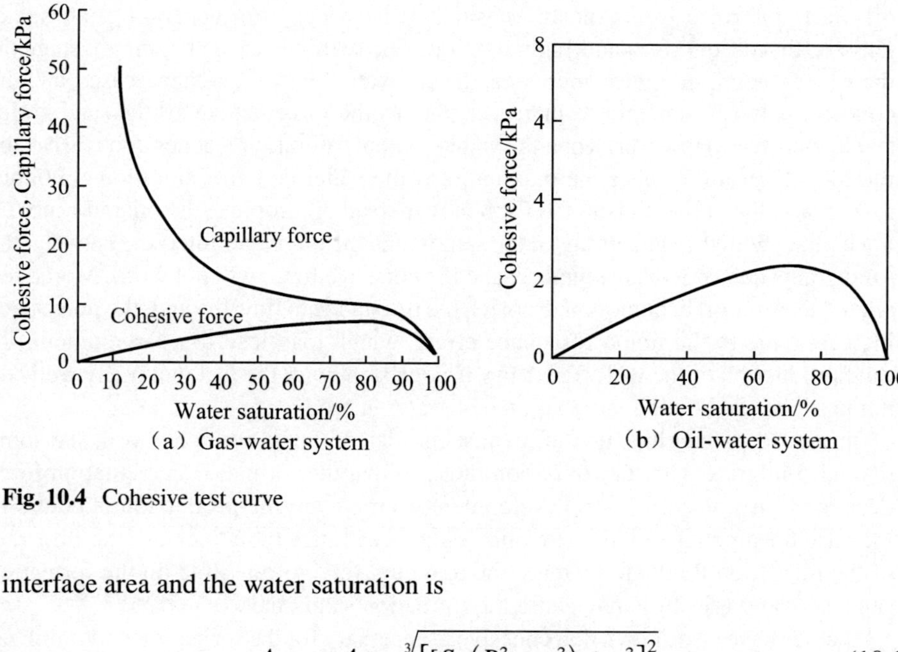


(a) Gas-water system (b) Oil-water system

**Fig. 10.4** Cohesive test curve

interface area and the water saturation is

$$A_{\text{o-w}} = 4\pi \cdot \sqrt[3]{\left[\left[S_{\text{w}}\left(R^3 - r^3\right) + r^3\right]^2\right.} \tag{10.1}$$

where the porosity is

$$\phi = 1 - \frac{r^3}{R^3} \tag{10.2}$$

Then, the relationship between oil–water interface area and porosity is.

$$A_{\text{o-w}} = 4R\pi \cdot \sqrt[3]{[S_{\text{w}}\varphi + 1 - \varphi]^2} \tag{10.3}$$

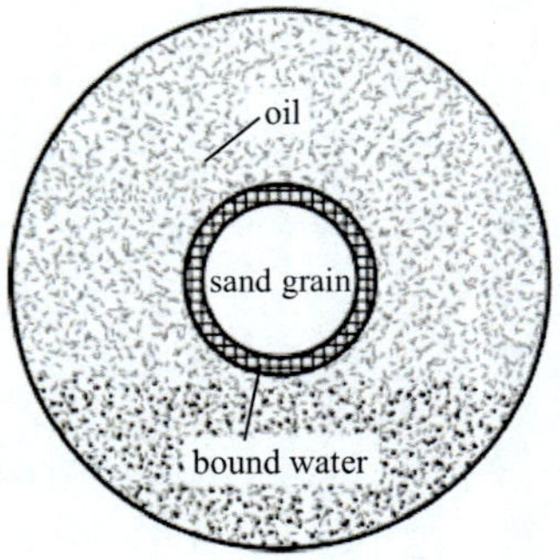


**Fig. 10.5** Cohesive analysis model

For hydrophilic sandstones, when the water content saturation is relatively low, an oil–water interface forms on the outside of the water film, generating interfacial tension. From Eq. (10.3) and Fig. 10.4, it can be seen that at the beginning stage, as the water content saturation increases, the oil–water interface area increases, and the cohesive between sand grains increases, that is, the force that keeps the sand grains stable increases. However, when the water content saturation reaches about 80%, the interfacial tension reaches the maximum; if the water content saturation continues to increase, the crude oil is in the form of dispersed oil droplets, the oil and water do not contact with the sand grains at the same time, the oil–water interface between the sand grains does not exist anymore, and the cohesive force will not work. Moreover, when the crude oil becomes oil droplets, the resistance to flow through the pore space increases due to the liquid resistance effect, which may lead to a large amount of sand production of the well. This view is consistent with the fact that many wells do not produce sand until a large volume of water is produced.

In addition, water injection may cause the clay of the reservoir to swell, and some may also migrate with the formation fluid, so that the formation cementation force decreases. In water injection development, to maintain the production is bound to increase the amount of fluid extraction, which will inevitably increase the flow rate of the formation fluid and increase the dragging force of the fluid on the formation sand, so water injection may cause the formation sand out.

Studies have also shown that cohesive is necessary for the formation and stabilization of sand arches, and that severe intrusion of water-based fluids usually results in the destruction of sand arches and causes sand emergence, so sand prevention requires reducing water intrusion and minimizing drilling fluid leakage, cement slurry water loss, and intrusion of injection, completion, and formation water.

#### 10.1.2.5 Impact of Gas Intrusion

It was found through experiments that the gas overflowing from the produced liquid has a great influence on the sand output, which increases 60 times when the gas–liquid ratio is 6.7; when the gas–liquid ratio is 9.4, the sand output increases 2000 times. This shows that the gas has a great influence on the sand output. The following is some exploratory analysis of the effect of dissolved gas on sand output (Zhang 2000).

**Oil and gas two-phase flow increases oil flow resistance**

The gas in the original formation contains both dissolved and free gas states. The free gas that is focused at the top of the gas is called gas-top gas; the gas in the dissolved state that is uniformly saturated within the oil is called dissolved gas. During the development of an oil field, the formation pressure decreases continuously due to the consumption of energy from the reservoir itself, and when the pressure at the bottom of the well falls below the saturation pressure, the gas originally dissolved in the crude oil near the bottom of the well separates out and a two-phase flow of oil and gas appears near the bottom of the well.

After the gas is separated from the oil, it starts as small free bubbles evenly dispersed in the oil, and the bubbles keep increasing as the pressure of the oil layer decreases. When the bubbles pass through the pore throat, the resistance increases due to the deformation of the oil–gas interface, and this phenomenon is called the Jamin effect. As shown in Fig. 10.6, the fluid resistance increases and the drag force on the sand particles increases, which creates the conditions for the second stage of sand emergence.

$$p_2 = p_1 - 2\sigma\left(\frac{1}{R_1} - \frac{1}{R_2}\right) \tag{10.4}$$

where $R_1$ and $R_2$ are the two different radii of curvature of the bubble, respectively.

Considering that the oil flows into the well through numerous such defiles, and that the bubbles in the oil are increasing in size, and that the bubbles undergo siltation, these resistances will add up to a substantial number. The increase in flow resistance is equivalent to giving impetus to the movement of the sand particles, resulting in an increase in the amount of sand coming out.

**Fig. 10.6** Jamin effect and bubble siltation

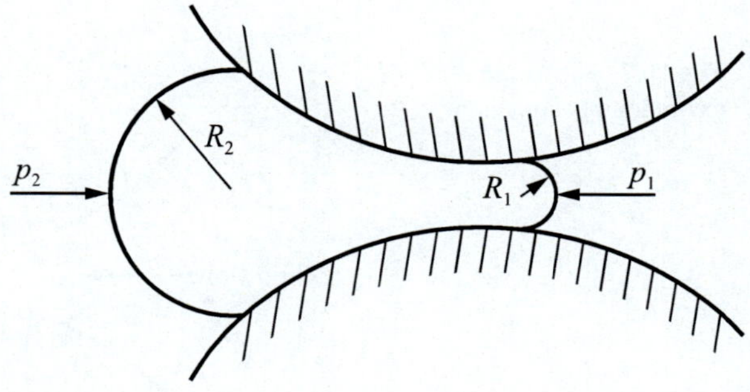

### Cavitation erosion of rock particles

The bubbles in the formation first scramble into the large pore channel, and due to the abnormal viscosity of the bubble film layer and the change in the radius of the pore channel, the large pore channel is blocked, the resistance keeps increasing, and the bubbles enter the smaller pore. In addition, the bubbles can adapt to the odd shape of the pores, and due to the antifoaming effect of the formation, the bubbles break before and after, so that the alternating stresses acting on the rock skeleton may cause fatigue damage.

In the local area where the liquid flows through, when the pressure is below a critical value, the liquid will cavitate. In the low-pressure region, the cavitated liquid carries a large number of bubbles to form a "two-phase flow" movement, as shown in Fig. 10.7, thus disrupting the macroscopic continuity of the flow, and the bubbles collapse when the produced liquid carries them through the higher pressure region. Since the bubbles generate a large instantaneous pressure when they collapse, when the collapse occurs near the rock surface, the repeated action of the high pressure generated by the continuously collapsing bubbles in the produced fluid will not only destroy the rock surface and cause "cavitation", but also the instantaneous high pressure generated by the bubble rupture may cause the sand particles to start and enter the borehole.

Due to the impact of the cavitation jet, the sand is initiated at a much higher rate than the flow rate of the produced fluid, which makes it easy for the sand to flow out on the one hand; on the other hand, it is equivalent to an abrasive jet, which will produce a greater impact on other sand, gravel fill or casing, thus destabilizing the fill and causing damage to the casing or certain mechanical anti-sand tools.

### Formation of an gas column, which reduces borehole pressure and accelerates shear damage to the well wall

During production, gas enters the borehole and as the pressure decreases a column of gas is formed, which further reduces the pressure in the well. As the pressure at the bottom of the well decreases, the well wall is prone to shear damage, thus laying the foundation for the first stage of sand production.

**Fig. 10.7** Formation of cavitation phenomenon

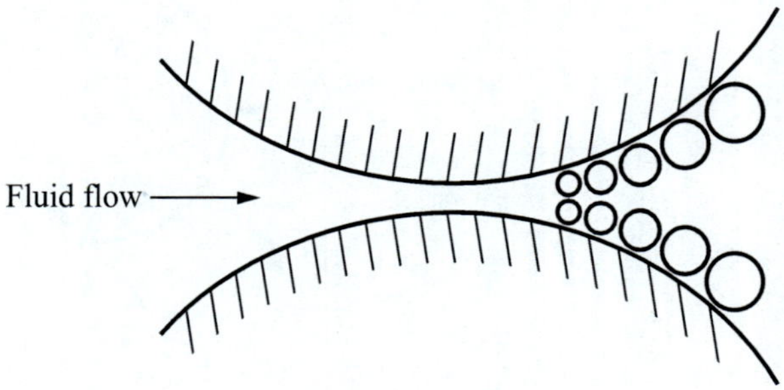

In addition, when the well is shut in, the natural gas slips off and rises due to its density being less than the produced fluid density and has a tendency to accumulate up through the produced fluid at the wellhead. Since the well is shut in, the natural gas cannot expand, so the volume of the natural gas does not change during the ascent, which allows the pressure of the natural gas to remain at its original bottom-of-well pressure value during the ascent. When the natural gas rises to the surface, this pressure is added to the producing fluid column and acts throughout the wellbore, causing excessive bottomhole pressure, which, if the well is opened for production, will produce large pressure fluctuations that will destabilize the formation or sand arch and increase the volume of sand coming out. For example, a large amount of sand emerges after shutting down a well and then producing from a weakly cemented natural gas reservoir in the Aragaska field, which may be caused by the above-mentioned reasons.

### *10.1.3 Hazards of Sand Emergence and Prevention*

Sand emergence from oil and gas wells is one of the major problems encountered in oil extraction. The hazards of sand production can be summarized into three categories: (1) wear and corrosion of downhole and wellhead oil recovery equipment. Sand in the production fluid makes various oil recovery pumps and pipelines subject to wear and tear, which greatly shortens their life span. For oil transmission pipelines, the corrosion rate is accelerated by abrasive wear. (2) Borehole stability problems. For example, the borehole is scrapped due to sand out, and the casing is squeezed and destroyed due to excessive sand out. Through the analysis of the information of a sparse sandstone block in Shengli oilfield found that: with the intensification of sand and the reduction of formation pressure, casing loss wells increase year by year, the sparse sandstone reservoir well casing damage is serious, the total number has accounted for more than 10%, individual blocks have reached more than 30%. (3) The problem of handling sand out of the borehole. If the sand comes out of the borehole, anti-sand measures have to be taken; the oil sand circulated to the surface has to be treated for environmental protection and other reasons, which increases the additional cost.

In view of the hazards of sand production, a variety of sand control methods have been adopted, which can be summarized into two major categories: first, natural completion method of sand control, i.e., using oil recovery parameters such as differential pressure and production rate to control sand emergence from the well; second, active sand control methods, including gravel filling, wireline screen tubing, chemical cementing and sand arch stabilization. Among them, gravel filling is one of the most important methods, which involves filling the gravel near the screen. There are two methods of gravel filling: gravel filling in the casing and bare eye gravel filling. The size of the gravel is determined empirically based on the size of the outgoing sand particles. Wire-wound screen pipe sand control is a sand control method that uses the gaps between the wires to filter the formation outgoing sand, and fluid passes through the gaps into the production pipe column, preventing most of the formation sand from entering the production pipe column. The chemical cementation method is a method of injecting a chemical substance near the borehole to increase its strength to prevent sand emergence, but this method reduces the permeability of the formation. The sand stabilization arch method or high-pressure filling method is a method of mechanically compacting the formation near the borehole with a hydraulic packer, which aims to improve the bridging and plugging ability between the formation particles. The common denominator of active sand control methods is that they are expensive (fabrication and maintenance) and can reduce the productivity of the well.

## 10.2 Analysis of the Sand Production Mechanism

Oil well sand emergence mechanism has received increasing attention as a theoretical basis for sand prevention and sand emergence prediction. There are two main approaches to the study of the sand emergence mechanism of oil wells, namely experimental studies and theoretical studies. The experimental study is to observe the sand emergence behavior under controlled conditions, so as to understand the sand emergence mechanism and the influence law of various production process parameters on sand emergence. At the same time, the sand emission simulation experiments can verify the theoretical model, and there are two main research methods: one is to conduct large-scale sand emission simulation experiments, which has the advantage of being able to simulate the actual sand emission phenomenon in a more realistic way, but the disadvantage is that the equipment is huge, the experimental cost is high, and it is not reproducible; the other is to conduct small-scale sand emission simulation experiments, which has the advantage of being able to simulate several factors that control the sand emission, and the experiments are flexible, simple and reproducible. The disadvantage is that it differs greatly from the actual sand emergence situation. The theoretical study of sand emergence is to assume the rocks around the borehole into a certain physical model, and then determine whether the rocks are damaged according to the mechanical intrinsic relationship of the materials, the force situation, combined with certain damage criteria.

### *10.2.1 Differential Production Pressure Required for Fluid Flow*

For high permeability reservoirs, increasing production means increasing the production pressure differential. And an increase in production pressure differential can increase the level of bias stress in the formation, which, if it exceeds the formation shear strength, can cause formation damage and lead to sand out of the well.

It is important to understand the relationship between production pressure differential and formation pressure drop in the near borehole zone. In addition to energy loss from Darcy flow, production pressure differential must overcome flow injuries such as (1) permeability injuries, radial flow convergence, and momentum effects due to changes in the stress field in the near-hole zone during drilling. (2) Borehole flow injuries, such as bottomhole imperfections, shot hole and surface skin injuries. (3) Drilling fluid solid phase injury. Production pressure differential not only perturbs the pore pressure, but also leads to stress concentration in the near-well zone. The rapid increase in effective stress in the near-well zone can increase the risk of formation damage in the early stages of production. Late production can also cause formation damage as formation pressures decay.

### *10.2.2 Stress in the Near-Wellbore*

Prior to development, in situ stresses within the reservoir are balanced and stable. Changes in in situ stresses, for whatever reason, affect the stability of the reservoir. In order to restore the stability of the formation itself, the stress field is automatically adjusted. It is well known that the in situ stress within a reservoir is represented by three principal stress components: the overburden pressure, the minimum horizontal in-situ stress, and the maximum horizontal in-situ stress. These stress components act within the reservoir, and in the vicinity of the borehole, their stress magnitudes change, creating a stress concentration zone. The stresses in the stress concentration zone may exceed the strength of the sparse sandstone and cause formation damage. The degree of stress concentration near the borehole and the mechanism of formation damage are controlled by the mechanical properties of the formation. Understanding the mechanical properties of the formation is critical in selecting an appropriate model to predict the formation damage mechanism. Formation damage can lead to the movement of formation sand and subsequent sand production out of the formation.

Formation sands exist in two forms: skeletal sands and loose sands. For skeletal sands, the stress concentration at the borehole may lead to shear damage and thus sand emergence, and this problem can be described by rock mechanics theory. However, for loose sands in weakly consolidated formations, it is difficult to describe accurately with this theory, and a more complex theory should be sought for analysis.

Drilling a circular borehole in a sandstone formation redistributes the stresses near the borehole, and the stress distribution pattern of the surrounding rock in a vertical borehole is given in Chap. 8. The stress state of the borehole confining formation and the reservoir pore pressure determine whether the borehole collapses or fractures.

Stress concentrations in the borehole confining generated by the drilling process may create damage zones near the borehole. Borehole damage is a natural phenomenon that mitigates stress concentrations when tangential stresses in the wellbore exceed the formation strength. Although borehole damage mitigates the stress concentration, it creates dilation phenomenon that creates the necessary conditions for sand emergence. Therefore, evaluating whether a well has produced damage during drilling is critical to the sand emergence analysis. If borehole collapse has occurred during drilling, the fine design of the perforation becomes critical.

Loose sands are generally non-adhesive and their mechanical properties are related to the state of stress. The stress state near the borehole may not conform to the principles of linear elasticity mechanics. A plastic yield zone may be generated near the borehole such that the difference between the maximum horizontal stress and the minimum horizontal stress converges to zero. The yielding of the formation within the plastic zone is the primary cause of sand production. As the yield zone increases, sand emergence continues. Loose sands exhibit nonlinear characteristics, and their sand-out evaluation methods should be different from those of skeletal sands. The mechanisms of sand emergence for nonlinear loose sands are dilation, capillary forces, and cohesive damage.

A borehole or perforation hole in a sparse sandstone formation typically produces a plastic zone, and damage to the formation is controlled by one or more of these mechanisms. Production and cyclic loading are the primary causes of plastic zone extension into the formation. In addition, the effect of reservoir pressure decay should be considered.

## *10.2.3 Mechanisms of Stratigraphic Damage*

A well should be analyzed for sand-out mechanisms prior to sand control treatment. From an engineering perspective, the two most common types of formation damage are borehole damage caused by stress action and borehole damage caused by chemical action.

### 10.2.3.1 Stress-Induced Borehole Damage

Opening the borehole will create a new stress distribution in the borehole confining. If the stress is high enough it may lead to damage to the well wall. Similarly, in drilling, completion, and treatments, the borehole is subjected to hydrostatic pressure and damage may occur. For the polar system, shear stress is a function of vertical, tangential, and radial stresses, and too large a value may cause formation sanding. As the bottomhole pressure increases, the radial stress increases and the tangential stress decreases. If the tangential stress changes from compressive to tensile, the wellbore will be damaged in tension. Most sedimentary rocks can withstand large compressive stresses, but damage may occur under very small tensile stresses. As the bottomhole pressure decreases (drilling fluid density decreases or production pressure differential increases), the tangential stress in the wellbore increases and the radial stress decreases. Larger production pressure differentials can result in shear stresses greater than the shear strength of the formation, creating a shear damage zone, when sand particles may flow with the fluid, producing massive sand outflow. Figure 10.8 represents a plot of shear stress versus effective normal stress, giving the

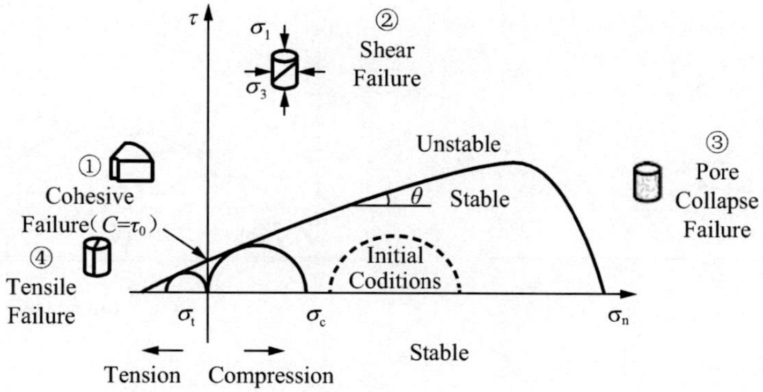


**Fig. 10.8** The four main mechanisms that cause sand out

stress locations corresponding to the four modes of damage. The diagonal straight line in the figure represents the shear strength of the formation, the slope is the internal friction coefficient of the formation, the intercept with the y-axis is the cohesion of the formation, and the intersection with the x-axis represents the tensile strength of the formation.

**Tensile damage**

For vertical wells, if the borehole fluid hydrostatic pressure is sufficiently high, the tangential effective stress in the wellbore will exceed the tensile strength of the formation and produce tensile damage in the direction of the maximum horizontal ground stress, which is determined by the following criterion.

$$\sigma_\theta - P_p = -\sigma_t \quad (10.5)$$

where $p_p$ is the reservoir pressure, $\sigma_\theta$ is the tangential stress at the wellbore,$\sigma_t$ is the tensile strength.

For weakly cemented sandstone reservoirs with high clay content, if the well is producing with a large production pressure differential, the radial stress on the wellbore wall will also be greater than the tensile strength of the formation and produce tensile damage, which is determined by the following guidelines.

$$\Delta p_w = \sigma_t \quad (10.6)$$

That is, if the production rate is very high, a very high pressure differential will be generated at the wellbore wall, separating the formation and causing tensile damage. Therefore, the production rate should be limited to a level such that the pressure differential does not exceed the tensile strength.

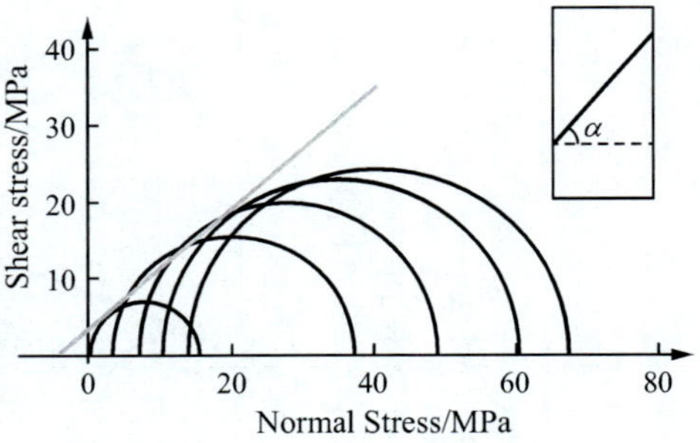


**Fig. 10.9** Damage molar circle, damage envelope and rupture angle for sparse sandstone

**Shear damage**

During the production of a vertical oil well, the pressure in the borehole is less than the formation pressure, and a strong stress concentration occurs near the borehole. For high-strength formations, the rock surrounding the borehole may behave elastically, and shear damage may occur if the formation shear strength is low. For weak sandstones, shear damage causes plastic yielding of the rock, resulting in borehole reduction; for brittle formations, shear damage can cause the borehole diameter to expand. Once shear damage is formed, the sandstone particles on the shear damage surface become free particles.

Determination of shear damage in the formation using the Mohr–Coulomb criterion.

$$\tau_f = \tau_0 + \sigma_n \tan\phi \tag{10.7}$$

Figure 10.9 gives the damage molar circles, strength damage envelopes and rupture angles obtained by triaxial strength experiments for a set of typical sparse sandstones in the Shengli oilfield.

**Cohesive damage**

Cohesive strength is the controlling factor in the erosion of formation free surfaces. The free surface of a formation can be the perforation channel, the borehole wall of a open-hole completion, the fracture surface or shear surface, the interface between formations, or the bond with the cement ring. As shown in Fig. 10.10, the shear strength of a formation consists of two components, the physical association between particles (cohesion) and the friction between particles. Cohesion is generated by two factors: the cementing force of the particles and the surface tension between the particles. If the drag force caused by fluid flow is large enough to exceed the cohesive strength of the formation (cohesion), wellbore wall instability or sand emergence begins to occur. Equation (10.8) gives an expression for the occurrence of cohesive damage at the surface of a borehole or perforation hole.

$$\frac{\mathrm{d}p}{\mathrm{d}r} = \tau_0 \tag{10.8}$$

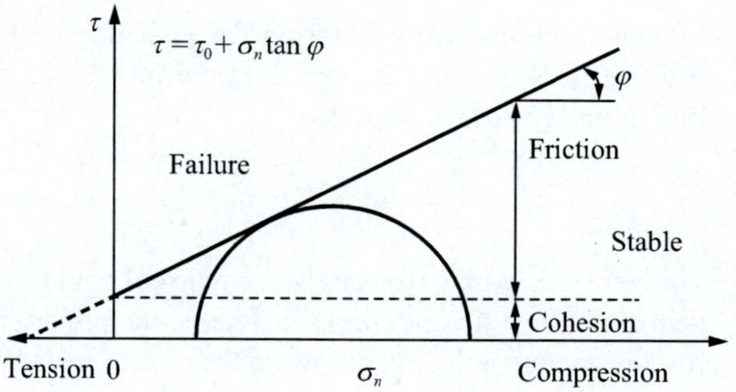


**Fig. 10.10** Two components of formation shear strength (friction and cohesion)

where $\tau_0$ is the cohesion of the formation, i.e., the intercept of the strength envelope. This type of damage produces fine powder yarns. The critical production rate for sand-free production can be found when the pressure drop is equal to the cohesive strength. It can be seen that when the pressure drop near the borehole is high, low adhesive strength is the cause of sand production.

## Pore collapse

For loose sandstone gas reservoirs, as the development time increases, the formation pressure decays severely, causing the effective stress in the formation to increase significantly. According to the definition of effective stress,

$$\sigma' = \sigma - \alpha P \tag{10.9}$$

The Biot coefficient $\alpha$ indicates the deformation of the rock skeleton and the pore pressure transfer efficiency, with values between 0 and 1. If the pore pressure transfer efficiency is low, the effect of pore pressure to offset the confing pressure will be reduced. According to the poroelastic theory, several experimental methods can be used to determine the Biot coefficient $\alpha$.

The stress state at a point within the formation can be represented by a molar circle. For highly porous and permeable reservoirs, the Biot coefficient $\alpha = 1$. If the formation pressure decay is severe, the Mohr circle characterizing the stress state at any point within the formation will shift to the right, as shown in Fig. 10.11.

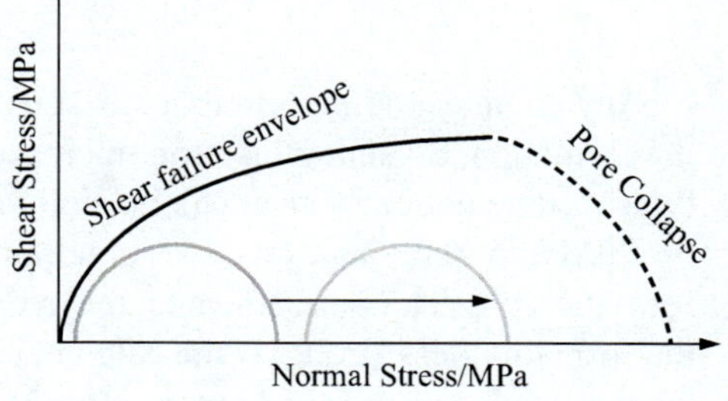


**Fig. 10.11** Pore collapse mechanism

The Coulomb-Mohr criterion for considering pore collapse is the envelope in Fig. 10.11. Where the diagonal straight line indicates shear damage and the dashed line indicates pore collapse.

$$\tau = \tau_0 + (\sigma_n - \alpha p_p) \tan \phi \tag{10.10}$$

Figure 10.11 gives the sandstone pore collapse mechanism, when the pore pressure decreases, the effective stress increases, the molar circle moves to the right, and the formation pore will collapse when the molar circle is tangent to the dashed line. Pore collapse is permanent damage and cannot be recovered. Once pore collapse occurs, the reservoir production will drop rapidly and even lead to the scrapping of the well.

#### 10.2.3.2 Chemically Induced Stratigraphic Damage

The dominant mechanism of stratigraphic damage may also be chemical. The absorption of water by clay in the formation can cause volume expansion or swelling pressure. This swelling pressure can shear the strength of the association between the particles and disintegrate the rock skeleton. The ability to produce volume deformation and swelling pressure varies by formation type. Water in the pores of a formation affects the strength of the formation in three ways: (1) by reducing internal friction; (2) by reducing the surface tension of the water-wetted formation; and (3) by reducing the cementation strength of the formation due to chemical action.

The results of rock mechanics experiments show that changes in water content affect the compressive strength and elastic parameters of the formation. The softening factor is defined as the ratio of the uniaxial compressive strength of dry rock to the uniaxial compressive strength of wet rock. The results of the softening coefficient for a typical loose sandstone specimen are given here.

$$F_{\text{soft}} = \frac{\sigma_c(\text{dry})}{\sigma_c(\text{wet})} = \frac{12.4\,\text{MPa}}{2.31\,\text{MPa}} = 5.3 \tag{10.11}$$

The cohesion of dry and wet rocks is

$$\begin{aligned} \tau_0(\text{dry}) &= 1.45\,\text{MPa} \\ \tau_0(\text{wet}) &= 0 - 0.14\,\text{MPa} \end{aligned} \tag{10.12}$$

Any chemical admixture that weakens the formation should not be injected into the formation. Scanning electron microscopy and X-ray analysis can determine the type of cementation, such as calcite, dolomite, illite, and imonite mixed layers. Based on SEM and X-ray analysis, two important insights have been developed: (1) If the formation is calcareous cemented, the hydrochloric acid in acid fracturing may make the formation less strong. If the calcium carbonate dissolves, the rock structure will be disrupted, causing sand outgrowth. (2) If the clay content of the rock is too high, a water-sensitive effect can occur. The swelling of the clay can affect the stability of the rock and cause sand outgassing.

## 10.3 Analysis of Sand Emergence Under Different Completion Methods

Different completion methods and different stress distribution patterns around the borehole channel lead to different critical states of sand emergence. Therefore, the study of sand emergence should be specific to the problem and should be carried out for different formations, different completion methods and different processes of sand emergence. At present, the main completion methods are open-hole completion method, through-hole completion method, liner tubing completion method, perforation completion method, screen tubing completion method, etc. For the first three completion methods, generally speaking, the formation is required to have a certain cementing strength, and the primary condition for sand emergence is the destruction of the formation destabilization; for perforation completion in a formation with a certain cementing strength, sand emergence is only possible if the perforation channel is destroyed. The key to sand prevention in both cases is to maintain the stability of the formation and the perforation channel. In the case of a perforation completion in a loose formation, sand will come out immediately or soon after the perforation, and a sand arch will be formed near the channel, so as long as the stability of the sand arch can be maintained, no sand production can be ensured. The optimization of sand control technology is the key to the situation where the stability of sand arch cannot be guaranteed.

### *10.3.1 Critical Sanding Conditions for Open-Hole Completions*

#### 10.3.1.1 Stress Analysis Around Vertical Well Walls

Assuming that the formation around the well wall is a porous elastic medium, the stress state around the wellbore can be solved by the following mechanical model: a circular borehole in an infinite plane is subjected to a uniform internal pressure $p_w$, while at an infinite distance from this plane it is subjected to two horizontal in-situ stresses $\sigma_H$,$\sigma_h$ and vertically to the gravity of the overlying rock layer, and the pore pressure $p_p$ within the formation is uniformly distributed, as shown in Fig. 10.12 (Zhang et al. 1999).

Considering the pressure transfer effect of pore pressure, the stress distribution in the surrounding rock of the wellbore is

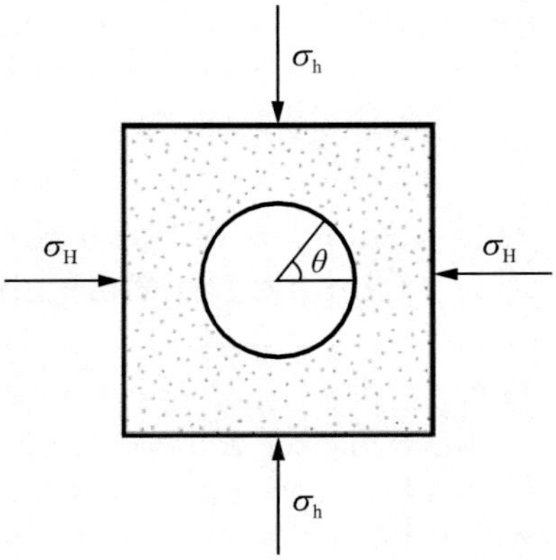


**Fig. 10.12** Mechanical model for production of bare boreholes in straight wells

$$\sigma_r = \frac{a^2}{r^2}P_w + \frac{1}{2}(\sigma_H + \sigma_h)\left(1 - \frac{a^2}{r^2}\right) + \frac{1}{2}(\sigma_H - \sigma_h)\left(1 + \frac{3a^4}{r^4} - \frac{4a^2}{r^2}\right)\cos 2\theta$$
$$+\delta\left[\frac{(1-\beta)(1-2\mu)}{2(1-\mu)}\frac{(r^2-a^2)}{r^2} - \phi\right](P_w - P_p) \quad (10.13a)$$

$$\sigma_\theta = -\frac{a^2}{r^2}p_w + \frac{1}{2}(\sigma_H + \sigma_h)\left(1 + \frac{a^2}{r^2}\right) - \frac{1}{2}(\sigma_H - \sigma_h)\left(1 + \frac{3a^4}{r^4}\right)\cos 2\theta$$
$$+\delta\left[\frac{(1-\beta)(1-2\mu)}{2(1-\mu)}\frac{(r^2+a^2)}{r^2} - \phi\right](p_w - p_p) \quad (10.13b)$$

$$\sigma_z = \sigma_v - 2\mu(\sigma_H - \sigma_h)\cos 2\theta + \delta\left[\frac{(1-\beta)(1-2\mu)}{1-\mu} - \varphi\right](p_w - p_p) \quad (10.13c)$$

$$\tau_{r\theta} = \frac{1}{2}(\sigma_H - \sigma_h)\left(1 + \frac{2a^2}{r^2} - \frac{3a^4}{r^4}\right)\sin 2\theta \quad (10.13d)$$

where $\sigma_H$ and $\sigma_h$ are the maximum and minimum horizontal in-situ stresses,$\sigma_v$ is the overburden pressure, $p_p$ is the original pore pressure of the reservoir, $p_w$ is the borehole pressure, $\beta$ is the volumetric compression factor of the rock, $\mu$ is the Poisson's ratio, $\phi$ is the rock porosity,$\delta$ is a function with $\delta = 1$ when the wellbore wall is permeable and $\delta = 0$ when the wellbore wall is impermeable.

For the vertical well production process, the radial stress is minimum and the tangential stress is maximum, therefore, the possibility of fracture from tensile damage in the tangential direction of the wellbore wall is small. In most cases, the sand production process consists of two stages: shear damage and tensile damage, i.e., shear damage occurs in the wellbore wall due to a drop in reservoir pressure or borehole pressure, and the rock becomes loose, and then tensile damage occurs due to the erosion of produced fluids to produce sand. Therefore, some wells start with high production and do not produce sand, but as production continues, the reservoir pressure decreases and low production occurs instead.

Since stress concentrations are generated near the borehole, making the stresses on the wall the highest, comparing the stresses on the wall with the strength criterion will determine whether the wall is stable.

The stress distribution at the wellbore wall is given by

$$\begin{cases} \sigma_r = p_w - \delta\phi(p_w - p_p) \\ \sigma_\theta = -p_w + \sigma_{\mathrm{H}}(1 - 2\cos 2\theta) + \sigma_h(1 + 2\cos 2\theta) + \delta\left[\frac{(1-\beta)(1-2\mu)}{1-\mu} - \phi\right](p_w - p_p) \\ \sigma_z = \sigma_v - 2\mu(\sigma_{\mathrm{H}} - \sigma_{\mathrm{h}})\cos 2\theta - \delta\left[\frac{(1-\beta)(1-2\mu)}{1-\mu} - \phi\right](p_w - p_p) \end{cases} \tag{10.14}$$

In Eq. (10.14), the circumferential and radial differential stresses are greatest along the direction of minimum horizontal in-situ stress when $\cos 2\theta = -1$, i.e., $\theta = \pm\frac{1}{2}$.

$$\begin{cases} \sigma_r = p_w - \delta\phi(p_w - p_p) \\ \sigma_\theta = -p_w + 3\sigma_H - \sigma_h + \delta\left[\frac{(1-\beta)(1-2\mu)}{1-\mu} - \phi\right](p_w - p_p) \\ \sigma_z = \sigma_v + 2\mu(\sigma_H - \sigma_h) - \delta\left[\frac{(1-\beta)(1-2\mu)}{1-\mu} - \phi\right](p_w - p_p) \end{cases} \tag{10.15}$$

#### 10.3.1.2 Development of the Sand Emergence Prediction Model

The most commonly used rock damage criterion is the Coulomb-Mohr criterion, which is expressed as

$$\sigma_1 = 2\tau_0 \tan\left(45^\circ + \frac{\varphi}{2}\right) + \sigma_3 \tan^2\left(45^\circ + \frac{\varphi}{2}\right) \tag{10.16}$$

where,

$\tau_0$ is the cohesion of the reservoir, MPa;
$\varphi$ is the angle of internal friction of the reservoir.

Assuming that the far-field reservoir pressure remains constant for a given production period, the relationship between wellbore wall stress and production pressure differential is

$$\begin{cases} \sigma_r = p_p + (1 - \delta\phi)\Delta p \\ \sigma_\theta = -\delta\left[\frac{(1-\beta)(1-2\mu)}{(1-\mu)} - \phi\right]\Delta p + \Delta p + 3\sigma_H - \sigma_h - p_p \\ \sigma_z = -\delta\left[\frac{(1-\beta)(1-2\mu)}{(1-\mu)} - \phi\right]\Delta p + \sigma_v + 2\mu(\sigma_H - \sigma_h) \end{cases} \tag{10.17}$$

Using Eq. (10.16) and 10.17), the critical production pressure difference $\Delta p$ for the reservoir not to produce sand can be determined.

$$\Delta p = \frac{p_p + 2\tau_0 \tan\left(45^\circ + \frac{\varphi}{2}\right) - 3\sigma_H + \sigma_h + p_p \tan^2\left(45^\circ + \frac{\varphi}{2}\right)}{(1+\phi\delta) - (1-\phi\delta)\tan^2\left(45^\circ + \frac{\varphi}{2}\right) - \frac{(1-\beta)(1-2\mu)\delta}{1-\mu}} \tag{10.18}$$

***Example 10.1*** A loose sandstone reservoir in an oil field with a well depth of 3600 m has a strength in accordance with the Coulomb-Moore criterion, a cohesion of 1.5 MPa, an internal friction angle of 20°, a reservoir porosity of 0.25, a modulus of elasticity of 1000 MPa, a Poisson's ratio of 0.32, a uniaxial tensile strength of 0.15 MPa, a formation pore pressure gradient of 12.5 kPa/m, and a borehole pressure at production of 40 MPa, if the overburden pressure is 80 MPa, the maximum horizontal in-situ stress is 68 MPa and the minimum horizontal in-situ stress is 60 MPa.

Using the open-hole completion method, the wellbore wall is well permeated ($\delta = 1$), try to determine if the reservoir is sandy when it is producing?

**Solution**

In situ stress component.

$$\sigma_v = 80\,\text{MPa}$$
$$\sigma_H = 68\,\text{MPa}$$
$$\sigma_h = 60\,\text{MPa}$$

Stratigraphic pore pressure.

$$P_p = 3600 * 12.5\,\text{KPa/m} = 45\,\text{MPa}$$

Reservoir volume compression factor.

$$\beta = \frac{3(1-2\mu)}{E} = \frac{3 \times (1 - 2 \times 0.32)}{1000} = 1.08 \times 10^{-3}(1/\text{MPa})$$

$$\tan\left(45^\circ + \frac{\varphi}{2}\right) = \tan\left(45^\circ + \frac{20^\circ}{2}\right) = 1.428$$

The critical production pressure difference for the reservoir without sand is

$$\Delta p = \frac{p_p + 2\tau_0 \tan\left(45^\circ + \frac{\varphi}{2}\right) - 3\sigma_H + \sigma_h + p_p \tan^2\left(45^\circ + \frac{\varphi}{2}\right)}{(1+\phi\delta) - (1-\phi\delta)\tan^2\left(45^\circ + \frac{\varphi}{2}\right) - \frac{(1-\beta)(1-2\mu)\delta}{1-\mu}}$$
$$= \frac{45 + 2 \times 1.5 \times 1.428 - 3 \times 68 + 60 + 45 \times 1.428^2}{(1+0.25) - (1-0.25) \times 1.428^2 - \frac{(1-1.08\times10^{-3})(1-2\times0.32)}{1-0.32}}$$
$$= 3.66\,\text{MPa}$$

The actual production pressure differential is

$$\Delta p' = 45 - 40 = 5\,\text{MPa}$$

thanks to $\Delta p' > \Delta p$

So the reservoir is producing out of sand.

### 10.3.2 Critical Sand-Out Conditions for Perforation Completions

At present, most oil and gas wells are completed by perforation completion, which starts with an elongated perforation path and can be simplified to the model shown in Fig. 10.13, where the perforation direction is arbitrary. For the convenience of the study, it is assumed that the hole is shot along the following two special directions, namely, along the direction of maximum horizontal in-situ stress and the direction of minimum horizontal in-situ stress (Cheng and Zhang 2001).

#### 10.3.2.1 Critical Production Pressure Differential for Perforation Along the Direction of Maximum Horizontal In-Situ Stress

As shown in Fig. 10.13a, the stress distribution around the perforation hole is shown by perforating along the direction of maximum horizontal in-situ stress, referring to Eq. (10.14), with the plumb direction as the x-axis and the direction of minimum horizontal in-situ stress as the y-axis.

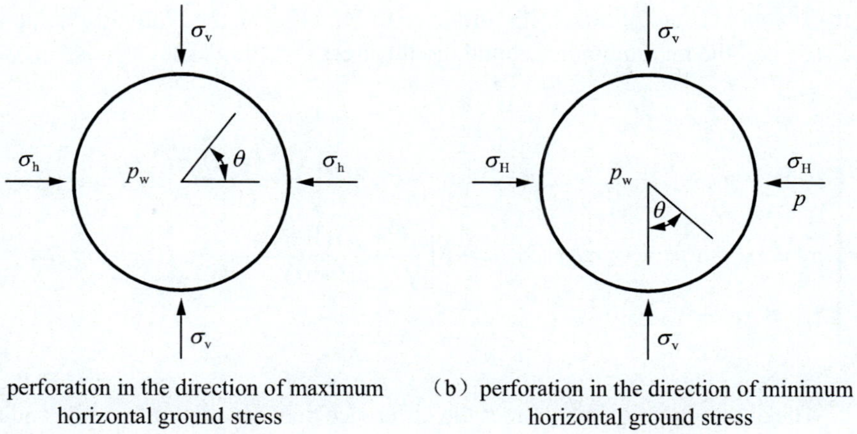


**Fig. 10.13** Shot hole borehole stress model

$$\begin{cases} \sigma_r = p_w - \delta\varphi(p_w - p_p) \\ \sigma_\theta = \sigma_h + \sigma_v - 2\cos 2\theta(\sigma_v - \sigma_h) - p_w - \delta\left[\frac{(1-\beta)(1-2\mu)}{(1-\mu)} - \varphi\right](p_w - p_p) \\ \sigma_z = \sigma_H - 2\mu(\sigma_v - \sigma_h)\cos 2\theta - \delta\left[\frac{(1-\beta)(1-2\mu)}{(1-\mu)} - \varphi\right](p_w - p_p) \\ \tau_{r\theta} = \tau_{\theta z} = \tau_{zr} = 0 \end{cases} \tag{10.19}$$

The above equation maximizes the difference between circumferential and radial stresses when $\cos 2\theta = -1$, i.e. $\theta = \pm\frac{\pi}{2}$, the direction of minimum horizontal in-situ stress, is reached. At this point.

$$\begin{cases} \sigma_r = p_w - \delta\varphi(p_w - p_p) \\ \sigma_\theta = -p_w + 3\sigma_v - \sigma_h + \delta\left[\frac{(1-\beta)(1-2\mu)}{(1-\mu)} - \varphi\right](p_w - p_p) \\ \sigma_z = \sigma_H + 2\mu(\sigma_v - \sigma_h) - \delta\left[\frac{(1-\beta)(1-2\mu)}{(1-\mu)} - \varphi\right](p_w - p_p) \\ \tau_{r\theta} = \tau_{\theta z} = \tau_{zr} = 0 \end{cases} \tag{10.20}$$

Using the Coulomb-Mohr criterion (10.16), the critical production pressure differential for sand exclusion from the well is

$$\Delta p = \frac{p_p + 2\tau_0 \tan\left(45^\circ + \frac{\varphi}{2}\right) - 3\sigma_v + \sigma_h + p_p \tan^2\left(45^\circ + \frac{\varphi}{2}\right)}{(1 + \phi\delta) - (1 - \phi\delta)\tan^2\left(45^\circ + \frac{\varphi}{2}\right) - \frac{(1-\beta)(1-2\mu)\delta}{1-\mu}} \tag{10.21}$$

#### 10.3.2.2 Critical Production Pressure Difference Along the Minimum Horizontal In-Situ Stress Direction of the Perforation

Similar to the above analysis, as shown in Fig. 10.13b, the stress distribution around the borehole is as follows, with reference to Eq. (10.14), the plumb direction as the $x$-axis and the maximum horizontal in-situ stress direction as the $y$-axis.

$$\begin{cases} \sigma_r = p_w - \delta\varphi(p_w - p_p) \\ \sigma_\theta = -p_w + \sigma_v + \sigma_H - 2\cos 2\theta(\sigma_v - \sigma_H) + \delta\left[\frac{(1-\beta)(1-2\mu)}{(1-\mu)} - \varphi\right](p_w - p_p) \\ \sigma_z = \sigma_h - 2\mu(\sigma_v - \sigma_H)\cos 2\theta - \delta\left[\frac{(1-\beta)(1-2\mu)}{(1-\mu)} - \varphi\right](p_w - p_p) \\ \tau_{r\theta} = \tau_{\theta z} = \tau_{zr} = 0 \end{cases} \tag{10.22}$$

The above equation maximizes the difference between circumferential and radial stresses when $\cos 2\theta = -1$, i.e. $\theta = \pm\frac{\pi}{2}$, the direction of maximum horizontal in-situ stress, is reached.

$$\begin{cases} \sigma_r = p_w - \delta\varphi\left(p_w - p_p\right) \\ \sigma_\theta = -p_w + 3\sigma_v - \sigma_H - \delta\left[\frac{(1-\beta)(1-2\mu)}{(1-\mu)} - \varphi\right]\left(p_w - p_p\right) \\ \sigma_z = \sigma_h + 2\mu(\sigma_v - \sigma_H) - \delta\left[\frac{(1-\beta)(1-2\mu)}{(1-\mu)} - \varphi\right]\left(p_w - p_p\right) \\ \tau_{r\theta} = \tau_{\theta z} = \tau_{zr} = 0 \end{cases} \tag{10.23}$$

Using the Coulomb-Mohr criterion (10.16), the critical production pressure differential is

$$\Delta p = \frac{p_p + 2\tau_0 \tan\left(45^\circ + \frac{\varphi}{2}\right) - 3\sigma_v + \sigma_H + p_p \tan^2\left(45^\circ + \frac{\varphi}{2}\right)}{(1 + \phi\delta) - (1 - \phi\delta)\tan^2\left(45^\circ + \frac{\varphi}{2}\right) - \frac{(1-\beta)(1-2\mu)\delta}{1-\mu}} \tag{10.24}$$

***Example 10.2*** Analyze the critical production pressure difference between the reservoir producing sand when the reservoir is shot along the maximum horizontal ground stress direction and when the reservoir is shot along the minimum horizontal ground stress direction for the Example 10.1 reservoir, which is extracted by the shot-hole completion method.

**Solution**

In-situ stress component

$$\sigma_{\text{v}} = 80\,\text{MPa}$$
$$\sigma_{\text{H}} = 68\,\text{MPa}$$
$$\sigma_{\text{h}} = 60\,\text{MPa}$$

Stratigraphic pore pressure.

$$P_{\text{p}} = 3600 * 12.5\,\text{KPa/m} = 45\,\text{MPa}$$

Reservoir volume compression factor.

$$\beta = \frac{3(1-2\mu)}{\text{E}} = \frac{3 \times (1 - 2 \times 0.32)}{1000} = 1.08 \times 10^{-3}(1/\text{MPa})$$

$$\tan\left(45^\circ + \frac{\varphi}{2}\right) = \tan\left(45^\circ + \frac{20^\circ}{2}\right) = 1.428$$

(1) According to Eq. (10.21), the critical production pressure difference when perforating in the direction of maximum horizontal ground stress is.

$$\Delta p = \frac{45 + 2 \times 1.5 \times 1.428 - 3 \times 80 + 60 + 45 \times 1.428^2}{(1 + 0.25) - (1 - 0.25) \times 1.428^2 - \frac{(1-1.08\times10^{-3})(1-2\times0.32)}{1-0.32}}$$
$$= 48.22\,\text{MPa}$$

(2) According to Eq. (10.24), the critical production pressure difference when shooting in the direction of minimum horizontal ground stress is

$$\Delta p = \frac{45 + 2 \times 1.5 \times 1.428 - 3 \times 80 + 68 + 45 \times 1.428^2}{(1 + 0.25) - (1 - 0.25) \times 1.428^2 - \frac{(1-1.08\times 10^{-3})(1-2\times 0.32)}{1-0.32}}$$
$$= 38.32\,\text{MPa}$$

Comparing Example 10.1 with Example 10.2, we can see that the critical production pressure differential for the bare-hole completion is much less than the critical production pressure differential for the shot-hole completion for the original ground stress state of this well.

### *10.3.3 Sand Arch and Its Stability Model*

For loose sandstones of much lower strength, sand emerges soon after the penetration hole and after a period of time, a sand arch forms around each borehole. The sand arch can provide stabilization to stratigraphic sands that have little or no cementation strength between the sand grains; to achieve this stabilization, the sand should be hydrophilic, but not too highly saturated with water. The surface tension of the system contributes to the formation and stabilization of the sand arch.

Figure 10.14 shows how the sand arch is formed and why it acts as a stabilizer for the formation. When the casing perforation is completed, hydrocarbons begin to produce (Fig. 10.14a). A certain volume of formation sand collapses down around each borehole and begins to flow into the borehole through the borehole. As it continues to collapse, more and more sand comes out, and a sand arch forms around the borehole right in the collapsed area (Fig. 10.14b). The combined forces of surface tension and friction between the produced fluid and the sand grains stabilize the sand arch, and if this combined force is overcome, the sand arch collapses. The frictional force is generated by the load of the overlying rock layer acting on the sand. As shown in Fig. 10.14c, a sand arch collapses when the stress on the sand layer is greater than the force that stabilizes the sand arch. If an area of the sand formation collapses, a large sand arch forms within the expanded sand formation, leaving the formation in a stable state again until a new collapse occurs.

In the following, the effect of fluid flow on the stability of the sand arch is investigated by analyzing the stresses in the sandstone (Bratli and Risnes 1981).

To study the stability of a sand arch, one must first know the stress distribution in the sandstone, which is generally a very complex problem. For simplicity, assume that the sand arch is symmetrical and spherical, the mechanical model as shown in Fig. 10.15, with an inner diameter of $R_1$, an outer diameter of $R_2$, a fluid pressure on the inner surface of $p_1$, and a fluid pressure on the outer surface of $p_2$.

Due to the plastic zone within the sand arch, the Coulomb-Mohr criterion should be satisfied.

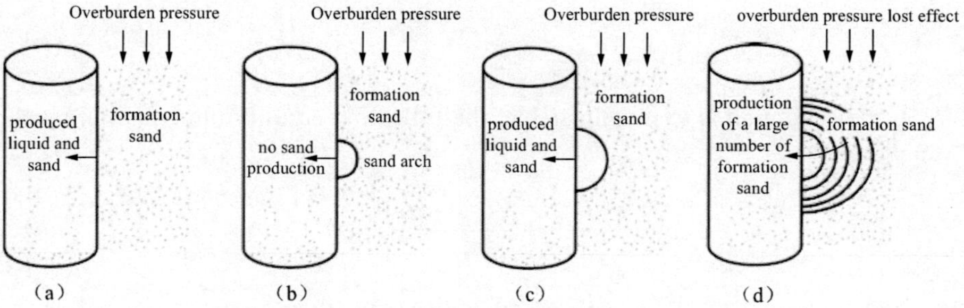


**Fig. 10.14** Sand emergence process for shot-hole completions in sparse reservoirs

**Fig. 10.15** Mechanical model of the sand arch

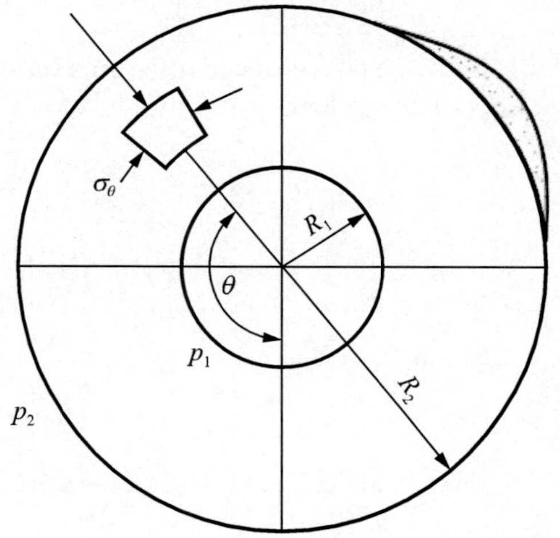


$$\sigma_\theta - \alpha p_p = 2\tau_0 \frac{\cos\varphi}{1-\sin\varphi} + \left(\sigma_r - \alpha p_p\right)\frac{1+\sin\varphi}{1-\sin\varphi} \tag{10.25}$$

For single-phase flow, the pore pressure distribution near the sand arch is

$$p_r = p_1 + \frac{q\mu}{4\pi K}\left(\frac{1}{R_1} - \frac{1}{r}\right) = p_2 - \frac{q\mu}{4\pi k}\left(\frac{1}{r} - \frac{1}{R_2}\right) = g_1 - \frac{g_2}{r} \tag{10.26}$$

where

$$g_1 = P_1 + \frac{q\mu}{4\pi k}\frac{1}{R_1} = P_2 + \frac{q\mu}{4\pi k}\frac{1}{R_2}$$

$$g_2 = \frac{q\mu}{4\pi k}$$

The pore pressure gradient is

$$\frac{dp_r}{dr} = \frac{g_2}{r^2} \tag{10.27}$$

If the sand arch is to remain stable, the following equilibrium equation must be satisfied.

$$\frac{d\sigma_r}{dr} + \frac{2(\sigma_r - \sigma_\theta)}{r} = 0 \tag{10.28}$$

Substituting (10.23) into (10.26), such that$\alpha = 1$, gives

$$\frac{d\sigma_r}{dr} = \frac{4}{r}\left[\frac{\sin\phi}{1-\sin\phi}\sigma_r + \frac{\cos\phi}{1-\sin\phi}\tau_0 - \frac{\sin\phi}{1-\sin\phi}p_p\right] \tag{10.29}$$

Since the pore pressure in the arch varies with radius, substituting Eq. (10.26) into Eq. (10.29) yields

$$\frac{d\sigma_r}{dr} = \frac{\frac{4\sin\varphi}{1-\sin\varphi}\left[\sigma_r + \frac{\tau_0}{\tan\varphi} - \left(g_1 - \frac{g_2}{r}\right)\right]}{r} \tag{10.30}$$

Solving the differential Eq. (10.30) yields.

$$\sigma_r = C_1 r^{\frac{4\sin\varphi}{1-\sin\varphi}} - \frac{\tau_0}{\tan\varphi} + g_1 - \frac{4\sin\varphi}{1+3\sin\varphi}\frac{g_2}{r}$$

Inner boundary conditions of the sand arch: $\sigma_\mathrm{r} = p_1$, $p_\mathrm{r} = p_1$, for $r = R_1$

$$\text{Got } C_1 = \left(\frac{\tau_0}{\tan\varphi} - g_1 + \frac{4\sin\varphi}{1+3\sin\varphi}\frac{g_2}{R_1} + p_w\right)R_1^{\frac{1-\sin\varphi}{4\sin\varphi}}$$

It further follows that.

$$\begin{aligned}\sigma_r = &-\frac{\tau_0}{\tan\varphi} + g_1 - \frac{4\sin\varphi}{1+3\sin\varphi}\frac{g_2}{r} \\ &+ \left(\frac{\tau_0}{\tan\varphi} - g_1 + \frac{4\sin\varphi}{1+3\sin\varphi}\frac{g_2}{R_1} + p_w\right)\left(\frac{r}{R_1}\right)^{\frac{4\sin\varphi}{1-\sin\varphi}}\end{aligned}$$

According to the definition of effective stress, the radial effective stress in the sand arch is

$$\sigma_r^{'} = \sigma_r - p_r$$

$$\text{Cause } \frac{d\sigma_r'}{dr} = \frac{d\sigma_r}{dr} - \frac{dP_r}{dr} = \mathrm{S}_a$$

where Sa is the sand flow initiation pressure gradient.

Rule.

$$\frac{q\mu}{4\pi k}\frac{1}{R_1} = \frac{4\tau_0 \cos\varphi}{1 - \sin\varphi} + R_1 S_a$$

That is, the critical pressure difference is.

$$p_2 - p_1 = \frac{4\tau_0 \cos\varphi}{1 - \sin\varphi} + R_1 S_a \tag{10.31}$$

The critical flow rate $q$ is.

$$q = \left(\frac{4\tau_0 \cos\varphi}{1 - \sin\varphi} + R_1 S_a\right)\frac{4\pi k R_1}{\mu} \tag{10.32}$$

Using Eq. (10.32) it is possible to determine the interrelationship of rock mechanics and flow parameters when the formation is to be kept stable.

The above analysis shows that for a formation with a certain cementing strength, shear damage is the first cause of sand production, whether by bare-hole completion or shot-hole completion. At this time, the fluid pressure only removes part of the sand in the damage zone, which is manifested as high production without sand out at the time of production, and with the decay of reservoir pressure and water intrusion, the amount of sand out increases significantly. Therefore, for this type of formation, the key to sand prevention is to prevent shear damage from occurring in the formation and to eliminate the possibility of sand production in the first stage of sand production. Although sand emergence is not necessarily caused immediately when the borehole pressure reaches the critical value for shear damage to occur, certain parameters such as borehole pressure and reservoir pressure must be controlled during production so that the critical value is not reached.

For loose sandstone, as long as the production will be accompanied by sand out, after sand out, sand arch will be formed outside the borehole, at this time to maintain the stability of sand arch is the key to sand prevention. And flow rate, i.e. production rate, is the most sensitive factor affecting the stability of sand arch, therefore, for this type of formation, sand prevention can be achieved by controlling production rate.

### *10.3.4 Effect of Pressure Depletion on Reservoir Critical Production Pressure Differential*

As the oil and gas field is exploited, the formation pressure in the reservoir section gradually decreases, the ground stress changes, the effective stress in the bedrock of the formation increases, and when the effective stress in the bedrock exceeds the compressive strength of the rock, it will produce crushing sand, and the critical production pressure difference of the oil and gas well will decrease. If the production pressure differential is kept constant, it will inevitably lead to sand emergence from oil and gas wells. In order to prevent sand emergence from oil and gas wells and affect the production of oil and gas fields, it is particularly important to study the effect of pressure decay on the critical production pressure differential.

Vaziri et al. (2002) proposed a prediction model for sand exit based on different rock damage mechanisms through extensive indoor experiments and comparative field tests. According to the tensile damage mechanism, the calculation model of the critical production pressure difference $\Delta p$ for sand emergence is shown in Eq. (10.33).

$$\Delta P = \left(CK + P_p\right) - \sqrt{\left(CK + P_p\right)^2 - 2CKP_p} \tag{10.33}$$

where, $K = \frac{4\cos\phi}{1-\sin\phi}$; $C$ is the cohesion of the formation rock, MPa; $\phi$ is the angle of internal friction, degrees.

The stretching model is able to predict the critical production pressure differential after reservoir pressure failure. The model requires few parameters for its calculations and is applicable to arbitrary well types.

Figure 10.16 gives the effect of reservoir pressure depletion on the critical production pressure difference in a gas field pattern. It can be seen that the critical production pressure difference of the reservoir decreases gradually with the depletion of formation pressure, the original pressure of this reservoir is 46.08 MPa, the critical production pressure difference is 32.6 MPa at the beginning of extraction, and the critical production pressure difference drops to 0 when the formation pressure depletes to 13.5 MPa.

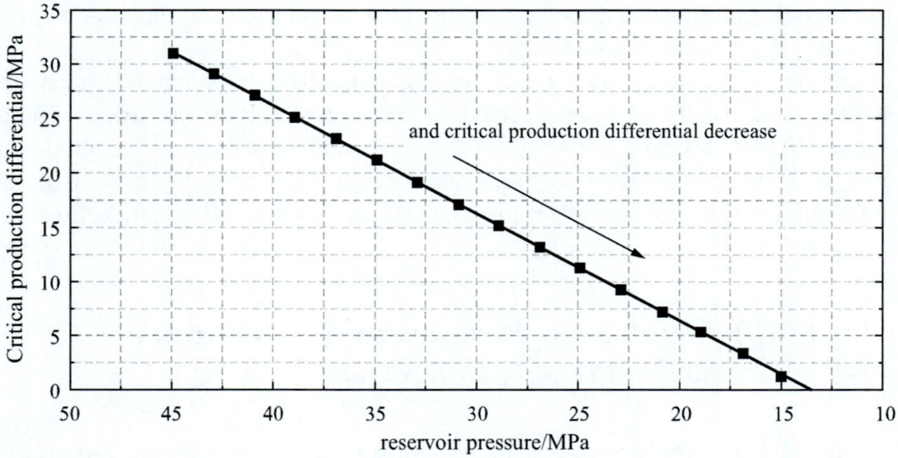


**Fig. 10.16** Law of reservoir pressure depletion on critical production pressure difference

## 10.4 Experimental Study of the Sanding Mechanism

From the analysis of the sand emergence mechanism, it can be seen that the plastic yielding of the borehole surrounding rock in the reservoir is a necessary condition to cause sand emergence from the well. Sand flow under a certain production pressure difference is a sufficient condition for sand emergence. The main influencing factors of well sand emergence include production pressure difference, sand flow initiation pressure gradient, in situ stress, geometry of sand arch, loading history, reservoir deformation and yielding characteristics parameters, perforation parameters, borehole profile parameters, reservoir liquid–solid surface tension, and strength weakening factor. Due to the complexity of the problem, it is difficult to establish the controlling equations and analytical solutions that finely describe the sand-out process. Therefore, the sand-out simulation experiment becomes one of the effective means to analyze the sand-out mechanism and obtain the sand-out control parameters.

The sand emergence experiment is a research method to obtain the main control factors of sand emergence and their evolution laws by simulating the borehole and its production environment, and then integrate with theoretical sand emergence models and actual production for sand emergence prediction.

The main indoor sand emergence experiments are: permeability force experiment, sand arch stabilization experiment and sand emergence process simulation experiment.

### *10.4.1 Permeability Force and Critical Pressure Gradients*

#### 10.4.1.1 The Concept of Permeability Force

Figure 10.17 shows a permeability force test setup with a sand body of length L and cross-sectional area A. A pressure measuring tube is installed at each end of the sand body and the pressure is provided by the hydrostatic pressure at heights $h_1$ and $h_2$, respectively, relative to the datum 0–0. When $h_1 = h_2$, the pressures at both ends of the sand body are equal and no seepage occurs.

If the interconnecting reservoir on the left side is lifted upwards so that $h_1 > h_2$, an upward seepage will occur in the sand due to the presence of a head difference. The head difference $\Delta h$ represents the energy lost as the seepage flows through a sand sample of $L$ length. Having an energy loss indicates that the sand grains give resistance to the seepage as the water percolates through the pores of the sand sample; conversely, the sand grains are necessarily subject to the reaction force of the seepage, which gives forces such as push and friction to each sand grain. For the convenience of analysis, the seepage force on a sand grain per unit volume of sand is defined as the permeability force, denoted by $j$.

To further investigate the magnitude and nature of the infiltration forces, a force analysis was performed on the sand sample subjected to steady seepage shown in

**Fig. 10.17** Schematic diagram of the permeability test

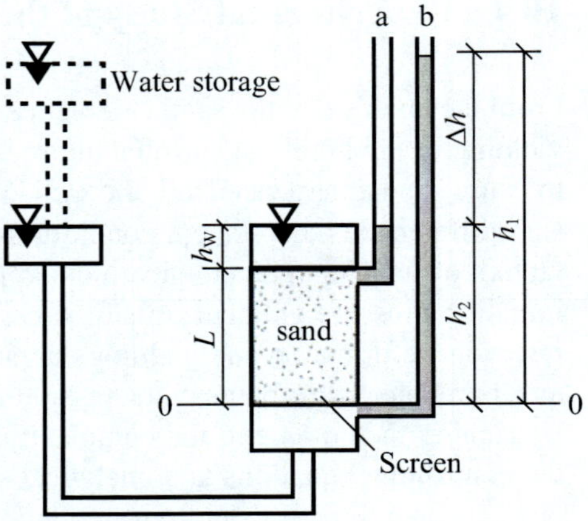


Fig. 10.17. The sand body (containing the sand skeleton and pore water) was taken as the isolated body, and the forces acting on the sand sample are shown in Fig. 10.18, including.

Total sand-water weight.

$$w = \rho_b g L = (\rho_m + \rho_w) g L \tag{10.34}$$

Water hydrostatic pressure at both ends of the sand sample.

$$p_1 = \rho_w g h_w \tag{10.35}$$

$$p_2 = \rho_w g h_1 \tag{10.36}$$

The support reaction force of the lower screen of the sand sample is *R*.

Under this condition, the forces between the sand grains and the water are internal and do not appear in the force analysis of the sand sample. The support reaction force R of the lower screen of the sand sample is an unknown quantity and can be found by the force balance condition of the sand sample in the vertical direction, i.e.,

**Fig. 10.18** Sand-water force analysis

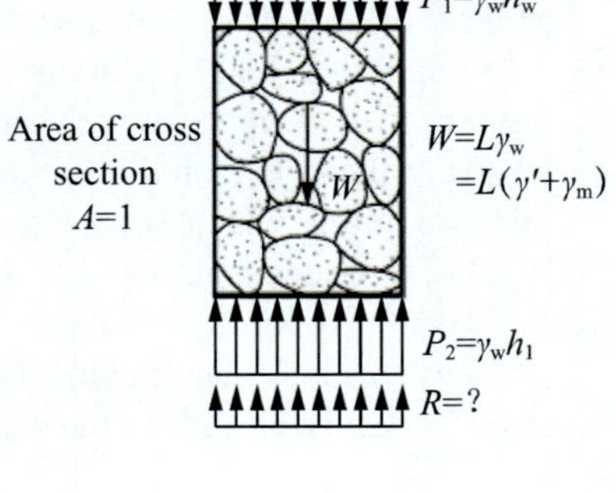

$$p_1 + w = p_2 + R \tag{10.37}$$

Therefore

$$\rho_w g h_w + (\rho_m + \rho_w) g L = \rho_w g h_1 + R \tag{10.38}$$

Collated from

$$R = \rho_m g L - \rho_w g \Delta h \tag{10.39}$$

As seen in Eq. (10.37), when there is no pressure difference between the two ends, the support reaction force of the lower screen of the sand sample $R = \rho_m g L$; when there is upward seepage, the screen support force is reduced accordingly $\rho_w g \Delta h$. In fact, this reduction is taken up by the permeation force $J$ acting on the sand skeleton as a whole, i.e. the total permeation force $J$ acting on the sand sample is

$$J = \rho_w g \Delta h \tag{10.40}$$

Thus, the seepage force on the sand particles per unit volume of sand, i.e., the infiltration force $j$, is

$$j = \frac{J}{V} = \frac{\rho_w g \Delta h}{1 \cdot L} = \rho_w g i \tag{10.41}$$

The above equation shows that the magnitude of the permeability force on the sand skeleton in the seepage field is proportional to the pore pressure gradient, and its direction of action is the same as the direction of the pressure gradient. The permeability force is a volumetric force whose magnitude is *F/V*.

The seepage force reflects the pushing and dragging force of the infiltrating water flow on the skeleton within the unit volume of sand in the seepage field. For isotropic sands, the direction of seepage flow velocity and the direction of pore pressure gradient are the same, at which time the direction of infiltration force action and the direction of seepage flow velocity are the same; for anisotropic sands, the direction of infiltration force action and the direction of seepage flow velocity are different at this time because the direction of seepage flow velocity and the direction of pore pressure gradient are not the same.

For a general sand body, the particle shape and the water flow in the pores are very complex, so that lift forces perpendicular to the direction of the water flow are usually present for individual sand particles. However, for isotropic sand bodies, there are a considerable number of particles contained within the characteristic volume under consideration, so that when the action on all particles is sought to be summed, the lift forces in the direction of the perpendicular water flow can cancel each other out so that their combined forces are in the same direction as the direction of the flow velocity. In the case of anisotropic sand bodies, the presence of factors such as the directional arrangement of the particles makes the lift forces in the direction of the vertical water flow velocity non-cancellable, and in this case, the direction of the infiltration force and the direction of the seepage flow velocity do not coincide.

#### 10.4.1.2 Critical Pressure Gradient

As seen in Eq. (10.39), under hydrostatic conditions, i.e., $\Delta h = 0$, the support reaction force on the lower screen of the sand sample $R = \rho_m gL$; and when there is upward percolation, i.e., $\Delta h > 0$, the support reaction force on the screen decreases accordingly $\rho_w g\Delta h$. If the reservoir at the left end of Fig. 10.17 is continuously raised, the$\Delta$ h gradually increases, and thus the permeability force acting on the sand body also gradually increases. When $\Delta h$ increases to a certain value, the upward infiltration force overcomes the downward gravitational force of the sand particles, the sand body has to undergo suspension or bulging, commonly known as flowing sand.

The pore pressure gradient when the sand body is in the critical state of flowing sand is studied. From Fig. 10.18, it can be seen that when flowing sand occurs, the pressure of the sand sample pressed against the screen $R = 0$. According to Eq. (10.39), it is obtained that

$$R = \rho_m gL - \rho_w g\Delta h = 0$$

Consequently

$$S_a = \rho_m g = \frac{\rho_w g\Delta h}{L} \tag{10.42}$$

where, $S_a$ is called the critical pressure gradient, which is the pressure gradient at which the sand body starts to undergo flow sand damage.

#### 10.4.1.3 Infiltration Damage of Sand

The deformation or damage of the sand body under the action of seepage is called infiltration deformation or infiltration damage, such as reservoir production out of sand, reservoir settlement or uplift. There are two main types of infiltration damage of sand, flowing sand and pipe surge.

**Flowing sand**

For production wells, the phenomenon of sand flowing on the surface of a bare borehole wall or on the surface of a shot hole borehole under the action of permeable water flowing in the direction of the borehole is called flowing sand. Any type of loose or yielding sand can be damaged by flowing sand as long as the hydraulic pressure gradient reaches a certain value.

**Tube surge**

Tube surge refers to the role of seepage, a certain gradation of sand in the sand body of fine sand particles, through the larger particles formed by the pore space to move, and eventually formed in the sand body and the borehole surface through the channel, thus triggering a large number of sand, commonly known as earthworm hole phenomenon, as shown in Fig. 10.19. The occurrence of pipe surge damage will generally show a gradual development of the process over time, is a progressive nature of the damage. First in the infiltration of water flow, the finer particles in the coarser particles formed in the pore space mobile loss; after that, the pore space of the sand body is expanding, the seepage rate is increasing, the coarser particles will also be carried away by the water; with the above process of scouring the continuous development, will be formed in the sand body through the seepage channel, and eventually may cause the sand body collapse.

The factors affecting the occurrence and development of infiltration damage in the sand body consist of two parts: endogenous and exogenous. The endogenous factors are the composition and structure of the sand body particles, i.e. geometric conditions; the exogenous factors are the hydraulic conditions, i.e. the magnitude of the permeability forces acting on the sand body skeleton.

At seepage overflows, flowing sand occurs as long as the infiltration pore pressure gradient is satisfied to be greater than the critical pore pressure gradient. Whether the sand body occurs tube surge, the first decision on the nature of the sand. Generally, when the clay content is high, only flowing sand will occur and no pipe surge will occur. In the clay content is low, the occurrence of pipe surge must have the corresponding geometric and hydraulic conditions.

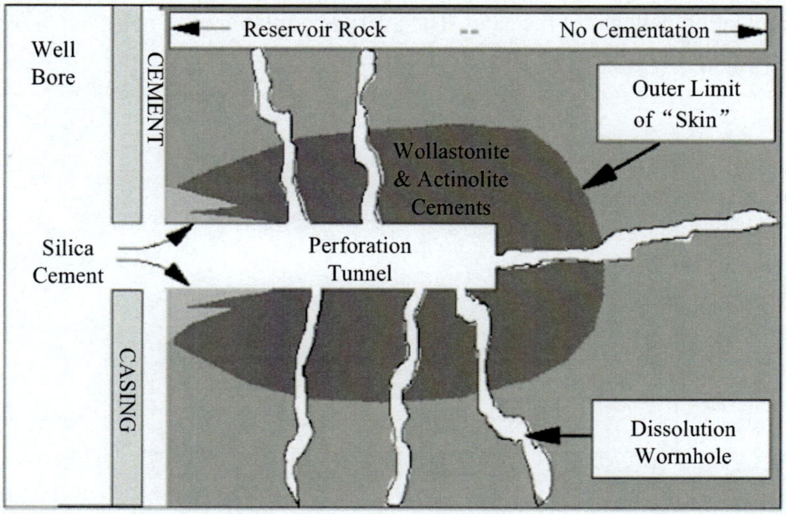


**Fig. 10.19** Schematic diagram of shot hole orifice pipe surge

The diameter of the pore formed by the coarse particles in the sand body must be larger than the diameter of the fine particles before it is possible for the fine particles to move in it, which is a necessary condition for the occurrence of pipe surge. For inhomogeneous coefficient $C_u < 10$ uniform sandy soil, particle coarse and fine difference is not large, coarse particles formed by the pore diameter is not larger than the fine particles, so the fine particles can not move in the pore, it is not possible to occur tube surge. For the inhomogeneous coefficient $C_u > 10$ of the inhomogeneous sandy soil, both may occur tube surge may also occur flow sand, depending mainly on the sand grading situation and fine particle content. The following two cases are discussed.

For sandy soils lacking intermediate grain size and with discontinuous grading, the form of permeable broken ring depends mainly on the fines content. The so-called fine material refers to the grain size below the horizontal section of the grading curve, such as the curve in Fig. 10.20 ①. The grain size below point *b* in Fig. 10.20. Experiments show that when the content of fine material in the following 25%, the fine material can not fill the pores formed by the coarse material, infiltration damage basically belongs to the pipe surge type; when the content of fine material in 35% or more, the fine material is sufficient to fill the pores formed by the coarse material, coarse and fine material to form a whole, enhanced seepage resistance, infiltration damage is flow sand type; when the content of fine material in 25–35%, belongs to the transition type, the specific form of damage also depends on the sand body Loose and dense degree. For uneven sand with continuous grading, as in the curve in Fig. 10.20 ②, it is difficult to find out the dividing line between the skeletal particles and the filling fines.

Some scholars have proposed that the average pore diameter $D_0$ of the sand body is compared with the diameter of the finest fraction of particles ds to discern the type of permeation damage of the sand. The average pore diameter of a sand body can be expressed by the following empirical formula.

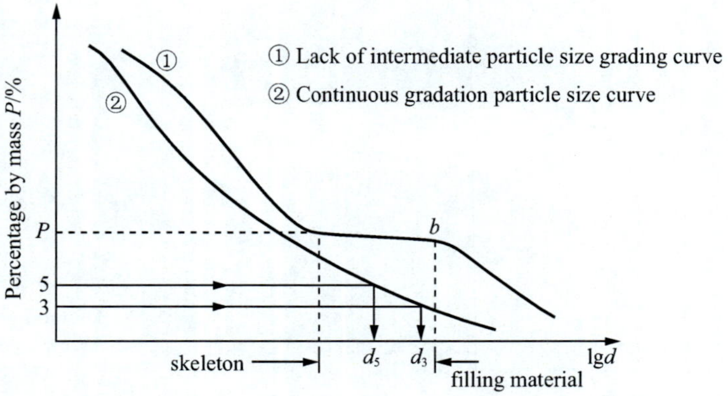


**Fig. 10.20** Grain size grading curve

$$D_0 = 0.25d_{20} \tag{10.43}$$

where, $d_{20}$ indicates that the mass of sand smaller than this particle size is 20% of the total mass. The experimental results show that when more than 5% of the fine particles in the sand are smaller than the average diameter of the pores of the sand $D_0$, that is, $D_0 > d_5$, the form of damage is tube surge; and when the content of the fine particles in the sand less than $D_0 < 3\%$, that is, $D_0 < d_3$, there are few particles that may be lost and no tube surge will occur, presenting flow sand damage.

Permeability can drive the fine particles in the pore space between the roll or move is the occurrence of the hydraulic conditions of the tube surge, available to occur in the tube surge critical hydraulic slope drop to express. But so far the critical hydraulic slope drop of the tube surge calculation method is not mature. Domestic and foreign scholars proposed more calculation methods, but the calculation results vary greatly, so there is no a recognized suitable formula to characterize. It is usually determined by the infiltration damage test.

### *10.4.2 Sand Arch Stability Experiments*

The arch is a curved structure that spans an orifice and supports the load by decomposing vertical stresses into horizontal stresses, and this force decomposition inevitably creates shear stresses within the structure. Because of the shear strength of the compressive sand, some of the load caused by the self-weight of the overlying sand can be transferred to the sand adjacent to the arch structure under the arch effect. In order to study the mechanism of sand exiting from unconsolidated sand reservoir shot-hole completion production, (Hall and Harrisberger 1970; McNulty 1965; Bratli and Risnes 1981; Tippie and Kohlhaas 1973) were the first to investigate the stability of sand arches in loose sands using a movable portal sand arch simulation experimental setup. The sand was first subjected to an enclosing pressure in the laboratory, and then the stresses applied to a small part of the sand were allowed to drop to zero. The sand in the small area around the unstressed sand will compact to form an arch structure. The arch structure establishes a static equilibrium between the stressed and unstressed sand. In experiments, the unstressed area is usually created by removing a small portion of the bottom end after stress is applied to the sand pile, called a "live-plate gate" experiment, as shown in Fig. 10.21. For loose sandstone perforation completion, the sand within the perforation hole is not stressed, whereas most of the reservoir sand will be stressed due to overburden loading. Sand arches are formed around each shot hole and have the effect of stabilizing the sand and preventing sand out. If these arches fail it will lead to sand out. Therefore, the study of the necessary conditions for the formation of these arches, the stability limits and the causes of their destruction is one of the important elements in the study of the sand-out mechanism.

To study the stability of sand arches, Hall and Harrisberger first investigated the triaxial strength characteristics of loose sands and determined the shear damage

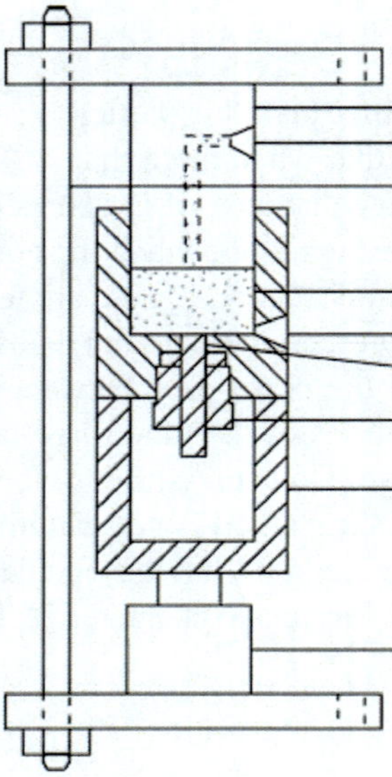


**Fig. 10.21** Schematic diagram of the sand arch stability experiment

envelopes of several loose sands. For loose sands, the damage envelopes are not straight lines, as shown in Fig. 10.22. The envelope curvature is determined by the mechanical properties of the sand when it breaks at different stress levels. Based on the failure characteristics, the damage envelope is divided into three regions. In region I, the stress level is low and the destruction of the sand is accompanied by volume expansion. The increase in volume causes the sand particles in the sand body to roll against each other, as in Fig. 10.23. In region III at high stress levels, the energy required to increase the volume of the sand body is greater than the energy required to crush the particles along the damage surface. Region III is primarily a zone of particle crushing, accompanied by a significant reduction in sand volume. Region II is a transitional region between swelling and crushing, where neither of these types of damage dominates. The size of the crushed particles is smaller than the original particles, so the rock becomes more dense. The swelling effect at low stress levels gradually disappears with increasing stress.

In Region I, the angle of internal friction is approximately constant and is a function of sand type. Given a positive stress at the damage surface, the higher the angle of internal friction, the higher the shear stress required for damage. The angle of internal friction is higher if the particles are poorly sorted, well angled and well compacted. The internal friction angle is related to the shear expansion angle. It is shown that sand arches are formed only in region I. The expansion is necessary for the formation of stable arches. If crushing damage occurs at any location, it leads to volume reduction and load transfer to adjacent locations, at which point the stresses exceed the level required for crushing damage and crushing damage propagates. In contrast, in the shear expansion zone, stress arch damage in the sand body does not

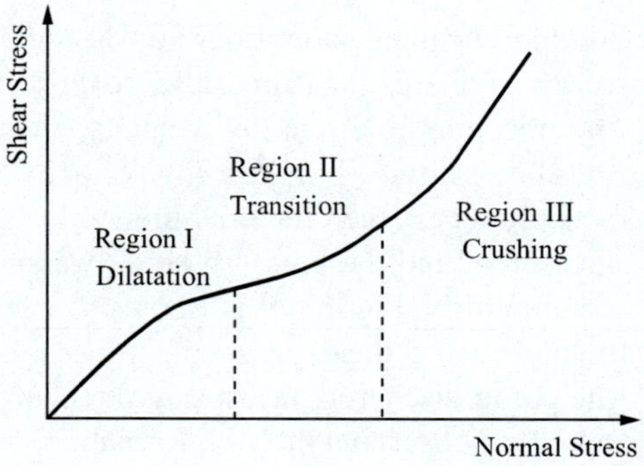


**Fig. 10.22** Loose sand shear damage envelope

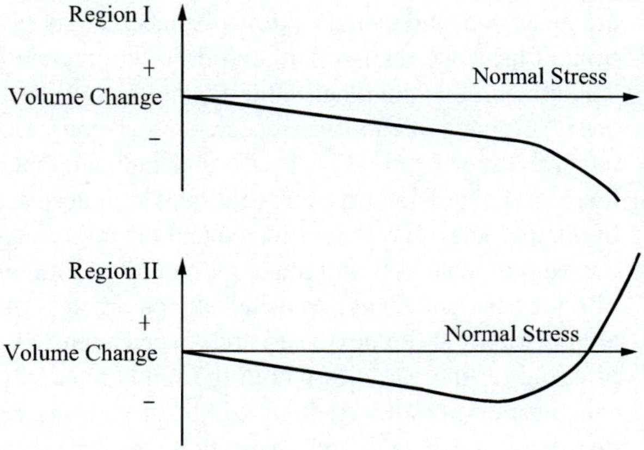


**Fig. 10.23** Loose sand volume strain characteristics at different stress levels

propagate and the damage is accompanied by shear expansion of the entire arch structure. As the particle density decreases, the angle of the particle prism increases, or the particle distribution becomes wider, the transition zone II narrows and appears at lower stress level locations, and these properties can be used as indicators for evaluating the stability of various sand arches.

Hall and Harrisberger (1970) simultaneously investigated the effect of various saturation conditions on the stability of sand arches. It was shown that the arch structure is not stable in either dry or saturated sand under pressure conditions. If two immiscible phases are present and the saturation of the wetted phase is less than the movable saturation (funicular), the arch structure will be stable. When the saturation of the wetting phase exceeds the movable saturation, the sand arch fails. This phenomenon was repeatedly observed for various fluid combinations as well as for wetting inversion conditions between fluids. Thus, the interfacial tension of the liquid–liquid-solid contact provides a cohesive force that is necessary for sand arch stabilization.

Tippie and Kohlhaas (1973) studied the pattern of fluid flow rate effects on the formation of sand arches and their stability. For water-saturated sands, sand arch simulation experiments were carried out by replacing water with oil in a pressure kettle to residual saturation. For the sand arch experiments, the initial morphology of the arch was measured after a short period of time by starting to pass through the injection hole at a low velocity; then the pressure kettle was reassembled and the flow started again at a low velocity with a gradual increase in rate until sand was produced; then the pressure kettle was removed, the geometry of the arch was measured, and the pressure kettle was reassembled, and the process was repeated continuously. Figure 10.24 shows a set of sand arch formation-destruction experimental curves, the experimental process of constant overburden pressure, increasing the pump discharge, measuring the change of the outer boundary pressure $p_e$ with time, and observing the arch formation, destruction phenomenon. It can be seen that the arrow in the figure points to the moment of sand arch destruction. After a certain stage of experimental sand arch destruction, the geometric parameters of the arch are measured, and then the pump displacement is slowly increased to the value at the time of arch destruction. No arch destruction occurred and the pump displacement was increased to a higher value for the arch to destroy, creating a larger size sand arch. This phenomenon was repeated in subsequent experiments. The $\Delta p / Q$ plots for the arch experiments (Fig. 10.25) also indicate that the permeability decreases with increasing repulsion time and that sand arch destruction occurs only after migration of fine particles. It is hypothesized that the migrating fines within the sand bridge the coarse pore channels and make the vicinity of the arch more dense. This migration of fines causes a change in the elastic parameters of the sand body. After each sand arch disruption, the next sand arch experiment will still show fines migration. The bridging of fine particles within the sand arch leads to a reduction in permeability, causing the pressure gradient within the arch to exceed the gradient required for disruption. Each sand arch disruption was accompanied by fine particle migration. When the infiltration channel is blocked, the fluid flow rate through a single pore channel increases ($\mathrm{d}p/\mathrm{d}r$ increases), creating greater flow stresses on the surface of the sand particles.

These experiments show that flow is a factor in determining sand arch size and stability, and that flow control has a significant effect on the production characteristics of sparse sandstone reservoirs.

Sieve analysis of the sand arch cavity sand and output sand from the sand arch experiment was performed and the results are shown in Fig. 10.26. The particle size distribution of the output and cavity sand was coarser than that of the base filling sand. The fine particles flow out before the sand arch is destroyed, which is responsible for the coarse grain size of the cavity sand and output sand at the time of sand arch destruction. As each new arch is formed, the sand grain size distribution within the arch is variable. The strength characteristics of the sand in the arch continue to change during the growth of the arch.

## 10.5 Predicting Models for Oil Well Sand Production

In order to effectively prevent oil well outgassing, it is necessary to understand well outgassing prediction techniques.

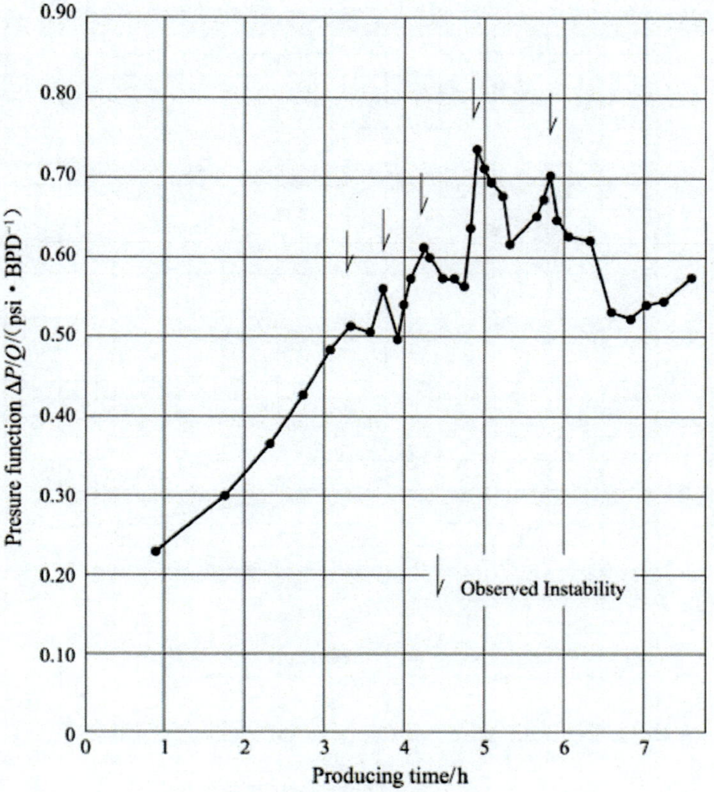


**Fig. 10.24** Sand arch formation—failure experiment curve

The development of sand-out prediction models has a relatively short history, and the first important technique was proposed by Nathan Stein et al. in 1972, which relates the sanding trend of a well to the shear strength of the formation and uses data from sonic logs and density logs to analyze the factors influencing sand-out. The limitations of this method are that wells must be tested for sand emergence after completion, requiring a large amount of sand emergence before reliable data can be obtained, and that the effects of reservoir pressure decay and produced water are not considered. Tixier et al. (1975) adopted a method similar to Stein's, using logging data to analyze and evaluate the sand emergence of wells in their current state, and proposed a threshold value for the intrinsic strength of the formation The threshold value of the formation strength, if the formation strength is higher than this value, the oil and gas formation will not come out. The main drawback of this method is that it cannot give quantitative results and can only determine whether the well is sandy or not, but cannot calculate the maximum sand-free oil production rate.

Another model was proposed by George R. Coates et al. in 1981, which, unlike the previous model, relates the sand-out trend to the stress in the reservoir rock near the borehole and calculates its compression factor using the mechanical property

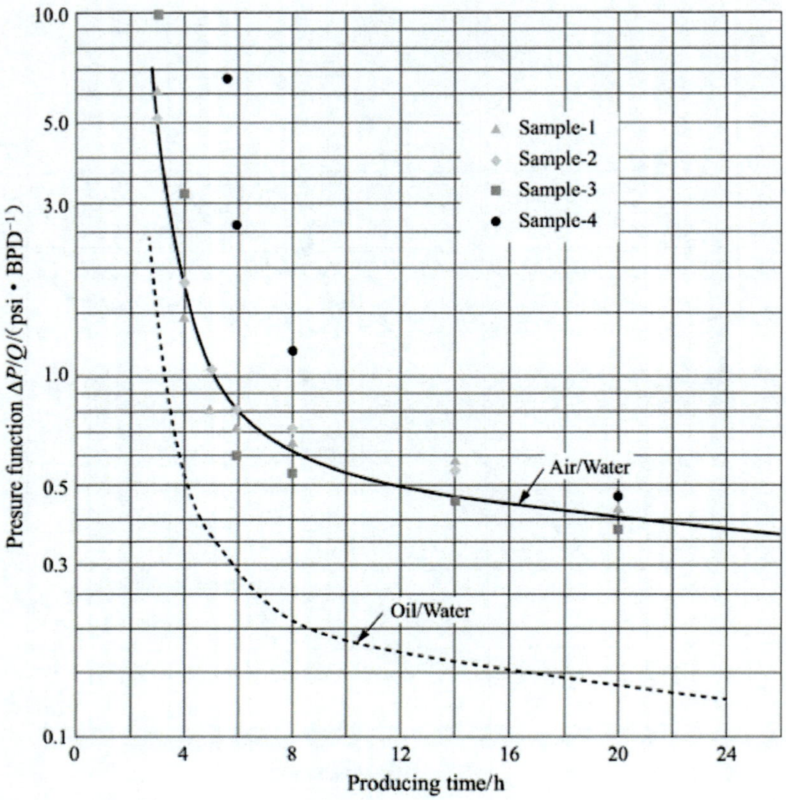


**Fig. 10.25** Arching experiment $\Delta p/Q$ curve

parameters obtained in the logs. The core of this method is the Moore's circle stress analysis method, which can be used for sand emergence prediction in wells that do not produce too much water.

In 1984, Barrow DC et al. used information from 16 Gulf Coast wells to analyze the reliability of predicting sand production in wells using Moore's circle analysis and concluded that sand control decisions could be made based on the predictions of the model when a safety factor of 200 psi (1.4 MPa) was given. Although the Moore's circle analysis method gives better prediction results when the well is not producing water, its limitation is that it cannot be used in the case of water production.

In 1989, Ghalambor A et al. considered the influence of multiple factors on sand discharge and proposed a method to calculate safe production pressure drop using multi-parameter linear regression, which has the disadvantage that too much data is required and it is difficult to promote its use in the field.

In summary, the following models exist for predicting sand emergence.

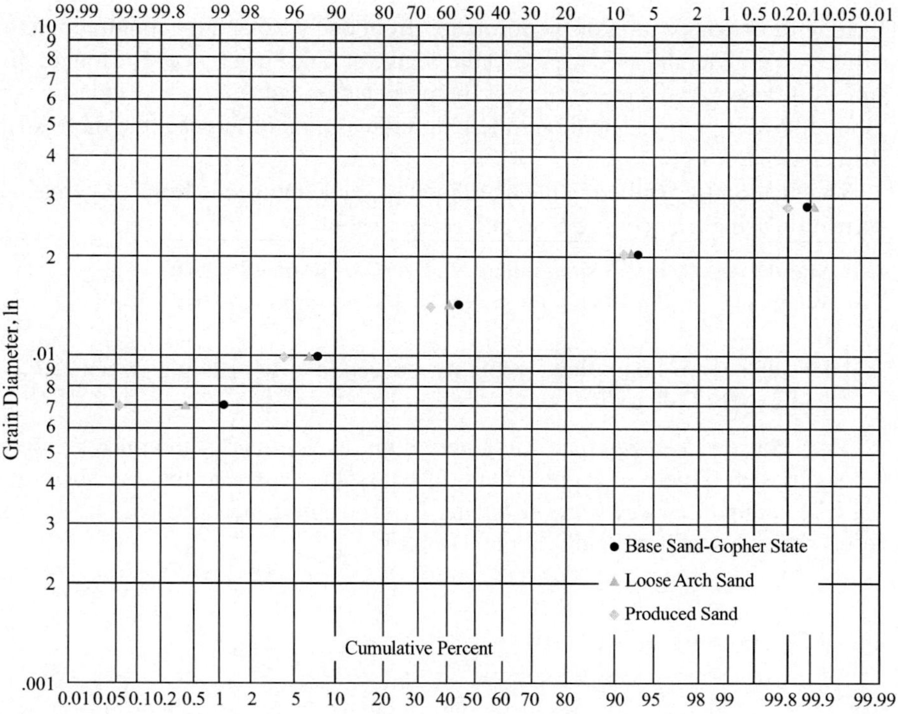


**Fig. 10.26** Particle size distribution pattern of stratigraphic and in-arch sands

## 10.5.1 Single Parameter Model

### 10.5.1.1 Porosity Method

The pore structure of the formation is related to the cementation strength of the formation, and the size of the cementation strength is closely related to the depth of burial of the reservoir, the type of cementation, the cementation method, the size of the formation particles, and the shape of the particles. Generally speaking, the deeper the formation is buried and the smaller the porosity, the higher the strength of the formation is. Research shows that the porosity of the formation is greater than 30%, which is very easy to produce sand; when the porosity is between 20–30%, the sand production slows down; the porosity is less than 20% basically does not produce sand.

### 10.5.1.2 Acoustic Time Difference Method

The sonic time difference method uses the time difference of sonic waves in a formation to predict the sand emergence of the formation. The sonic time difference reflects the degree of compaction of the formation from one side, and the longitudinal time difference is generally used to predict the degree of sand emergence. The longitudinal time difference $\Delta t_\mathrm{p}$ is the inverse of the propagation velocity of the acoustic wave along the well profile. The larger the longitudinal time difference, the more sparse the cementation.

The method for predicting the likelihood of sand emergence from a hydrocarbon formation using the acoustic time difference method is.

(a) Stabilization without sand when $\Delta t_p < 311.7\,\mu\text{s/m}$ (95 μs/ft)
(b) When 311.7 μs/m (95 μs/ft) $\le \Delta t_p \le$ 344.5 μs/m (105 μs/ft), sand may come out.
(c) Unstable sandstone that is highly susceptible to sanding when $\Delta t_p >$ 344.5 μs/m (105 μs/ft)

The acoustic time difference judgment limit used in this method takes different values in each oilfield. A large amount of field data statistics show that the critical value of longitudinal wave time difference for sand emergence from sparse sandstone reservoirs in Shengli oilfield is about 310 μs/m.

#### 10.5.1.3 Sanding Index Method

The sanding index method is a method to calculate the rock mechanical parameters using the relevant data such as sound velocity and density from the logging data, and to calculate the sanding index of the formation using the combined modulus method, and then to make sand emergence prediction. The dynamic elastic modulus and Poisson's ratio of the formation can be calculated based on the longitudinal velocity, transverse velocity and bulk density from the full-wave column acoustic logging and density logging data.

$$E_d = \frac{\rho_r V_s^2\left(3V_p^2 - 4V_s^2\right)}{V_p^2 - V_s^2} \times 10^{-3}\,\text{MPa} \tag{10.35}$$

$$\mu_d = \frac{V_p^2 - 2V_s^2}{2\left(V_p^2 - V_s^2\right)} \tag{10.36}$$

where $E_\mathrm{d}$ is the modulus of elasticity derived from rock acoustic logging and density logging, $\mu_\mathrm{d}$ is the Poisson's ratio derived from rock acoustic logging, $\rho_\mathrm{r}$ is the formation density, $V_\mathrm{p}$ is the longitudinal wave velocity, and $V_\mathrm{s}$ is the transverse wave velocity.

Halliburton Oilfield Services, Schlumberger Petroleum Services used the combined modulus method to calculate the sanding index of the formation. The formation strength is well correlated with the shear modulus $G$ and bulk modulus $K$ of the rock, and both are functions of parameters such as sonic, density, well diameter, and mud content in the logging data, and the relationship between the sanding index and the combined modulus of the rock is.

$$B = K + \frac{4G}{3} = \frac{E_d}{3(1 - 2\mu_d)} + \frac{2E_d}{3(1 + \mu_d)} \tag{10.37}$$

where $B$ is the sand-out index, $K$ is the bulk modulus of elasticity of rock, and $G$ is the shear modulus of elasticity of rock.

The smaller the value of the sanding index, the lower the strength of the rock and the more sand-prone the formation. The criteria for determining this are:

- Normal producing oil and gas wells are sand free when the sanding index $B \geq 2 \times 10^4$ MPa
- Slightly sandy oil and gas reservoirs when the sanding index is. $1.4 \times 10^4$ MPa $\leq B \leq 2 \times 10^4$ MPa
- When the sanding index $B \leq 1.4 \times 10^4$ MPa, the oil and gas formation is severely sandy.

#### 10.5.1.4 Schlumberger Sanding Index Method

The Schlumberger sanding index method is a prediction method to determine the sand potential of a formation by calculating the Schlumberger sanding index of the rock. The Schlumberger sanding index is equal to the product of the shear modulus of rock and the bulk modulus of elasticity, or is calculated directly from logging data as follows.

$$SR = K \cdot G = \frac{E_d}{3(1 - 2\mu_d)} \cdot \frac{E_d}{2(1 + \mu_d)} \tag{10.38}$$

where *SR* is the Schlumberger sand-out index, $MPa^2$.

The larger the *SR* value, the stronger the rock, the better the stability, and the less likely to produce sand. *SR* value can be calculated from logging data, when *SR* is greater than $3.8 \times 10^7$ $MPa^2$, the formation will not produce sand. And the China Petroleum Exploration and Development Institute recommends $SR > 5.9 \times 10^7$ $MPa^2$ when the formation is not sandy.

#### 10.5.1.5 Combined Modulus Method

The combined modulus method is a method of calculating the elastic combined modulus of rocks based on sonic and density logging data, which leads to sand emergence prediction. The combined modulus of the stratigraphic rocks is.

$$E_c = \frac{9.94 \times 10^8 \times \rho_r}{\Delta t_c^2} \tag{10.39}$$

where $E_c$ is the modulus of elastic combination of formation and $\Delta t_c$ is the longitudinal wave time difference of the formation obtained from acoustic logging.

In general, the smaller the $E_c$ value, the greater the likelihood of sand emergence from the formation. The determination of whether sand is present or not is as follows.

(a) $E_c \geq 2.0 \times 10^4$ MPa, no sand at normal production.
(b) $1.5 \times 10^4$ MPa < $E_c$ < $2.0 \times 10^4$ MPa, slight sand out at normal production.
(c) $E_c \leq 1.5 \times 10^4$ MPa, severe sand out during normal production.

#### 10.5.1.6 Uniaxial Compressive Strength Method

Empirical field data show that when the production pressure difference is less than half of the uniaxial compressive strength of the reservoir, it can keep the reservoir from producing sand at the initial stage of extraction. Indoor core experiments can directly measure the uniaxial compressive strength of the formation, while for formations that are not cored, analytical processing of log data can be used to derive the strength parameters of the formation, and thus the critical production pressure difference for sand emergence from the formation.

$$\Delta p = UCS/2 \tag{10.40}$$

where $\Delta p$ is the reservoir critical production pressure difference and *UCS* is the uniaxial compressive strength of the reservoir rock.

### *10.5.2 Multi-parameter Models*

The multiparameter model considers depth, acoustic propagation time, production rate, production pressure difference, production index, mud content, and water content among the sand emergence factors. Veeken et al. (1991) defined the sum of reservoir pressure decay value ($\Delta p_{\mathrm{de}}$) and production pressure difference ($\Delta p_{\mathrm{dd}}$) as the total pressure difference ($\Delta p_{\mathrm{td}}$) and developed a two-parameter sand emergence prediction model using the field sand emergence data, as shown in Fig. 10.27.

As can be seen in Fig. 10.27, there is a wide risk zone between sand emergence and non-sand emergence, which is a result of other factors. For this reason, Ghalambor and Asadi (2002) proposed the problem of predicting sand emergence from gas wells after water intrusion, and developed a model for predicting sand emergence from gas reservoirs after water intrusion using on-site gas well data and reservoir property parameters obtained through logging data using a statistical approach.

$$\begin{aligned}\Delta \mathrm{P} = {} & 0.18457 P_{ws} - 1153.97 V_{sh} + 0.00557 G \cdot V_{wc} - 20.29912 P_{\mathrm{M}} \\ & + 13036.1 \Phi_e \cdot V_{wc} - 23655.6 C_b \cdot G + 0.015832 \Phi_1 \cdot G + 15177.4\end{aligned} \tag{10.41}$$

where $\Delta p$ is the ultimate differential production pressure, psi; $p_{\mathrm{ws}}$ is the static pressure at the bottom of the well, psi; $V_{\mathrm{sh}}$ is the volume fraction of mud shale; $G$ is the

**Fig. 10.27** Total drawdown versus transit time for intervals with and without sand problems

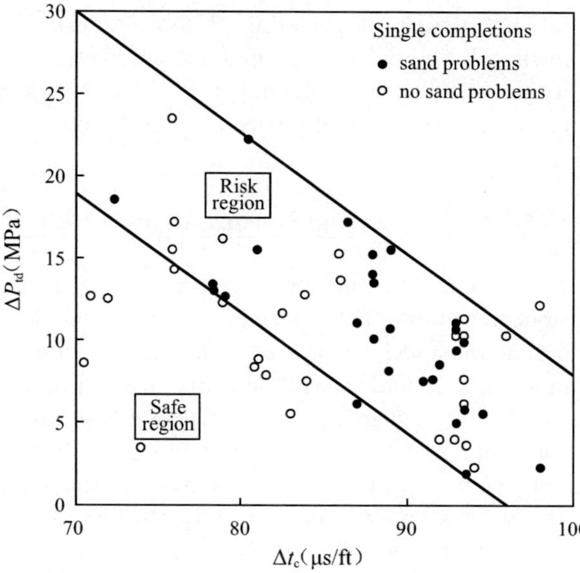


shear modulus, psi; $V_{wc}$ is the volume of wet clay, ft$^3$; $p_M$ is the differential pressure predicted by the molar circle, psi;$\phi_e$ is the effective porosity;$C_b$ is the volume compression factor; $\phi_1$ is the total porosity.

However, the results of linear regressions using sand production data from different wells may confound the actual effects of reservoir and production process parameters. Moreover, the effect of well water production tends to mask the effect of reservoir pressure decay; in fact, a reduction in reservoir pressure will lead to shear-damaged sand emergence, and such sand emergence problems often occur in a relatively short period of time, and the amount of sand emergence is large, sudden and catastrophic.

In general, the use of less-parametric models is more common because multi-parametric models require a large amount of data, which is very difficult to record and monitor out of the sand over time.

## 10.5.3 Engineering Forecasting Method

In practical engineering, sand emergence predictions for developed fields are based on empirical relationships and fitting of historical data, while for new development areas, especially offshore fields, well completion design must take into account sand prevention predictions. However, there are no reliable similar wells, and because many factors affect well sand emergence, the optimized development plan should accurately discern the requirements and characteristics of geology, rock mechanics, logging, production process and reservoir engineering.

K.W. Weissenburger et al. (1987) established an engineering system analysis method for completion method and parameter design based on core analysis and logging data, with well sand free as a criterion and reservoir management and production process requirements as constraints.

#### 10.5.3.1 Engineering Systems Analysis Method

The goal of the engineering systems analysis is to evaluate alternative completion strategies prior to field implementation. Among other things, the sand arch stability evaluation model can be used to design the optimal shot hole size and phase and the pressure drop and production ceiling over the entire reservoir life; reservoir simulation can calculate the daily production of a full or selective shot hole section and compare it to the daily production of a gravel-filled completion. Capacity forecasts can then be evaluated against reservoir management planning.

The basic logic of an engineered system for evaluating completion strategies based on the amount of sand produced is shown in Fig. 10.28. The system starts with core and logging data. Initially, low-intensity formations (physical sand control is required in all cases), high-intensity formations (physical sand control is not required in any case), and moderate-intensity formations are identified. Rock mechanics experiments and logging rock mechanics interpretation are then carried out. Potential well sections for natural completions or other completions are selected based on calibrated logging data. If the simulated production of the selected completion scenario is consistent with the reservoir management plan, the appropriate completion scenario is preferred. If the simulations do not agree with the reservoir management plan, the engineering systems approach must be repeated until a suitable completion option is selected, or all completion options are unsuitable, at which point the reservoir management plan itself must be modified.

#### 10.5.3.2 Optional Completion Methods

This engineered system is used to predict "sand-free" production conditions for natural completions. Under reservoir geometry and other constraints, selective shot-hole drilling in stronger formations and adherence to operational guidelines is an effective sand control method. If the risk or predicted operational constraints of selective injection are unacceptable, alternate options in the engineered system are "aggressive" sand control methods such as gravel filling, fracture filling, or chemical cementing. Reservoir constraints, equipment constraints and management objectives will determine which sand control techniques are used. The advantages and disadvantages of these four sand control techniques are briefly described below.

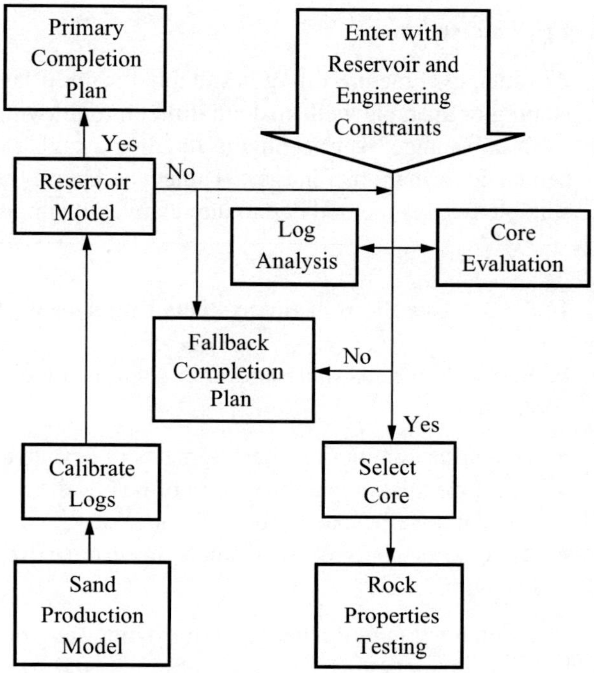


**Fig. 10.28** Flow chart of reservoir management method for oil well out of sand

## Selective shot hole

Benefits: Does not cause formation damage. Acceptable light sand-out can significantly increase production. For thick reservoirs, selective shot hole may be effective unless weakly cemented well sections dominate.

Disadvantages: It is important to follow operating guidelines, and operating errors may result in sand out. Cores need to be taken from several wells to calibrate mechanical property logs to select shot hole intervals.

## Gravel filling

Benefit: Gravel filling is an active sand control technique.

Disadvantages: Optimum daily production requires good infill construction. Weakly cemented, highly permeable and thick reservoir sections are susceptible to formation injury during gravel filling. Flowing fine particles or waxy crude oil may form deposits near the wellbore or within the gravel fill, reducing daily production rates.

## Chemical sand control

Advantages: this method is relatively economical and does not require changes to the wellbore production tool set.

Disadvantage: formation consolidation is only applicable to thin reservoir sections.

**Fra-Pac treatment**

Advantages: Fracture filling is suitable for multi-layers, thin reservoir sections and sloping or unstable wells that are difficult to fill with gravel to prevent sand.

Disadvantage: The technique requires a high strength barrier near the weakly cemented formation of interest. Thinner and poorly cemented well sections are more suitable for this method due to the relatively high inflow in the fractures.

#### 10.5.3.3 Key Steps of the Systems Engineering Approach

Engineered systems and sand arch stability models are suitable for a variety of applications.

- Predicting completion characteristics of designed wells in reservoir planning.
- Optimize the completion design of new wells by predicting the sand emergence trend and well capacity for different design options.
- Analyze the historical sand emergence from existing wells and predict future sand behavior.

Predicting the future behavior of existing sand-out wells can be the most difficult because attempts must be made to estimate the size, shape, and connectivity of the peroration cavity. For new well sand-out predictions, the initial geometry of the perforation is better known and can be managed through production to keep the perforation cavity size within a limited range.

**Reservoir geology and core testing**

Formation rock mechanics logging interpretation and drill stem testing are the primary means of predicting sand strength during the initial phase of production. However, experience has shown that formation strength logging and drill stem testing are not yet able to accurately predict sand emergence throughout the development life of the reservoir. Cores provide a qualitative calibration method for formation rock mechanics logging, i.e., further quantitative calibration is possible through rock mechanics property testing and sand arch cavity stability modeling analysis. Advances in drilling engineering have greatly improved the success rate of coring weakly cemented formations, thereby making cores more readily available than in the past.

The geological description of the reservoir regarding trends in formation strength provides the framework for modeling the shot hole cavity. Therefore, it is important to consider vertical trends, lateral trends and saturation trends in formation strength.

**Logging**

As in geology, in conventional reservoir descriptions, logging is used for reservoir stratification and lithology evaluation, which in turn assesses the strength of the sand. In this engineering system, cable logging is critical in sand strength prediction applications, providing a continuous profile of rock mechanical property parameters along the wellbore, which allows for continuous observation and analysis of sand strength.

However, mechanical property logs need to be calibrated due to the indirect nature and simplicity of the interpretation of the mechanical properties of the logs. In older blocks of development, mechanical property logs can be calibrated based on production history. In newer blocks, logs can be calibrated to some extent based on drill stem testing, but most effectively based on core strength testing and shot hole cavity stability model results.

**Rock mechanics experiments**

Rock mechanics is at the heart of the engineering system for sand emergence prediction. Compression and tension experiments on reservoir cores provide the damage and deformation parameters needed to model the stability of the shothole cavity. Where possible, it is desirable to test each reservoir well section of interest. Well sections for rock strength testing are usually selected based on geologic description and strength logging results.

**Sand emergence prediction model and data collection**

Develop a complex numerical simulation model for quantifying the stability of sand arches to predict the sand exit characteristics of wells. Determine the safe operating envelope from the shear damage limit and the tensile damage limit. The model should consider multiple sand emergence mechanisms simultaneously, including shear damage and tensile damage due to fluid seepage forces, boundary loads, and residual stresses.

Dimensional analysis was performed to determine important variables for cavity stability. Production pressure differential, horizontal and vertical ground stress differential and completion shot hole skin are all related to reservoir pressure, so the model is able to assess the effect of reservoir pressure depletion or pressure maintenance on sand arch stability during the life of the reservoir.

Fluid percolation force coefficients combine the effects of flow rate, viscosity and relative permeability. For example, fluid seepage force effects allow shot-hole cavity stability models to study the effects of water breakthrough and two-phase flow on sand arch stability during the life of the reservoir. Deformation and damage parameters from stress/strain tests are used in the cavity stability model to describe complex rock mechanical behavior (i.e., nonlinear stress/strain relationships). Other control parameters in the model include cavity geometry (e.g., shot hole diameter, length, and phase), well slope angle, and the angle between the shot hole channel/cavity and horizontal ground stress, as well as loading factors for the cumulative weakening effects of multiple cycles of start-up and shutdown production.

In-situ stress and sandstone stress/strain deformation and damage characteristics are key parameters in the numerical analysis of the outgoing sand. Reservoir stresses are derived from direct measurements at the beginning of development and can also be derived from later pressure transient test wells. Ground stress requires density logging rock mechanics analysis, leakage testing/rupture pressure testing, and elastic strain recovery measurements.

The model allows the effects of shot hole diameter, length and phase to be evaluated when predicting the stability of new wells. As a rule of thumb, completion pressure losses (epithermal effects) must be assigned to the planned shot hole. Calculation of flow rates and fluid permeability factors requires relative permeability, fluid viscosity and compressibility, and cavity geometry parameters.

Rock mechanics is at the heart of the engineering system for sand emergence prediction. Reservoir sandstones are complex mechanical materials with nonlinear stress/strain relationships. Compression and tension tests of reservoir cores provide the damage and deformation parameters required for cavity stability models. The tests should be performed at four or more peritectic pressures, with the cores grouped subjectively over a range of formation strengths. Calibrate formation strength logs with the results of rock mechanics experiments and cavity stability analysis.

The cumulative weakening effect and stress relief will be accompanied by repeated cycles of flow opening and closing in the shot hole region. The loading repetition factor is another parameter that must be evaluated empirically.

#### 10.5.3.4 Reservoir Management

Reservoir management refers to the overall operating philosophy of the reservoir. Sand emergence prediction (SPP) is only one of the tools used by completion engineers. SPP inputs need to consider other technical construction constraints such as equipment availability and data availability. Reservoir management will provide other constraints in terms of production, cash flow or other targets. The completion engineer will coordinate planning with the reservoir engineer and production engineer. For example, a typical management goal for an offshore reservoir may be a high initial rate to offset a large capital investment in a production facility. For medium-strength reservoirs, high initial production options via natural completions may be compared to gravel-fill options. Management will weigh whether a decline in production after water breakthrough is acceptable for high initial production rates from natural completions. With the reservoir simulator, our engineering system provides a quantitative assessment of sand-free production over the life of the reservoir.

## Exercises

1. Briefly describe the hazards of sand emergence from oil well production and the basic process of sand emergence from oil wells?
2. Briefly describe the main mechanisms of sand emergence from oil wells and analyze the main causes of sudden sand emergence from oil wells during high water content periods?
3. Consider the basic principles of circular arch self-stabilization and analyze the main role of circular arch stability for sand control?
4. A weakly cemented sandstone reservoir in the Shengli oilfield is buried at a depth of 2200 m, the reservoir pressure is 30 MPa, the cohesion is 3 MPa, the

angle of internal friction is 30°, the modulus of elasticity is 3000 MPa, and the Poisson's ratio is 0.26. What is the critical production pressure difference for a well completion without sand production by the shot-hole completion method?

## References

Barrow DC, Lasseigne CA. A field evaluation of the Mohr's circle technique for predicting sand strength. SPE-13087-MS; 1984.

Bratli RK, Risnes R. Stability and failure of sand arches. SPEJ. 1981;21(2):236–248.

Cheng YF, Zhang JG. Foundation of Sand Predicition Model in Perforated-hole Completion. Petroleum Drilling Technol. 2001;29(6):41–3.

Coates GR, Denoo SA. Mechanical properties program using borehole analysis and Mohr's circle. The SPWLA 22nd Annual Logging Symposium, Mexico City, Mexico; June 1981.

Durrett JL, Golbin WT, Murray JW et al. Seeking a solution to sand control. JPT. 1977;29(12):1664–1672.

Ghalambor A, Asadi M. A study of relevant parameters to predict sand production in gas wells. SEP Drilling & Completion 2002;17(02):88–99.

Ghalambor A, Hayatdavoudi A, Alcocer CF et al. predicting sand production in U.S. gulf coast gas wells producing free water. JPT. 1989;41(12):1336–1343.

Hall CD Jr, Harrisberger WH. Stability of sand arches: a key to sand control. J Pet Tech. 1970;821–829.

McNulty JW. An experimental study of arching in sand. Urbana: University of Illinois; 1965.

Stein N, Kelly J, Baldwin WF et al. Sand production determined from noise measurements. JPT. 1972;24(07):803–806.

Tippie DB, Kohlhaas CA. Effect of flow rate on stability of unconsolidated producing sands. Las Vegas SPE. 1973;4533.

Tixier MP, Loveless GW, Anderson RA. Estimation of formation strength from the mechanical-properties log (includes associated paper 6400). JPT. 1975; 27(03):283–293.

Vaziri H, Xiao Y, Palmer I. Assessment of several sand prediction models with particular reference to HPHT wells. SPE. 78235; 2002.

Veeken CAM, Davies DR, Kenter CJ, et al. Sand production prediction review: developing an integrated approach. In: SPE22792;1991.

Weissenburger KW, Morita N, Martin AJ, et al. Engineering approach to sand production prediction. SPE16892-MS; 1987.

Zhang JG, Cheng YF, Cui HY. Development of a model for predicting sand out of bare-hole completions. Petroleum Drilling Technology. 1999;27(6):39–41.

Zhang JG. Effect of gas intrusion on sand emergence. Oil Drilling Technol. 2000;v28N2:39–40

# Glossary

## Chapter 2 Stress and Strain

**Outside load** The load acting on the object surface is called outside load.

**Concentrated force** The force acting on a point is called concentrated force.

**Distributive force** The force acting on a surface or a line is called distributed force.

**Volume force** refers to the force in unit volume, such as gravitational centrifugal force.

**Normal stress** The stress acting in the normal direction of any inclined section inside the object is called normal stress.

**Shear stress** The stress acting in any inclined section plane inside the object is called shear stress.

**Stress state** It can be seen that only six of the nine stress components on the micro-unit are independent, namely $\sigma_x, \sigma_y, \sigma_z$ and $\tau_{xy}, \tau_{yz}, \tau_{zx}$. If these six stress components are known, then the stress components on any azimuthal section past that point can be found. Therefore, the stress components on the micro-unit near that point can represent the stress state at that point.

**Principal direction and the principal stress** It can be proved that through the point there are always three special planes perpendicular to each other and with no shear stress through this point, the section is called the principal plane, the direction of its outer normal is called the principal direction of the point, the normal stress acting on the principal plane is called the principal stress.

**Invariants of the stress tensor** The invariants of the stress refers to the coefficient of the characteristic equation of the principal value of the stress, including $I_1$, $I_2$ and $I_3$.

$$I_1 = \sigma_x + \sigma_y + \sigma_z$$

$$I_2 = \sigma_x\sigma_y + \sigma_y\sigma_z + \sigma_z\sigma_x - \tau_{xy}^2 - \tau_{yz}^2 - \tau_{xz}^2$$

$$I_3 = \sigma_x\sigma_y\sigma_z + 2\tau_{xy}\tau_{yz}\tau_{zx} - \sigma_x\tau_{yz}^2 - \sigma_y\tau_{zx}^2 - \sigma_z\tau_{xy}^2$$



Y. Cheng et al., *Foundations of Rock Mechanics in Oil and Gas Engineering*,
https://doi.org/10.1007/978-981-99-1417-3

$I_1 = I_2 = I_3 = \sigma$ are called the first, second and third invariants of the stress tensor.

**Maximum shear stress** is half the difference between the maximum principal stress and the minimum principal stress.

**Hydrostatic stress state** When $\sigma_1 = \sigma_2 = \sigma_3 = \sigma$, the principal shear stress is zero regardless of the values of *l*, *m* and *n* and the normal stress is $\sigma$. This situation is usually referred to as a hydrostatic stress state. In this stress state, only volumetric strain can be induced in the micrometeoroid without any change in shape.

**Deviatoric stress state** is the portion remaining after hydrostatic stresses are deducted from the stress state.

**Invariants of the deviatoric stress** The deviatoric stress invariant refers to the coefficient of the characteristic equation of the principal value of the deviatoric stress, including $J_1$, $J_2$ and $J_3$.

$$J_1 = s_1 + s_2 + s_3 = \sigma_x - \sigma_0 + \sigma_y - \sigma_0 + \sigma_z - \sigma_0 = 0$$

$$\begin{aligned} J_2 &= s_1 s_2 + s_2 s_3 + s_3 s_1 \\ &= \frac{1}{6}\left[\left(\sigma_x - \sigma_y\right)^2 + \left(\sigma_y - \sigma_z\right)^2 + (\sigma_z - \sigma_x)^2 + 6\left(\tau_{xy}^2 + \tau_{yz}^2 + \tau_{zx}^2\right)\right] \end{aligned}$$

$$J_3 = s_x s_y s_z + 2 s_{xy} s_{yz} s_{zx} - s_x s_{yz}^2 - s_y s_{zx}^2 - s_z s_{xy}^2$$

**Mohr Circle** The circle drawn by the two principal stresses σ1 and σ2 in the σ and

τ coordinate systems is called a Mohr Circle.

**Strain** The strain includes the positive strain, which refers to the change of unit length before and after the deformation of an object, and the shear strain, which refers to the change of Angle between two linear elements before and after the deformation of an object.

## Chapter 3 Rock Composition and Physical Properties

**The most common rock-forming minerals are 16 minerals** which are quartz, feldspar (orthoclase, plagioclase), mica (black mica, white mica), chlorite, hornblende, pyroxene, olivine, calcite, dolomite, gypsum, hard gypsum, rock salt, pyrite, and graphite.

**Magmatic rocks** Magmatic rocks are formed when molten magma cools at the surface or underground.

**Sedimentary rocks** Sedimentary rocks are formed by the deposition of sediments and have a complex history.

**Metamorphic rocks** Metamorphic rocks are rocks formed by the recrystallization of other rocks at high temperatures and pressures.

**Structure of intact rocks** refers to the size, shape, surface characteristics, quantitative relationships and interconnectedness characteristics of their constituent units (individual grains, aggregate grains, glassy).

**Types of cementation in clastic sedimentary rocks** According to the interrelationship between cement and debris can be distinguished as (1) basal cementation, where debris particles are immersed in cement without contacting each other; (2) pore cementation, where cement fills the pores between particles in contact with each other; (3) contact cementation, where cement is distributed only on the contact points between particles and debris; (4) mosaic cementation, where under the consolidation effect during the diagenesis, especially when the pressure-solution effect is obvious, the clastic particles in sandy sediments come into closer contact, and the particles develop from point contact to line contact, concave-convex contact, or even form suture-like contact. p12.

**Pore space of the rock** The space in a rock that is not filled with mineral particles, cement or other solid material is called **the pore space of the rock**.

**Pores and throats** The main components of rock pore space are pores and throats, and the larger space surrounded by rock particles is generally called a pore, while the narrow part connected only between two particles is called a throat.

**Tectonics of rocks** here refers to structural surfaces such as beddings, cleavages, joints, and faults.

**A fault** is a type of rupture tectonics in the earth's crust in which a body of rock is significantly displaced along a rupture plane.

**Bulk density of rocks** The mass per unit volume of rock (including the volume of pores within the rock) is called the bulk density of the rock.

**Porosity of the rock** The porosity of a rock is the ratio of the total volume of rock pores to the total volume of the rock and is often expressed as a percentage.

**Permeability of the rock** The property of a rock to permit the passage of fluids under pressure is called the permeability of the rock.

**Specific surface area of a rock** is the total surface area of particles within a unit volume of rock. The units are expressed in $cm^2/cm^3$.

## Chapter 4 Strength and Deformation Characteristics of Rocks

**Rock mechanical properties** mainly refer to the deformation characteristics of rocks and the strength of rocks.

**The whole stress-strain curve** The stress-strain curves obtained from rock compression tests conducted with a rigid testing machine include the deformation characteristics after the stress level reaches the peak strength.

**Confining pressure** The pressure acting on the circumference of the cylindrical sample represents the circumferential constraint on the underground rock.

**Dilation** In high stress level, the rock specimen continuously produces micro-rupture as well as slip within or between grains, which is the apparent inelastic deformation that the rock has before it is damaged, and this phenomenon is called dilation.

**Tangent modulus, secant modulus, and mean modulus** The tangent modulus $E_t$ is generally chosen as the slope of the tangent line of the stress–strain curve at the origin or at the point where the stress is half the strength of the rock sample. The secant modulus $E_s$ is mostly used as the ratio of stress to strain when the stress is half the strength of the rock sample, i.e., $E_{50}$, also known as the deformation modulus of the rock. The mean modulus $E_{av}$ is the slope of the approximately straight line portion of the stress–strain curve.

**Brittle failure and ductile failure** For rocks, a pre-damage strain of less than 3% can be considered as brittle damage, more than 5% as ductile damage, and 3–5% as a transitional stage.

**Pore pressure and the effective stress** The concept of pore pressure and effective stress was first introduced by Terzaghi (1933) in his analysis of saturated soils: when a load is applied, the compressive stress $\sigma$ in the soil is carried by two components, the effective compressive stress $\sigma'$ at the point of contact of the particles and the pore pressure $p_p$ generated by the saturated water in the pores (assuming that the pore water cannot drain freely).The effective stress is $\sigma' = \sigma - p_p$.

## Chapter 5 Characterization and Indoor Determination of the Strength of Rocks

**Compressive strength of rock** Uniaxial compressive strength is referred to as compressive strength, for uniaxial compression test, when the pressure reaches damage, the specimen damage stress is called the compressive strength of rock.

**Tensile strength of a rock** is the ultimate stress at which the specimen reaches damage under uniaxial tensile conditions.

**Shear strength** is generally defined in two ways: one is the maximum shear stress on the shear breaking surface of the rock when the specimen is subjected to a normal load; the other is defined as the maximum shear stress on the shear breaking surface in pure shear (i.e., without a normal load). The former takes into account the cohesion and internal friction contained in the rock during shear damage; the latter depends only on the cohesion.

## Chapter 6 Rock Strength Failure Criterion

**Fracture angle** The angle between the normal direction of the shear failure plane and the axial stress ($\sigma1$) is called the fracture angle.

## Chapter 7 In-Situ Stress States

**In-situ stress** Rocks buried deep in the ground are subjected to stresses prior to engineering disturbances, and this stress is generally referred to as in situ stress.

## Chapter 8 Mechanics of Wellbore Stability

**Formation collapse pressure** The critical wellbore fluid hydrostatic pressure when the shear failure occurs on the wellbore is called formation collapse pressure.
**Formation fracture pressure** The critical wellbore fluid hydrostatic pressure when the tensile failure occurs on the wellbore is called formation fracture pressure.

## Chapter 9 Mechanics of Hydraulic Fracturing

**Net pressure** The difference between treatment pressure and minimum horizontal in-situ stress is referred to as the net pressure.
**Average fracture width** is defined as the fracture volume divided by the fracture face area, or the single flank volume divided by the single flank area.
**Fracturing efficiency** is defined as fracture volume divided by injection volume.

## Chapter 10 Mechanics of Oil Well Sand Production

**Production pressure difference** The difference between the pore pressure of the reservoir and the wellbore pressure of the production well.